Our Earth's Changing Land

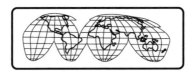

International Geographical Union
COMMISSION ON LAND USE/COVER CHANGE

This work was supported by the International Geographical Union's Commission on Land Use/Cover Change (IGU-LUCC)

The Commission involves a network of geographers with research interests in land use/cover change. Its objectives are:

- To promote geographical research on land use/cover changes, at scales ranging from the local to the global

- To stimulate the production and the use of land-use information bases of both the present and the past

- To coordinate the comparative study and the model study of land use/cover changes and their driving forces in different regions.

Our Earth's Changing Land

An Encyclopedia of Land-Use and Land-Cover Change

Volume 2: L–Y

Edited by Helmut Geist

GREENWOOD PRESS
Westport, Connecticut · London

Library of Congress Cataloging-in-Publication Data
Our earth's changing land : an encyclopedia of land-use and land-cover change / edited by
  Helmut Geist.
    p. cm.
    Includes bibliographical references and index.
    ISBN 0–313–32704–1 (set : alk. paper)—ISBN 0–313–32783–1 (v. 1 : alk. paper)—ISBN
0–313–32784–X (v. 2 : alk. paper)   1. Landscape changes—Encyclopedias.   2. Land use—Encyclopedias.   3. Landscape ecology—Encyclopedias.   I. Geist, Helmut, 1958–
    GF90.O87 2006
    333.7—dc22   2005019212

British Library Cataloguing in Publication Data is available.

Library of Congress Catalog Card Number: 2005019212

ISBN: 0–313–32704–1 (set)
      0–313–32783–1 (v. 1)
      0–313–32784–X (v. 2)

First published in 2006

Greenwood Press, 88 Post Road West, Westport, CT 06881
An imprint of Greenwood Publishing Group, Inc.
www.greenwood.com

Printed in the United States of America

The paper used in this book complies with the
Permanent Paper Standard issued by the National
Information Standards Organization (Z39.48–1984).

10 9 8 7 6 5 4 3 2 1

Contents

List of Entries

Guide to Related Topics

Analysis of Land Change
Aerial Photography
Auto-Correlation
Cadastre
Change Detection
Conversion
Cumulative Change
Decision-Making
Disequilibrium Dynamics
Driving Forces
Feedback
Georeferencing
Hot Spot Identification
Human Appropriation of Net Primary
    Production
Initial Conditions
Land Cover Classification System
Land Evaluation
Land-Use History
Land-Use System
Land-Use Transition
Mediating Factor
Meta-Analysis
Metabolism
Mode of Interaction
Multiplier Effect
Paleo Perspectives
Pattern Metrics
Pattern to Process
Pixelizing the Social
Proximate Causes
Remote Sensing
Scale
Socializing the Pixel
Spatial Diffusion
Syndrome
Systemic Change
Thermal Band Analysis
Tradeoffs
Transition Matrix
Validation

Biomass Burning
Advanced Very High Resolution
    Radiometer
Along Track Scanning Radiometers
    World Fire Atlas
Alternatives to Slash-and-Burn (ASB)
    Programme
Amazonia
Atmosphere-Land Interlinkages
Biodiversity
Biomass
Boreal Zone
Brazilian Cerrado
Canada
Carbon Cycle
Climate Impacts
Cumulative Change
Degradation Narrative
Disequilibrium Dynamics
Ecological Colonization
Fire
Global Burnt Area 2000
Global Observation of Forest and
    Land Cover Dynamics
GLOBCARBON
Grassland
Great Plains
Hot Spots of Land-Cover Change
Human Impact on Terrestrial Ecosystems
    (HITE) Initiative
LANDFIRE Project
Modification
Mountains
Nutrient Cycle
Paleo Perspectives
Pasture
Pristine Myth
Remote Sensing
Russia
Savannization
Secondary Vegetation
Slash-and-Burn Agriculture
Southern Africa
Southern Cone Region
Taiga
Threshold

Tropical Humid Forest
Turnover
United States of America
Vegetation

Case Studies

Amazonia
Boreal Zone
Brazilian Cerrado
Canada
China
Coastal Zone
East-Central Europe
Europe
Great Plains
Himalayas
Japan
Latin America
Meta-Analysis
Miombo
Mountains
River Basin
Russia
South Africa
South America
Southeast Asia, Mainland
Southeast Europe
Southern Africa
Southern Cone Region
Sudan-Sahel
Taiga
United States of America
West-Central Europe
Western Europe

Causes

Access
Agricultural Intensification
Agriculture
Agro-Industry
Arena of Land Conflict
Cattle Ranching
Colonization
Consumption
Contract Farming
Conversion
Cultural Factors
Cumulative Change
Decision-Making
Deforestation
Degradation Narrative
Desertification
Driving Forces

Economic Growth
Economic Liberalization
Economic Livelihood
Economic Restructuring
Endogenous Factor
Exogenous Factor
Extensification
Fire
Forest Degradation
Forest Transition
Global Environmental Change
Globalization
Green Revolution
Industrialization
Institutions
Investments
Land Abandonment
Land Concentration
Land Degradation
Land Fragmentation
Land Rehabilitation
Land Rent
Land Rights
Land Tenure
Land-Use Policies
Land-Use Transition
Leapfrogging
Mineral Extraction
Mode of Interaction
Modification
Pasture
Population Dynamics
Poverty
Property
Proximate Causes
Reforestation
Regrowth
Remittances Landscape
Slash-and-Burn Agriculture
Soil Degradation
Soil Erosion
Subsistence Agriculture
Suburbanization
Swidden Cultivation
Syndrome
Systemic Change
Technological-Scientific Revolution
Tragedy of Enclosure
Tragedy of the Commons
Transhumance
Urbanization
Urban Sprawl

Conventions and Environmental Agreements

Agenda 21
Conservation
Cropland Change
Forest
Institutions
Kyoto Protocol
Public Policy
Land Evaluation
Land-Use Planning
Land-Use Policies
Millennium Ecosystem Assessment
Ramsar Convention on Wetlands
United Nations Convention on Biological
    Diversity
United Nations Convention to Combat
    Desertification
United Nations Environment Programme
United Nations Food and Agriculture
    Organization

Cropland Change

Agricultural Intensification
Agricultural Revolution
Agriculture
Agrodiversity
Agro-Industry
BIOME 300 Project
Biotechnology
Cash Crops
Cassava
Coffee
Colonization
Consultative Group on International
    Agricultural Research
Consumption
Contract Farming
Conversion
Cotton
Cumulative Change
Decision-Making
Desertification
Disintensification
Domestication
Economic Livelihood
Eutrophication
Extensification
Fallow
Food Crops
Food Security
Globalization
Green Revolution

Hot Spots of Land-Cover Change
Human Appropriation of Net Primary
    Production
Initial Conditions
International Food Policy Research
    Institute
Land Abandonment
Land Concentration
Land Degradation
Land Rent
Land Rights
Land Tenure
Landholding
Land-Use History
Land-Use Legacy
Land-Use Policies
Land-Use System
Maize
Modification
Monoculture
Nutrient Cycle
Property
Proximate Causes
Remittances Landscape
Rice
Slash-and-Burn Agriculture
Soil Degradation
Soil Erosion
Spatial Diffusion
Species Extinction
Subsistence Agriculture
Swidden Cultivation
Tobacco
Urban-Rural Fringe
Wheat

Data

Aerial Photography
BIOME 300 Project
Cadastre
Center for International Earth Science
    Information Network
Center for Sustainability and the Global
    Environment
Change Detection
Continuous Data
Coordination of Information on the
    Environment (CORINE) Database
Discrete Data
Earth Resources Observation Systems
    (EROS) Data Center
Forest
Geographical Information System

Suburbanization
Sustainable Land Use
Urban Sprawl
Urbanization
Vegetation
Watershed
Wetlands

Land-Cover Products

AFRICOVER
Along Track Scanning Radiometers
    (ATSR) World Fire Atlas
BIOME 300 Project
Coordination of Information on the
    Environment (CORINE) Database
Earth Resources Observation Systems
    (EROS) Data Center
European Remote Sensing (ERS-1/-2)
    Satellites
Global Burnt Area 2000
Global Land Cover Map of the Year 2000
Global Land Cover Network
Global Terrestrial Observing System
GLOBCARBON
GLOBCOVER
GLOBICE
GLOBSCAR
GLOBWETLANDS
History Database of the Global
    Environment
Hot Spots of Land-Cover Change
IGBP-DIS Global 1-Km Land Cover
    Data Set
Sahel Land Cover
Tropical Ecosystem Environment
    Observations by Satellite (TREES)
    Project
United States Geological Survey (USGS)
    Program

Land-Use Dynamics

Access
Agricultural Frontier
Agricultural Intensification
Agricultural Revolution
Agrodiversity
Agroforestry
Agro-Industry
Arena of Land Conflict
Biomass
BIOME 300 Project
Biotechnology
Cattle Ranching

Colonization
Common-Pool Resources
Community Involvement
Conservation
Consumption
Contract Farming
Conversion
Corporate Strategies
Cotton
Cultural Factors
Cumulative Change
Decentralization
Decision-Making
Deforestation
Degradation Narrative
Desertification
Disequilibrium Dynamics
Disintensification
Domestication
Driving Forces
Ecological Colonization
Economic Growth
Economic Liberalization
Economic Livelihood
Economic Restructuring
Exotic Species
Extensification
Fallow
Feedback
Fire
Food Crops
Food Security
Forest Anomalies
Forest Degradation
Forest Transition
Forestry
Globalization
Green Revolution
Green Wall Project
Human Appropriation of Net Primary
    Production
Human Impact on Terrestrial Ecosystems
    (HITE) Initiative
Indigenous Knowledge
Industrial Revolution
Industrialization
Initial Conditions
Institutions
Investments
Land Abandonment
Land Concentration
Land Privatization
Land Reform

Rangeland Change

Satellite Imagery

Urban-Industrial Change

L

Land Abandonment. The withdrawal of land from production, which often follows marginalization processes in agriculture and rural areas. Land abandonment leads to changes in **vegetation** and landscapes (most of the land would turn into **forest** if left unmanaged). It may provide environmental benefits, in particular if pollution by agricultural chemicals were reduced. However, abandonment may also entail losses of **biodiversity** in cases where much of the land was **grassland** with high nature value for botanical interest, as habitats for birds or traditional landscapes (mainly semi-natural grasslands and other areas important for birds).

Agricultural marginalization is a process that is driven by a combination of social, economic, political, and environmental factors, by which in certain areas **agriculture** ceases to be viable under an existing **land-use system**. Although marginalization of agricultural land is not new, its extent and speed of growth are evident most clearly in remote areas under harsh climatic and soil conditions, but also observed more locally. **Driving forces** behind marginalization of land can be found in a range of factors at various spatial **scales**. The consequences operate on the local and regional scales, making their appearance in a change of the socio-economic position of the rural population in areas in question. However, **vulnerability** to marginalization depends on the local or regional, social, economic, political, and environmental conditions.

Competitive advantages of production may therefore result from a range of socio-economic factors like **access** to markets and the availability of proper **institutions**, skilled labor, credits available to farmers, infrastructure, international politics, **cultural factors**, and human choices. A major driving force behind vulnerability of rural areas (or marginalization of agriculture) therefore is economic marginality. In areas marginal for agriculture, agricultural productivity is often low because of climatic constraints, poor soils, and poor accessibility of agricultural lands or traditional low-input agriculture. Declining trends in agricultural production could result from farming becoming less viable, getting marginalized, and land eventually becoming abandoned. Marginalization has sometimes far-reaching consequences—social, economic, and ecological. Marginalization and abandonment of agricultural land have become a serious threat to **biodiversity** and landscape features as well. The effects of marginalization and abandonment on flora and fauna can vary considerably; the ecological consequences can be complex, site-specific, and far from uniform, even in a relatively small geographical region. Marginalization does not necessarily lead to progressively more extensive farming systems followed by abandonment. There may be a combination of **agricultural intensification** and **extensification** within the same farm or region, or the restructuring of **landholdings** or new land uses such as **forestry**. Both ecological and socio-economic aspects of marginalization could be solved through the provision of payments in support of biodiversity **conservation** and as rewards for income loss due to application of specific land management practices.

See also: **African Trypanosomiasis; Agricultural Frontier; Biodiversity; Coffee; Economic Growth; Food Crops; Forest Transition; Hot Spots of Land-Cover Change; Investments; Mountains; Pasture; Pristine Myth; Reforestation; Remittances Landscape; Secondary Vegetation.**

Further Reading

Baldock, D., Beaufoy, G., Brouwer, Floor, and F. Godeschalk, *Farming at the Margins: Abandonment or Redeployment of Agricultural Land in Europe*, London, UK, the Hague, NL: Institute for European Environmental Policy, Agricultural Economics Research Institute, 1996.

FLOOR BROUWER

Land Concentration.   Amassing of large amounts of land in the hands of a few individuals, corporations, or **institutions**. It involves processes that result in a particular structure of bimodal land distribution in which a large number have very little land and a very few have a great deal of land. Land concentration involves two processes: one that leads to the acquisition of land by some, and another that leads to its loss by others. The dynamics of land concentration are largely mediated by economic, social, and political factors. Land concentration can take place in several contexts. In some cases, it occurs in situations in which customary rights are negated to local people and land usurpation takes place by national or local elites through land encroachment. In other situations, particularly in those in which market economies are dominant, land transactions may tend to favor the concentration of land in few hands. Finally, land concentration is a common phenomenon in situations in which there are still open **access** areas, particularly in frontiers for agricultural expansion.

The three different situations often have taken place in tropical contexts where there are strong pressures for forest **conversion** to agricultural uses. In tropical areas, particularly of **Latin America**, land concentration of public land has occurred in different historical contexts, mainly in those where there were increasing economic interests for land as a means of production, or to support speculative purposes. This process, rather than being stopped, was stimulated by political patronage mechanisms in support of national or local elites who concentrated the lands, and who often held the power to legitimate or enforce land ownership structures by legal or semi-legal means. Land concentration has also taken place in areas already occupied by indigenous communities, or small farmers. Land encroachment has been the most-used mechanisms for ranchers and loggers to usurp communal lands, in contexts in which the lack of **property** rights and the absence of legal mechanisms to protect such rights both constituted the norm rather than the exception.

In areas dominated by small-scale farmers, land concentration tends to take place through land transactions. While the reasons that large landholders have to buy land, or to amend new areas to existing rural establishments, are strongly linked to economic expectations, the motivations that land sellers have to sell their land are somewhat ambiguous. On one hand, large landholders (i.e., large-scale agricultural producers or ranchers) buy the land expecting to increase their economic benefit from using such land on productive activities (i.e., raising cattle or producing grains), or prompted by some sort of speculative interest by which they expect either to increase their profits by selling the land at some time in the future, or to obtain some institutional rent from holding, such as **land rights** (i.e., cheap credit). On the other hand, small farmers tend to sell their lands in two different circumstances. First, they sell due to their inability to sustain productive activities in precarious conditions. Second, they sell in cases in which the resources they are offered for the land exceed the estimated benefits they will obtain from working on the land. The two dynamics lead to higher land concentration.

See also: **Agricultural Frontier; Cattle Ranching; Land Fragmentation; Landholdings; Land Privatization; Land Rent; Land Tenure; Mediating Factor; Tragedy of Enclosure.**

PABLO PACHECO

Land Cover Classification System.   Land-cover assessment and **monitoring** of its dynamics are essential requirements for the sustainable management of natural resources and for environmental protection. Land-cover data are also essential for environmental, **food security**, and humanitarian programs of many United Nations, international, and national institutions. At present there is no internationally accepted land-cover classification system. Some are more universally applicable than others, but none have been accepted as the international standard. The increasing number of regional and global initiatives involving land-cover mapping and monitoring requires a close interaction among them, in order to ensure their benefits. Since land cover has also become a basic data layer in **geographical information systems**, the synergy among land-cover mapping projects has a high priority.

To solve this problem, the land cover classification system (LCCS) was developed and implemented by the **United Nations Food and Agriculture Organization** (FAO) and the **United Nations Environment Programme** (UNEP) with the financial support of the government of Italy. LCCS was developed during the **AFRICOVER** project in 1994, to allow the harmonization of different nomenclatures, legends, and classifications of twelve African countries. It is a comprehensive methodology for the description, characterization, classification, and comparison of land-cover data, identified anywhere in the world, at any **scale**. LCCS allows the description of different land-cover features in a standardized way.

Classification Methodology

Classification systems exist in hierarchical and non-hierarchical formats. Most systems are hierarchically structured. They start with broad-level classes, which allow further systematic subdivision into more detailed subclasses. The classification process can be done *a posteriori* or *a priori*. A *posteriori* classification is based upon the definition of classes after the field samples have been collected. Although it is flexible and adaptable, it is often developed for a specific area being described; the classes therefore often need to be adapted for each new location. For such a classification, it is therefore difficult to define standardized classes. In an *a priori* classification system, all the classes are defined before the data is collected. The main advantage is that a standard set of classes and methodology is created, which is independent of the geographic area. The disadvantage is that an enormous number of predefined classes have to be created to allow an accurate description of all the possible land-cover types occurring in the world.

The classification system used in LCCS is an *a priori* classification system, but LCCS is unique as it provides a scale-independent method of classifying land cover. The approach supports all types of land-cover monitoring and enables a comparison of land-cover classes regardless of data source, sector, or country. This is achieved by the use of a set of universally valid classification criteria that uniquely identify the land-cover classes worldwide. The system works on two levels. On the first level, eight major land-cover types are defined. At the second level, land-cover classes are created by the combination of sets of predefined classifiers. These classifiers have been carefully defined to describe different land-cover variations present inside each major land-cover type and also to avoid inappropriate combinations of classifiers. The number of classifiers used determines the level at which the land cover is classified. The system is highly flexible but allows land-cover classes to be clearly characterized, thus providing internal consistency. The system is truly hierarchical and applicable to a variety of mapping scales and in any geographic location. It can be used as a reference standard system because its diagnostic criteria allow correlation with existing classifications and legends.

Software

A user-friendly interface has been built around the set of databases holding the numerous classifiers and attributes. The software facilitates the user and prevents errors in creating the required land-cover classes. The LCCS software contains three main modules. The classification module allows the user to define the land-cover class using two main phases. In the initial dichotomous phase, the major land-cover type is defined by selecting between the two main land-cover types, "primarily vegetated" and "primarily non-vegetated," and then selecting the more detailed classification fields. After having defined the main land-cover type, the user can set up a generic class using only the results of this initial phase; or can proceed to the modular-hierarchical phase, where more detailed land-cover classes can be created by combining different sets of predefined classifiers. The classifiers available depend on the main land-cover type and the classifiers previously selected (see figure).

Basic Classifier Modifier

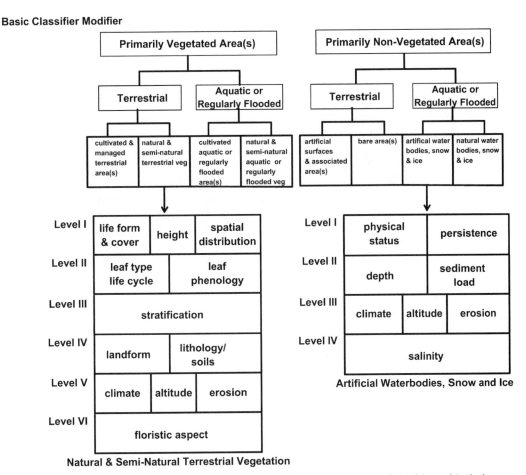

The LCCS Classification Module—the initial dichotomous phase and the modular-hierarchical phase.

The main purpose of the legend module is to take the land-cover classes identified and store them in a hierarchical structure that groups the classes according to the main land-cover type. The legend usually contains only a subset of the classification, namely those classes that are applicable in the area (to be) interpreted or mapped. In addition to providing a hierarchical structure, the legend also offers the ability to display, edit, and add user-defined attributes to a land-cover class. Standard descriptions are provided for the classes identified and the classifiers used, and all this information can be exported in various formats. In the legend section, mixed mapping units can be used, which comprise two, three, or more classes.

The translator module allows the translation of existing classifications and legends into LCCS. The reference base that is created can be used to compare classes of translated classifications and legends and their attributes, including the direct comparison of the individual land-cover classes.

Use of LCCS

The LCCS concept is rapidly being adopted by both national and international initiatives. Organizations that have already adopted LCCS include

FAO, the **Global Land Cover Map of the Year 2000** initiative of the **Joint Research Centre (JRC) of the European Commission**, the Global Land Use Cover Classification (GLUCC) working group of the **International Geosphere-Biosphere Programme**, and UNEP. Through the AFRICOVER project, a number of countries have also adopted LCCS, including Burundi, the Democratic Republic of the Congo, Egypt, Eritrea, Kenya, Rwanda, Somalia, Sudan, Tanzania, and Uganda. Numerous other countries from other regions of the world are now also adopting LCCS to implement their national mapping programs.

In addition, LCCS is one of the main tools of the **Global Land Cover Network** (GLCN), which has been developed by FAO and UNEP to increase the reliability of land-cover information for a large user community. GLCN provides direction, focus, and guidance for harmonization of land-cover mapping based on LCCS. AFRICOVER and ASIACOVER are the first regional outreach programs that are supported by GLCN, but additional networks are now being developed in the Caribbean, Central America, Central Asia, **east-central Europe**, **Latin America**, and the Middle East.

Future Developments

At the beginning of 2005, a new version of LCCS (version 2.0) was released with many developments and improvements, including a new user interface, modified class names and class descriptions, increased availability of classifiers and environmental attributes, and increased flexibility in class creation. The software and user manual of this new version of LCCS will be available in a number of languages including Arabic, English, French, and Spanish.

An LCCS Web Site has now been created with free information, documentation, and software. In addition, online assistance is provided to facilitate users in the development of their own LCCS initiatives and projects. A forum on LCCS has also been created to promote discussions among LCCS users. To further assist users, regional and national training workshops in LCCS and other mapping and GIS tools have, and are, being organized by GLCN with the support of UNEP and the governments of Italy and the Netherlands. In addition, self-tuition tools will soon be available to allow individuals to acquire the skills to use the LCCS software in their own initiatives and programs.

LCCS has the potential to become accepted as the international land-cover classification standard because of its inherent flexibility, applicability in all climatic zones and environmental conditions, and compatibility with the existing classification systems. LCCS has been submitted for approval to become a standard of the International Organization for Standardization (ISO). ISO is the world's largest developer of standards, with a network of 147 national standards institutes with a central secretariat in Geneva, Switzerland.

See also: **Conservation; Remote Sensing.**

Further Reading

Di Gregorio, Antonio, and Louisa J. M. Jansen, *Land Cover Classification System*, Rome: United Nations Food and Agriculture Organization, 2001; Di Gregorio, Antonio, and Louisa J. M. Jansen, *Land Cover Classification System: Classification Concepts and User Manual* [Online, October 2004], The United Nations Food and Agriculture Organization Web Site, www.fao.org/WAICENT/FAOINFO/SUSTDEV/EIdirect/EIre0062.htm; International Organization for Standardization [Online, November

2004], The ISO Web Site, www.iso.org; Global Land Cover Network; The Land Cover Classification System [Online, December 2004], The LCCS Web Site, www.glcn-lccs.org.

SOPHIA LINKE, REUBEN SESSA

Land Degradation.   Loss of life-supporting functions due to land use that is coupled with natural processes, which will remain an important global issue for the twenty-first century because of its adverse impact on agronomic productivity, the environment, and its effect on **food security** and the quality of life. The livelihoods of more than 900 million people in some 100 countries are now directly and adversely affected by land degradation due to complex biophysical and socio-economic interactions and regionally varied mechanisms. Unless the current rate of land degradation is slowed and reversed, the food security of humanity will be threatened—and the number of people adversely affected is expected to double to 1.8 billion until 2025.

Concept and Definition

There are biophysical (e.g., land management), socioeconomic (e.g., income and **land tenure**), and political (e.g., incentives and political stability) forces that determine causes and extent of land degradation. Land degradation has no single attribute, but comprises a bunch of processes. Processes such as **soil erosion**, loss of **biodiversity**, or **eutrophication** take place on different time **scales**. Often the degradation effects caused by human activities are delayed, and therefore people are not aware of the potential impacts on land. Apart from degraded ecosystems and deteriorated natural resources (e.g., soil, wood, and species), land degradation has socio-economic impacts (e.g., productivity decline and loss of income) on people.

The main function of land is to support human life, that is, the production function of land is one **perception** of defining land degradation. Apart from the production function there are other important functions of land, however, that sustain human life, such as the recreation function, cultural function, habitat function, and regulation function. An anthropocentric definition of land degradation that integrates the different perceptions of land (use) and keeps a multifunctional perspective on land in mind is the following: land degradation is the loss of a sustained economic, cultural, or ecological function due to human activity in combination with natural processes.

Land degradation results from a "mismatch" between functional capability of land and its utilization by people. Land degradation is therefore closely related to **resilience**, sensitivity, and carrying capacity of land, as well as **vulnerability** of people living on and from these lands.

Causes

The most frequently expressed main causes of land degradation are overgrazing of rangelands, overcultivation of cropland, waterlogging and **salinization** of irrigated land, **deforestation** and overlogging, and pollution and industrial causes. These **proximate causes** include factors such as **agricultural intensification**, cropland expansion, livestock extension, fuelwood extraction, infrastructure extension, settlement extension, non-adapted technologies, habitat fragmentation, and introduction of invasive species.

The underlying **driving forces** comprise elements such as population growth and density, migration patterns, **land-use policies**, tenure rights, institutional capability, education, gender policy, conflicts, and **poverty**. Without a comprehensive and regional specific analysis of proximate causes and underlying **driving forces**, it will be not possible to deal adequately with sustained solutions of land-degradation problems.

Main Processes

The main mechanisms that initiate land degradation include physical, chemical, and biological processes. Important among physical processes are unsustainable use of natural resources, removal of functional landscape elements, and a decline in soil structure, all leading to crusting, compaction, erosion, desertification, anaerobism, and environmental pollution.

Significant chemical processes include acidification, leaching, salinization, decrease in cation retention capacity, and fertility depletion. Biological processes include reduction in carbon stocks in **vegetation** and soils as well as a decline in species diversity. Many of these processes are related to eutrophication of surface water, contamination of groundwater, and emissions of trace gases (CO_2, CH_4, N_2O, NO_x) from terrestrial/aquatic ecosystems to the atmosphere.

The Dynamic of Land Degradation

Land degradation is a phenomenon with a history of centuries or even millennia (e.g., medieval soil erosion and **forest degradation** in Central Europe or historical salinization of arable land in Mesopotamia). Given a long history of land degradation, the threat and the dynamic of land degradation have considerably increased during the last sixty years.

Global **wetlands** have decreased about 50 percent since 1900. The majority of the remaining wetland is degraded or under threat of degradation. The intensity of these problems is closely related to the intensity of human activity in and around the wetlands. Further pressure to drain land for **agriculture** is highest in tropical and subtropical areas of Asia, Africa, and **South America**.

More than 39 percent or about 553 million hectares of primary **forests** are threatened with degradation or extinction by human actions such as logging, mining, clearing for agriculture, and overhunting. In 1992, the World Resources Institute stated that since 1950, about 580 million hectares of fertile land have been lost due to soil erosion caused by deforestation. Overgrazing caused damage on 20 percent or 685 million hectares of the world's **pasture** and rangelands in the last decades. Recent losses have been most severe in Africa and Asia.

According to the **United Nations Environment Programme**, about 19.5 million hectares of different land-cover types are assessed to be degraded due to **urbanization** and industrial processes. Often highly productive agricultural land is lost.

Soil degradation was assessed to have increased by about 2 billion hectares since 1945, which accounts for nearly 18 percent of the earth's vegetated land. As a consequence, productivity has been significantly reduced on roughly 60 percent of these 2 billion hectares. The current annual impact of land degradation on agricultural land is assessed to be about 21 million hectares of land that became uneconomic for crop production, and an additional 6 million hectares of land that are irreversibly lost for production.

Soil erosion is one of the major causes of reduced potential for food production in both developed and developing countries. For example, the **United States of America** has lost approximately one-third of its topsoil since farming began less than 300 years ago, and it continues to lose twelve tons per hectare per year. However, in certain regions of Africa, India, and Central America, soil erosion reaches up to 100 tons per hectare per year, with severe impacts on food security and the regional economy.

On average, the impact of soil degradation on agricultural productivity in recent decades was assessed to be about 0.5 percent per year, which to a large degree foils yield increases as a result of technical development.

The dry areas of the world are the origin of a large number of globally important cereals and food legumes, such as barley, **wheat**, fava beans, and lentils. In these dry areas, biodiversity is seriously eroding through the degradation of natural habitats, the intensification and expansion of cultivation, and overgrazing in natural rangelands.

A Global Overview

Available statistics show a large variation in the extent and rate of land degradation because of different definitions and terminology. A study, for example, by Harold E. Dregne and Nan-Ting Chou focused on land degradation (soil degradation and degradation of vegetation) in drylands, which is known as **desertification**, while the report "Global Assessment of Human Induced Soil Degradation" (GLASOD) by L. Roel Oldeman and colleagues emphasizes different aspects of soil degradation. The tables "Estimates of all degraded lands in dry areas" and "Assessment of soil degradation" give an overview of both studies, while the table "Regional soil degradation of main land-cover types" highlights regional soil degradation of main land-cover types according to the GLASOD assessment. Both assessments, however, are qualitative assessments carried out by regional experts and do not refer to a quantitative and reliable database. Until now, such a global database of harmonized quantitative data focusing on soil and land degradation has been lacking.

According to calculations performed by Harold Dregne and Nan-Ting Chou, about 3.6 billion hectares of dryland areas are degraded, an area that comprises about 70 percent of all global drylands (see table "Estimates of all degraded lands in dry areas"). Drylands cover a large part of the developed and the developing world, but the countries of the developing world are most seriously affected because 1.4 billion people live on or from degraded land, of which most are less-favored areas, including marginal agricultural areas and woodland

Estimates of all degraded lands (in million ha) in dry areas (Dregne and Chou, 1994)

Continent	Total Area	Degraded Area[†]	Percent Degraded
Africa	1,433	1,046	73
Asia	1,881	1,342	71
Australasia	701	376	54
Europe	146	94	65
North America	578	429	74
South America	421	306	73
Total	5,160	3,592	70

[†] Comprises land and vegetation.

Assessment of soil degradation (in million ha)
(Oldeman et al., 1992)

Continent	Total Area[†]	Degraded Area	Percent Degraded	Share of Seriously* Degraded Area
Africa	2,234	494	22	65
Asia	3,771	748	20	61
Australasia	787	103	13	6
Europe	949	219	23	72
North America	1,810	95	5	83
South America	1,993	306	15	65
Total	11,544	1,965	17	62

[†] Without wastelands (e.g. deserts, bare land).
* Seriously degraded area comprises degradation classes from moderate to extreme (Oldeman et al., 1992).

areas. The marginal drylands of the world experience enormous pressure on their environments, caused by human utilization and recurrent droughts. The resilience of dryland resources (i.e., soil and vegetation) is usually low, and this is why drylands are particularly susceptible to degradation. The considerably higher amount of degraded area calculated in the study by Harold Dregne and Nan-Ting Chou compared to the GLASOD assessment results from consideration of vegetation degradation.

In the GLASOD study, 2 billion hectares are assessed to be prone to soil degradation. From these 2 billion hectares, around 562 million hectares belong to arable land, 685 million hectares are permanent **pastures**, and 719 million hectares are forests or woodlands (see table "Regional soil degradation of main land-cover types"). As can be seen in the table "Assessment of soil degradation," land degradation is not evenly distributed. **Europe** and Africa show the highest amount of degraded area, while in North America and **South America**, soil degradation is limited to 5 percent and 15 percent of the total land area, respectively. However, these numbers reflect only one aspect of the picture of soil degradation and could be misleading. Looking at the soil degradation of arable land (see table "Regional soil degradation of main land cover types") reveals that both Africa and South America face considerable degraded area. Apart from the extent of soil degradation, the GLASOD assessment describes the severity of soil degradation, which is related to adverse effects on site productivity. Rosamund Naylor, based on the GLASOD assessment, stated that about 38 percent of the degraded land is estimated to be

Regional soil degradation of main land-cover types (in million ha) according to Oldeman et al. (1992)

Continent	Arable Land			Permanent Pasture			Forests/Woodlands		
	Total	Degraded	%	Total	Degraded	%	Total	Degraded	%
Africa	187	121	65	793	243	31	683	130	19
Asia	536	206	38	978	197	20	1,273	344	27
Australasia	49	8	16	439	84	19	156	12	8
Europe	287	72	25	156	54	35	353	92	26
North America	236	63	26	274	29	11	621	4	1
South America	180	92	51	572	78	14	962	137	14
Total	1,475	562	38	3,212	685	21	4,048	719	18

lightly degraded (leading to a loss in agricultural productivity of up to 10 percent, yet retaining potential for full recovery), 46 percent moderately degraded (leading to a 10–25 percent reduction in agricultural productivity and capable of being restored only through considerable financial and technical investment), 15 percent severely degraded (exhibiting a 25–50 percent loss in agricultural productivity and reclaimable only with substantial international assistance), and 0.5 percent extremely degraded (incapable of supporting agriculture and irreclaimable). The last column in the table "Assessment of soil degradation" shows the regional share of seriously degraded soils (moderately to extremely degraded soils) with an agricultural productivity decline of at least 10 percent.

A Regional View

Africa

In Africa, **conversion** of marginal land, forests, and wetlands due to expansion of agriculture is a major driver of land degradation. Land degradation is linked to rural poverty and stems to a large degree from colonial imbalances in land distribution, lack of incentives for **conservation**, insecure land tenure, and the failure to provide for diversified rural production systems. Livelihoods of pastoralists and wildlife as well as habitats and biodiversity are threatened by draining wetlands for agriculture. Loss of natural habitats has reduced vegetation cover and exposed soils to wind and water erosion. Soil erosion reduces the productivity of land, requiring farmers to apply more and more fertilizers, which causes additional economic pressure. Soil erosion also causes **siltation** of reservoirs and rivers, and increased risk of flooding.

Asia

In Asia, land degradation problems are directly related to land-use practices, particularly agricultural expansion and intensification. Biological degradation, soil fertility depletion, soil erosion, and soil acidification are of particular concern. Large-scale clearance of forestland has caused a decline in soil structure and substantial loss in biodiversity. Due to a large extent of steep land, severe water erosion occurs in the **Himalayas**, Central Asia, and **China**. Agricultural mismanagement—for example, the application of acid fertilizers—has led to substantial soil degradation in Cambodia, Malaysia, Thailand, and Vietnam. Irrigation farming is responsible for salinization in parts of northern India and Bangladesh. Asia comprises large areas of drylands in Central Asia, south Asia, and northeast Asia, which are seriously affected by wind erosion. In south and southeast Asia, soil contamination from lead and arsenic is a prevalent problem.

Australia

Land degradation is one of the most important environmental issues in Australia. More than half of Australia's farmland requires treatment for land degradation. Large-scale mining causes the loss of vegetation cover and the lowering of the groundwater water table. Half the major forests and about 35 percent of Australia's woodlands have been cleared or severely modified.

Various combinations of soil erosion, salinity, acidification, soil structure decline, and waterlogging affect much of Australia's cropping land. Apart from losses in agricultural productivity, the impacts of soil degradation include damage to water resources, biodiversity, pipelines, houses, and roads.

Europe

In Europe, **urban sprawl**, an increasing transportation network, and highly intensive agriculture are major reasons for land degradation. The increase in sealed surfaces together with a decrease in forest cover has increased the frequency and size of storm runoff, causing flooding, mudflows, and landslides. Soil erosion in Europe is mainly caused by water and is largely a result of unsustainable agricultural practices, clearcutting of forests, and overgrazing. Soil erosion causes particular problems in the Mediterranean region of Europe. Soil contamination occurs throughout Europe, and is particularly severe in urban areas due to industrial activities and inadequate waste disposal, as well as in areas with a long tradition of heavy industry, mining and military activities, and accidents. In eastern Europe, large irrigation and hydroelectric projects together with poor water management have resulted in salinization and waterlogging of large areas, especially in Azerbaijan, Belarus, the Russian Federation, and Ukraine.

North America

In North America, land degradation is associated with agricultural expansion, intensification, and **industrialization**. One of the key issues of agricultural land use is the use of chemical pesticides, which have contributed to increased food production but have also had important environmental and **human health** effects. North America accounts for 36 percent of world pesticide use. In **Canada**, the land area treated with chemical pesticides increased 3.5 times between 1970 and 1995. Some of the direct pressures leading to degradation due to erosion as well as chemical and physical degradation have been agricultural expansion, intensification, and overgrazing in arid lands. However, conservation measures have led to significant declines in erosion over the past thirty years. The expansion of farmland and urban areas, especially into wetlands, destroyed habitats and contributed to a substantial loss of biodiversity. North America's forests are becoming increasingly fragmented and biologically deteriorated due to large-scale logging by the timber industry.

South America

In South America, the agricultural expansion and the intensified use of natural resources have triggered land degradation. Soil erosion is the main cause of land degradation, while nutrient depletion is also a serious issue, largely driven by **agricultural intensification**. Further nutrient depletion is closely linked to rural poverty, which, in turn, has contributed to greater environmental degradation and land deterioration. Large-scale farming leads to erosion and compaction due to mechanization, as well as salinization because of improper irrigation and chemical pollution. Expansion of permanent pastures into previously forested areas is still the main source of deforestation in **Amazonia**. However, smallholder-farming systems increase deforestation as well, and lead to erosion and loss of soil fertility because they are used intensively without allowing for adequate **fallow** periods.

Institutional Framework and Policy Response

Important global conventions related to land degradation are the **United Nations Convention to Combat Desertification**, the **United Nations Convention on Biological Diversity**, and the United Nations Framework Convention on Climatic Change. To bring these conventions into action, an infrastructure is needed to cover funding mechanisms, research and development centers, and coordinated international programs. On the international level, organizations such as the **United Nations Food and Agriculture Organization** (FAO), the **United Nations Environment Programme** (UNEP), and the United Nations Development Programme are the main institutions coping with land degradation. These organizations have gained significant expertise and experience with land degradation on both the regional and global scales. One example of a joint international effort on land degradation is the **Land Degradation Assessment in Drylands (LADA) Project**. This project is carried out in partnership by international organizations such as FAO and UNEP to develop and test an effective assessment methodology for land degradation in drylands.

International funding mechanisms such as the Global Environment Facility (GEF) and organizations can support national governments, but land degradation needs to be tackled at the local and regional scales, with direct intervention in the affected areas taking into account competition among concurrent uses of land and soil (e.g., food production, living space, infrastructure and industrial production, and conflicts between private and public use of land resources). Particularly in developing countries, land degradation is about rural people and their **institutions**. To avoid further land degradation, these people need both sectoral and macroeconomic policies that positively affect farm incomes as well as independent local institutions that take care of future land management.

See also: **Access; Arena of Land Conflict; Blaikie, Piers; Brookfield, Harold Chilingworth; Carbon Cycle; Common-Pool Resources; Decision-Making; Degradation Narrative; Disequilibrium Dynamics; Exotic Species; Extensification; Hot Spots of Land-Cover Change; Land Rights; Marsh, George Perkins; Nutrient Cycle; Private Property.**

Further Reading

Barrow, Christopher J., *Land Degradation: Development and Breakdown of Terrestrial Environments*, Cambridge, UK: Cambridge University Press, 1991; Blaikie, Piers, and Harold Brookfield, *Land Degradation and Society*, London: Methuen, 1987; Carman, Bill, ed., Challenges and Opportunities for Integrated Land Management [Online, May 2003], The International Development Centre Web Site, web.idrc.ca; Daily, Gretchen C., "Restoring Value to the World's Degraded Lands," *Science* 269 (1995): 350–4; Dregne, Harold E., and Nan-Ting Chou, "Global Desertification Dimensions and Costs," in *Degradation and Restoration of Arid Lands*, eds. Harold E. Dregne, Lubbock, TX: Texas Technical University, 1994, pp. 249–81; Middleton, Nick, and David Thomas, eds., *World Atlas of Desertification*, 2nd ed., London: Arnold, 1997; Naylor, Rosamund N., "Energy and Resource Constraints on Intensive Agricultural Production," *Annual Review of Energy and Environment* 21 (1996): 99–123; Oldeman, L. Roel, "The Global Extent of Land Degradation," in *Land Resilience and Sustainable*

Land Use, eds. Dennis J. Greenland and Ivan Szabolcs, Wallingford, UK: CAB International, 1994, pp. 99–118; Pagiola, Stefan, *The Global Environmental Benefits of Land Degradation Control on Agricultural Land*, Washington, DC: The World Bank, 1997; Stuip, Mishka A. M., Baker, James C., and Willem Oosterberg, The Socio-Economics of Wetlands [Online, April 2004], The Wetlands International Web Site, www.wetlands.org; United Nations Environment Programme, GEO: Global Environment Outlook 3: Past, Present and Future Perspectives [Online, July 2002], The UNEP Web Site, www.unep.org; Watson, Robert T., Dixon, John A., Hamburg, Steven P., Janetos, Anthony C., and Richard H. Moss, Protecting Our Planet, Securing Our Future: Linkages among Global Environmental Issues and Human Needs [Online, November 1998], The World Bank Group Web Site, www-esd.worldbank.org/planet; World Resources Institute, *World Resources 2000–2001: People and Ecosystems: The Fraying Web of Life*, Washington, DC: WRI, 2000.

GERALD BUSCH

Land Degradation & Development. An international journal that seeks to promote recognition, **monitoring**, control, and rehabilitation of degradation in terrestrial environments. The journal focuses on the character, causes, impacts, and scale of **land degradation**. It addresses the history, current status, and future trends of land degradation (including the potential for **sustainable land use** or management); and discusses avoidance, mitigation, and control of land degradation, as well as remedial actions to rehabilitate or restore degraded land.

See also: **Arena of Land Conflict; Deforestation; Desertification; Driving Forces; Land Rehabilitation; Mineral Extraction; Perception; Proximate Causes; Soil Erosion; Species Extinction; Urban Sprawl.**

Further Reading
Wiley Publishers, Land Degradation & Development Journal [Online, December 2004], The Publishers Web Site, www.wileyeurope.com.

HELMUT GEIST

Land Degradation Assessment in Drylands (LADA) Project. One of the most recent attempts, funded by the Global Environmental Facility, to assess the magnitude of **desertification** and to provide a baseline for **monitoring**. In earlier attempts associated with the **United Nations Convention to Combat Desertification**, among others, case studies were collected from national governments, focusing on randomly selected sites rather than applying a well-conceived framework for analysis. This effort has been improved for pilot analysis and case studies related to the Land Degradation Assessment in Drylands (LADA) project, with national case studies stemming from Argentina, Egypt, Mexico, Uzbekistan, Senegal, Kenya, **South Africa**, and Malaysia. Different from the analysis of **hot spots of land-cover change** associated with the **Millennium Ecosystem Assessment**, LADA is a programmatic rather than a scientific effort of the **United Nations Food and Agriculture Organization** and other partners. It responds to the need to strengthen support to combat **land degradation**. LADA aims to generate up-to-date ecological, social, and economic and technical information, including a combination of traditional or **indigenous knowledge** and modern science, to guide integrated and cross-sectoral **land-use planning** and land management in drylands. LADA's main objective is foremost to develop and implement strategies, tools, and methods to assess and quantify the nature, extent, severity, and impacts of dryland degradation on ecosystems, **watersheds**, and **river basins**, and carbon storage in drylands at a range of spatial and temporal **scales**. The project will also build national, regional, and global assessment capacities to enable the design and planning of interventions to mitigate land degradation and establish **sustainable land use** and land management practices.

See also: **Carbon Cycle; United Nations Environment Programme.**

Further Reading

United Nations Food and Agriculture Organization, Land and Water Development Division, Land Degradation Assessment in Drylands [Online, February 2005], The LADA Web Site, www.fao.org/ag/agl/agll/lada.

HELMUT GEIST

Land Evaluation.   An assessment of the present and potential suitability of land for different land uses (**agriculture, forestry**, or other) in order to promote specific land-use patterns. Land evaluation is a necessary and initial step in the process of **land-use planning**. This understanding of the **land-use system** is also called analysis or diagnosis. Since the 1950s, the evolution of the agricultural system approach and the development of **geographical information systems** (GIS) as well as of dynamic simulation models allowed a more extensive coverage of spatial and temporal variations in land typologies. Non-soil, economic considerations and **indigenous knowledge** are also progressively taken into account. For over thirty years, the **United Nations Food and Agriculture Organization** (FAO) has been applying a land-planning method that is quite largely used. It is not only a soil survey interpretation, but also an evaluation of climatic, landform, and socio-economic conditions. The FAO land evaluation framework is a semi-quantitative method that compares the performance of different land uses. The method involves a consultation of indigenous people to formulate land-utilization types, that is, the relevant land uses and their ecological requirements. The next steps of the method are compilation or cartographic design of a land-use database; matching each land-mapping unit in terms of a suitability assessment (taking into account the productivity potentials under different uses, the environmental and economic impacts, and a social analysis); and development planning. Since 1976, computerized databases and **modeling** programs have been interfaced with GIS and transform the qualitative matching in a quantitative and automated one, which is called agro-ecological zoning.

See also: **Integrated Assessment; Land Quality; Land Rehabilitation.**

Further Reading

Verheye, Willy, Parviz Koohafkan, and Freddy Nachtergaele, The FAO Guidelines for Land Evaluation, in *Encyclopedia of Life Support Systems*, developed under the auspices of the UNESCO, Eolss Publishers, Oxford, UK [Online, September 2004], The EOLSS Web Site, www.eolss.net.

NATHALIE STEPHENNE

Land Fragmentation.   Land tenure structures in developing rural economies are often bimodal, which means that most of the land is owned or retained by a few people, and large populations have only small plots of land to make their livings. In this context, land fragmentation of individual **landholdings** takes place mainly through two main mechanisms. First, it occurs when the landholding **property** is divided among different family members according to specific inheritance rules for **land rights** transference that varies among regions; and second, when a portion of the land is sold in the land market. The two processes mentioned above constitute forces contributing to land fragmentation.

The first trend is somewhat straightforward due to which land rights are inherited after the person who held the property rights passed away, or decides to transfer his/her land rights. In this case, intensity of land fragmentation depends on the density of family members with potential rights over such pieces of land, as well as the inheritance rules' specificities. The second trend responds to broader economic determinants. In this case, large-scale landholders tend to fragment their landholdings in a situation in which they may want to achieve a medium-size scale that could meet some economies of scale that would fit better their intended activity (e.g., semi-intensive beef cattle production, intensive milk production, grain production, etc.). In other cases, landowners divide their establishments in the absence of any other valuable assets to sell to recover from economic crisis, or to invest the obtained capital in non-agricultural activities whose profits are higher than those that originated in **agriculture**. These latter cases often take place in rural economies in which land rights are relatively secure and there are no legal constraints to transfer them.

Land fragmentation is not commonly seen in tropical frontier areas. It takes place only in the so-labeled consolidated frontiers in which land tenure rights are guaranteed legally. The latter is in part the outcome of the **land tenure** evolution in the agricultural frontiers by which, after the initial rush driven by last speculation interests, original landholders begin to value the occupied land, either by legal or illegal means, through more developed land markets. The evolution of **agricultural frontiers** leads to growing investment interest from outsiders to buy a piece of land, and this in turn tends to increase the land prices. Larger **investments** in the frontiers along with trends to landholdings fragmentation tend to accelerate the economic activities in the frontiers, which leads to higher **deforestation**. The latter, in the Latin American context, often is an outcome of the transition from large-scale extensive **cattle ranching** toward semi-intensive systems of cattle ranching.

Finally, there is a third dynamic not mentioned above of induced land fragmentation. This results from the implementation of redistributive agrarian reforms. Though in the past, administrative agrarian reforms were applied through expropriation of the land rights mechanism to medium- and large-scale landholdings, currently there are sponsored mechanisms—with strong involvement of the World Bank—toward negotiated agrarian reforms assisted by the land markets. The underlying notion is to pulverize large landholdings by allocating land property rights and helping small landholders to negotiate and buy the land from large landholders through land credits.

See also: Amazonia; Arena of Land Conflict; Biodiversity; Land Concentration; Land Reform; Pattern Metrics; South America.

PABLO PACHECO

Land Privatization. A shift in **property** from state or other collective entities toward private actors. This shift can occur by way of expanding the types of **land rights** accorded to land users, or by granting them **private property** (ownership) or a bundle of rights close to private property. Land privatization can also take other forms. Reducing the control rights held by the state or other collective entities can enhance the protection of individual land users from interference into their land rights. Privatization can also expand the duration of land rights or strengthen them by way of their recognition and enforcement through the state. Land privatization programs have covered croplands and land used for urban and industrial purposes around the world. Their scope has been limited for **forests** and **grasslands**. Privatization has often been motivated by the presumed effects of private property on the efficiency of land management. Other primary motivations have been political and social, such as the use of land titles as signifiers of political participation and social equity.

Africa has witnessed a vivid debate about the benefits of land privatization. The debate was first stirred up by land registration programs in the 1980s and is now being revived in the wake of heightened attention to **institutions** and governance in development. Until today, rural people across much of Africa held use rights to land that were overlapping and not registered with the state. African land rights have been perceived as constraints on agricultural performance and causes of unsustainable land management. In reaction, the national governments, with support by the World Bank and other international donors, set out in the 1980s to demarcate, register, and certify land held so far under customary law. Land titling was conceived as a panacea to overcome agricultural stagnation and rural **poverty**. Proponents of land registration expected it to create secure land rights by clarifying local property relations. Secure rights would motivate landowners to invest in agriculture and perform land **conservation** practices. Land titles would also serve as collateral for landowners to obtain bank credit. Furthermore, land registration would create the basis for the development of land markets to facilitate the efficient allocation of land, in particular **land concentration**, in the hands of those with **access** to capital and modern technology.

Land registration has not delivered the expected rise in agricultural production and improvement in rural living standards, however. First, its effects on the security of land rights have been limited because insecurity originates from many other sources than a lack of clarity. For example, land users may have insecure land rights because of abuse of power by local elites and encroachment by outsiders. Land registration may even decrease the security of land rights by adding another layer of social control legitimating claims on land. Second, the lack of agricultural **investment** and conservation efforts is much more often due to inappropriate technology than insufficient security. Even in the absence of registration, local people adjust **land tenure** to emerging conservation requirements, granting investors long-term land rights. Third, land registration often does not increase the use of credit because that is constrained by factors independent of land rights. Farmers are reluctant to use land as collateral, as they are afraid to lose their most important productive resource in case of repayment problems. Banks, in turn, are reluctant to accept land titles as collateral because banks often cannot foreclose on land for political reasons. In addition, credit supply to agriculture tends to be very limited, as investments in other sectors yield higher returns. Finally, land registration rarely contributes to the development of land markets, as customary land tenure already includes mechanisms for land transfers despite the absence of ownership, and recognizes land sales when **agriculture** becomes more commercialized. In addition, the effects of land markets on agricultural productivity are dubious, as smaller **landholdings** are often more productive than large ones. Overall, the debate indicates that the benefits of land privatization are specific to political and economic contexts. Its effects on agricultural production often do not outweigh the high costs of land registration programs and negative consequences on equity, as women and other marginalized groups become excluded from access to land. More broadly, experience with land registration demonstrates that property is not an exogenous variable to be defined by states and international donors. The effects of land privatization on actual property relationships depend on the broader social dynamics of resource use.

Land privatization came to the fore again when previously socialist countries around the world embarked on massive property reforms after 1989. Land privatization programs in post-socialist countries transfer land rights to private entities by way of distribution and restitution. Distribution refers to policies that grant agricultural workers or the rural population equitable shares in the available land. Restitution implies the return of land to the historical owners (or their heirs), typically using as a reference point the period between the **land reforms** conducted by the new socialist governments after World War II and agricultural collectivization. There is great variation in the land privatization programs instituted by post-socialist governments. This variation reflects the highly political nature of land legislation, as many kinds of political actors have sought to influence the new land laws. The actors have asserted different goals for land privatization, often emphasizing notions of historical justice, political participation, and social equity more than more-narrow objectives of land management.

Land privatization has effected a massive shift toward private land rights in **west-central**, **east-central**, and **southeast Europe**. Today, most agricultural and urban land is no longer under the management of state units or collective entities but under private management. At the same time, closer looks at actual property relationships reveal high levels of heterogeneity and complexity. Local implementation of land privatization has witnessed intense political negotiations among villagers, urban residents, agricultural managers, ethnic groups, and emerging entrepreneurs. The political struggles have resulted in distributions and kinds of land rights that are specific to local political economic conditions and are often quite different from national legislation. For example, agricultural associations and large tenants control land markets and agricultural production in some regions, limiting the property rights held by the new landowners in practice; or, landowners often do not comply with the obligations tied to their land rights, such as the requirements to cultivate the land and protect valuable **biodiversity**. In addition, serious constraints limit the exercise of land rights, such as the lack of machinery, agricultural credit, inputs, and competitive product markets. As for land use, land privatization is associated with radical changes in land use across central and eastern Europe. Agricultural output dropped drastically during the first half of the 1990s. Cropping structures have changed in response to changes in farm structures and agricultural markets. There is also some evidence suggesting a link between land privatization and overgrazing, the destruction of irrigation systems, **soil erosion**, **deforestation**, and **urban sprawl**.

Land privatization has taken a different form in East Asia and the former Soviet Union. The shift toward private property has not been as complete as in central and eastern Europe. States tend to retain formal ownership to land, granting use rights only to private entities. Individual property rights include very limited rights of alienation only. As for policy, the emphasis has been on distributing land to agricultural workers previously employed in state and collective farms. In many places, collective entities have retained control rights over land, exercising their authority by way of redistributions of land and restrictions on allowable land use. Land privatization has contributed to strong increases in agricultural output in **China** and Vietnam, in stark contrast to the general output decline observed in central and eastern Europe. Cropping structures in East Asia have changed from **rice** toward other crops, such as vegetable and fruit trees.

See also: Cadastre; Economic Restructuring; Exogenous Factor; Land-Use Policies; Public Policy; Transhumance.

Further Reading

Bruce, John W., and Shem E. Migot-Adholla, eds., *Searching for Land Tenure Security in Africa*, Dubuque, IA: Kendall/Hunt Publishing House, 1993; University of Wisconsin–Madison, Land Tenure Center [Online, October 2004], Land Tenure Center Web Site, www.wisc.edu/ltc/; Verdery, Katherine. *The Vanishing Hectare: Property and Value in Postsocialist Transylvania*, Ithaca, NY: Cornell University Press, 2003.

THOMAS SIKOR

Land Quality. The condition of land relative to the requirements of land use, that is, **agriculture**, **forestry**, **conservation**, and environmental management. Indicators of land quality (LQIs) are needed to raise awareness of agricultural and environmental issues, and as instruments for **monitoring** progress toward or away from **sustainable land use**. The World Bank has been leading an international coalition to develop a program on land quality that is probably the most advanced effort to address the integration of socioeconomic (land management) data with biophysical information in the definition and development of appropriate indicators, and how to scale and aggregate indicators from local to global **scales**. In this regard, one of the requirements for basic or core LQIs is that land quality is to be assessed for specific agro-ecological zones (AEZs) in a country, allowing for a further breakdown by specific types of land use and land management. This requires the development and application of spatially located and georeferenced temporal databases of variables that are still under construction.

See also: **Agrodiversity; Georeferencing; Global Terrestrial Observing System; Land Evaluation; Mediating Factor; Proximate Causes.**

Further Reading

Dumanski, Julian, and Christian Pieri, Land Quality Indicators (LQI): Monitoring and Evaluation, in *Encyclopedia of Life Support Systems*, developed under the auspices of UNESCO, Eolss Publishers, Oxford, UK [Online, October 2003], The EOLSS Web Site, www.eolss.net.

H E L M U T   G E I S T

Land Reform. A politically, socially, and economically desirable or urgent need to change ways of **land tenure** or its ownership, and/or the redistribution of agricultural and **forest** lands owned by landlords and other big owners among farmers and landless persons. In communist countries, land reforms (LRs) involved the forcible joining of farmers' land into big cooperatives or state estates. A less-frequent definition concerns such innovations as improvement of farms (by reallocations of plots via exchange, sale, or lease), large-scale operations, mechanization, and so on, in order to change the field patterns optimal for economic efficiency.

The history of land reforms began in antiquity and had some common features with historical development during feudalism, capitalism, and communism. LRs have been connected particularly to the transition from feudalism to capitalism in **Europe** and America for the last three centuries. The process climaxed in the period of bourgeois revolutions; first in the Netherlands and England in the seventeenth century, in France at the end of the eighteenth century, in almost the whole rest of Europe in the years 1848–1849, and in the **United States of America** during the Civil War in 1861–1865.

The major goals of LRs have been to solve social tensions in the rural areas and to decrease relative overpopulation through parceling of large **landholdings**, in order to at least partially satisfy the common greed of small farmers for land. In contrast, the so-called enclosures in England from the late fifteenth to the nineteenth centuries brought an expropriation of peasant land and common fields, and then conversion of small farmers into wage-earning labor forces in agriculture, or predominantly enabled them to move to the cities. This has led to a rise of very big estates enclosing big tracts of **pastures** or arable land.

Land reforms have been proclaimed usually by the state or by some interest groups, mostly under political pressure and with the aim to remove survivals of the feudal system, which became a brake in the development of capitalism in **agriculture**, realized through the payments as monetary fixed tributes by serfs—hereditary tenants to feudal noblemen, or state or church landlords; tithes to clergy; or manorial labor for big landowners (the amount of time for such labor differed according to social and economic situations in the country or territory). LRs have introduced payments based on real land **property** of all, that is, payment of land taxes to the state, or tributes for leased land in order to farm larger homesteads or estates.

Land reforms have greatly influenced land use and land cover. In most cases, they have brought about an increase of agricultural land (except permanent **grasslands**) and primarily arable land. Small farmers who gained permanent grasslands usually ploughed them up for better securing of their self-sufficiency or for the market. Former landless peasants (workers in big farms or in industry) became part-time farmers for self-sufficiency. This situation was typical until the 1950s, mostly in mountainous regions in **west-central**, **east-central**, and **southeast Europe**. This has led to a higher share of arable land in some of these regions, which, as a consequence of **urbanization** and the **technological-scientific revolution** in agriculture, resulted in afforestation again in the second half of the twentieth century. These processes have enforced the disappearing of small farms and their joining with bigger farms and ranches. Consequently, the permanent decrease in the number of farms and ranches triggered a concentration of landownership or land tenure. It lasted until the present as a general feature of agricultural development, not only in Europe and North America.

Capitalism has eliminated feudal survivals that tied peasants to land, for example, through the abolishment of serfdom in the Austrian monarchy in 1848, and of slavery in the United States in 1863; and made land goods on a free market and enabled its free inheriting. Already in 1841, the so-called "federal land fund" started selling federal land or free lands to farmers at very cheap prices on the basis of laws in the United States. The first "Homestead Act," from 1862, was one of the most important from those laws. A colonist could gain about 195 hectares (480 acres) for a mere $200. The fund was abolished to ensure nature **conservation** on about 24 percent of U.S. territory (minus Alaska) as late as 1935. Thus, it led to extreme use of agricultural land, such as in the **Great Plains** in the United States, as well as to considerable wind erosion there (the so-called Dust Bowl in the dry first half of the 1930s). Water erosion prevailed in humid regions, where the layer of mold has been reduced, for example, by nearly one half in Illinois. A process comparable to the Dust Bowl event took place in the Soviet Union after tilling up of prairies in the 1970s in former Soviet Central Asia (Kazakhstan).

The following ways and goals of land reforms give evidence of the application of various so-called "ways of development of capitalism in agriculture." The "American way" has specifically created economically efficient large farms oriented toward market production. After the French Revolution in 1789, the "French way" realized political and social goals by abolishing serfdom, when the land of clergy and emigrated nobility was expropriated by state and sold to farmers. This led to an increase of arable land area, and above all to the crumbling of agricultural land among small farms, which currently burdens the agricultural policy of the European Union (EU).

The so-called "Prussian way" of land reform occurred on the borders of the former Prussia in 1939, and also in the Austrian monarchy. After the revolution of 1848–1849, big estates of nobility were not parceled or expropriated but transferred to capitalistic ways of farming based on the use of free labor as waged workers and particularly using the elements of the **Agricultural Revolution** and the technological-scientific revolution in agriculture. The capital obtained by noblemen from deliberated farmers markedly helped them to realize such innovations. In order to become owners of land cultivated by farmers in serfdom, these farmers had to pay to their former landlords a part of the value of land, which they had only in hereditary tenure or lease before revolution. Land reform in **Russia** was similar to Prussian and Austrian reform; but serfdom was abolished later, in 1861. Only homesteads became hereditary property of farmers; the land they had in tenure became common land, and peasants had to pay redemption to village authorities to receive it in ownership. The state compensated about 80 percent of the allotment land-market value to landlords.

After the First World War, further land reforms aimed to economically strengthen small farmers toward large estates, accomplished in some succeeding states of the abolished Habsburg monarchy and in Romania, for example. The government redeemed a part of large-estate land and sold it to peasants who mostly tilled it. Similar land reforms were realized in many countries in North Africa and the Middle East, usually following independence or political revolution.

Land reforms in communist countries had a very specific character. The aims were particularly political and only then economic. Individual character of farming didn't satisfy collectivism. Large estates and small estates were expropriated without any compensation. Small peasants were forced by political and sometimes even police pressure to put their land into cooperatives, become its members, and farm collectively. Thus, large farms were created, the areas of which became a comparative advantage through the development of mechanization and modernization. They farmed big track-fields that brought about simplification of the rural landscape and **soil erosion**. Another desirable result of such reforms was the release of many peasants as new labor forces for the extensive growing industry.

The fall of communism has been followed by restitution of land property "back" into the state before so-called socialist collectivization. Former members of cooperatives or their descendants received back their land (if they asked for it). After capital transformation of most cooperatives, they usually farm the same area of land, since their size enables greater effectiveness of production. They still farm the land of "old-new" owners, paying them very small tribute. These owners are usually not interested in doing business in agriculture because their farms are very small; or because they live in the cities or have jobs in other sectors of the economy, and thus have lost their affinity for farming.

See also: **Agro-Industry; Cadastre; Common-Pool Resources; Industrialization; Land Privatization; Land Rent; Land Rights; Private Property; South Africa.**

Further Reading

Ginter, Donald E., *A Measure of Wealth: The English Land Tax in Historical Analysis*, Montreal, Kingston, JM, London: McGill-Queen's University Press and Hambledon Press, 1992; Stanis, Vladimir E., *Socialist Transformation of Agriculture: Theory and Practice*, Moscow: Progress Publishers, 1976; Tuma, Elias H., *Twenty-Six Centuries of Agrarian*

Reform: A Comparative Analysis, Berkeley, CA: University of California Press, 1965; Turnock, David, ed., *Privatisation in Rural Eastern Europe: The Process of Restitution and Restructuring,* Cheltenham, UK, Northampton, MA, USA: Edward Elgar Publishing, Inc., 1998.

LEOŠ JELEČEK

Land Rehabilitation. Amelioration of land, by engineering or biological restoration measures, that has been destroyed by mining or quarrying, land-fills, occupation, and pollution in the course of production and construction; or demolished by natural disasters such as flood, coastal invasion, mud-flow, and sandstorm to a state where land can be used for farming or other purposes. It is often executed after good land has been demolished through industrial activities like mining, tile making, or coal-based electricity generation. In a much broader sense, land rehabilitation also includes readjusting and rearranging land-use structures by engineering or biological measures to increase the area of arable land; improve the quality and productivity of land; and finally amend the general conditions of production, living, and the environment. Land rehabilitation needs to follow the principles of pursuing integrated benefits in economy, society, and ecology and boosting harmony among socio-economic development, rational resources use, and **conservation**. Its operation should be in concordance with **land-use planning** and rely upon the advancement of technical methods. Moreover, the operation should also take into consideration market mechanisms. The main characteristics of land rehabilitation can be summarized as large **investment** and long-term return. Therefore, land rehabilitation is a large and complex project involving many factors of economy, society, and technology. Above all, it is an important way to complement farmland or offer construction land for countries with large populations and limited land resources, such as **China** with its limited cultivated land resources. Land rehabilitation is an effective means to keep the balance between population and land resources through resolving the conflict between them. In China, land rehabilitation intends to increase the area of farmland, to improve the quality of farmland, to optimize structures of land use, to ameliorate the conditions of production and living in rural regions, to protect the environment, and to likewise promote **agricultural intensification**. Land rehabilitation includes four courses of action. First, level off land, consolidate scattered plots, build terraces or fix breeding surfaces, and merge villages by engineering or biological measures. Second, build agricultural engineering works such as roads, wells, canals, slopes management, and forest shelter. Third, control sandy land, **salinization** and polluted land, meliorate soils, and renew **vegetation**. Fourth, ensure secure **land rights** as well as proper land use and make regular land-change investigations and registrations. In China, the policy of land rehabilitation follows the rule that who destroys land has to find answers for land rehabilitation. In the process of practice, land rehabilitation planning is scientifically worked out and functions as the head of all operations. Land and resources bodies are in charge of confirming the land-use purpose after rehabilitation. In 2001, a project about land consolidation and rehabilitation and **sustainable land use**, supported by the **United Nations Development Programme**

(UNDP), was implemented in China. The main items of this project include policy research, ability improvement, demonstration works, and information systems construction.

See also: **Arena of Land Conflict; Forestry; Green Wall Project; Land Evaluation; Land Reform; Reforestation; Technological-Scientific Revolution.**

QI LU

Land Rent. Payment for a plot of land. The price for a plot of agricultural or urban land derives from its real economic output and the average rate of deposit interest, with usually no land transfers happening if interest rates are higher than economic output. Land rent is a basic economic category that expresses and influences land-use/cover change, valid for both capitalist and communist systems. Differences in land characteristics, such as location and productivity, and/or differences in social power around the ownership of land, usually determine the precise level of rent on any specific tract of land, thus producing what is called differential (land) rent.

Two forms of differential rent can be distinguished. The extensive differential rent, or differential rent I (DR I), represents an extra profit that arises because all plots of land have equal amounts of capital **investment** but are different in terms of natural fertility and in their position to the market of agricultural products. DR I relates to the extensive growth of **agriculture** based on the extension of cultivated, especially arable land. Though losing its basic impact and importance, historically, advancing **industrialization** and **urbanization** created new location conditions, including growing demand for food. The intensive differential rent, or differential rent II (DR II), represents an extra profit that is created by unequal capital investments into plots of land with same degrees of natural fertility. Thus, DR II relates to **agricultural intensification**, that is, more effective use of capital (artificial fertilizer, machinery, genetic engineering) creating more fertile land, but also to **land abandonment** in areas with low prospects for intensification, such as those at high or remote locations. In both cases, the differential rent is determined by the difference in cost between any productive plot of land and the so-called marginal plot. The marginal plot is land so unproductive that the cost, for example, of producing a crop there is equal to its market price, thereby leaving no room for rent. There is an equivalent to the land rent model developed by **Johann Heinrich von Thünen**. While calculated in a similar manner, differential rent is interpreted differently in neoclassical and Marxist land economics. In the neoclassical interpretation of Thünen, for example, it is portrayed as an index of natural scarcity whether of locations or differentially fertile lands, while in the Marxist account it is an expression of the monopoly power of capital as a whole.

See also: **Agricultural Revolution; Land Reform; Land Rights; Technological-Scientific Revolution.**

Further Reading

Marx, Charles, *Capital: A Critique of Political Economy*, vol. III/2, Section IV, New York: International Publishers, 1967; Thünen, Johann Heinrich von, *Von Thünen's*

Isolated State: An English Translation of Der Isolierte Staat, Oxford, New York: Pergamon, 1966.

LEOŠ JELEČEK

Land Rights. Enforceable claims on land. The holders of land rights are able to make decisions about land (legal rights) and receive a share of the benefits derived from land (economic rights). Land rights and obligations together constitute **land tenure**, or **property** in land. Land rights involve complex social relationships among actors, including many different types. The rights pertaining to one parcel of land are typically many and may not be bundled in the hands of one single actor, instead belonging to several different actors. As land rights are embedded in broader social relations, they display a tremendous diversity around the world. Simple classifications of land rights as private, state, or communal do not capture the diversity. This diversity also makes clear statements about the relations between land rights and land use impossible. Land rights and land use are generally thought to influence each other. The nature of their relationship is highly dependent on contextual factors, however.

Like property rights in general, land rights are of many types. A simple distinction is between use rights (usufruct), such as rights of cultivation, passage, grazing, and collection; and control rights, such as the rights to exclude other claims, transfer land, make major alterations, and sell land. Land rights are often specific to particular resources associated with land. For example, the right to cultivate land does not automatically imply a right to the trees standing on the land. Another distinction is between private ownership (or freehold) and leasehold. Land ownership, or **private property** in land, accords landholders far-ranging rights to land, including rights of alienation. In contrast, the rights pertaining to land held under leasehold are more restricted and are often tied to more extensive obligations. Common forms of leasehold are land rentals, under which tenants pay landlords a fixed rent; and sharecropping, under which tenants pay landlords a fixed share of the harvest. Nevertheless, there are many mixed forms combining elements of ownership and leasehold. For example, **contract farmers** typically retain ownership rights to land, yet their contracts with agribusiness severely limit the exercise of the rights.

Land rights can be further distinguished by the social actors (legal entity) holding them. Land rights are public if they belong to organized political collectivities at the international, national, or local level. In most cases, states exercise public land rights on behalf of the political entities they represent. Land rights are private if they belong to natural or legal persons, such as individuals, families, corporations, and partnerships. Communal land rights are more complex. They combine control rights held by an organized collectivity with individual use rights allocated to their members. The members' rights are use rights, possibly including inheritance rights but not a right to sell the land. The degree of collective control may vary among different types of land. For example, communities may allocate arable land to individual members but keep **grasslands** open to all members for joint use. Again, there are also many mixed forms of property combining public, private, and communal elements.

Land rights differ from each other by the nature of authority backing up and enforcing claims. Land rights specified in state laws and regulations are called statutory. In contrast, customary land rights derive their legitimacy from social conventions and norms. In many settings, statutory and customary laws coexist side by side, emanating from states, religious bodies, international agreements, local customs, etc. Common is the juxtaposition of statutory rights based on European systems of law and customary rights, having their origin in attempts by colonial regimes to maintain indigenous legal systems. Land rights in such situations of legal pluralism are especially complex, as the rights legitimated by different legal systems rarely match. The rights are often hybrids, deriving their legitimacy from multiple legal orderings. Examples of such hybrid land rights are *Adat* laws in Indonesia and Indian reservations in the **United States of America**.

Land rights are key elements of social organization. As with property, land rights find attention in the social and cultural norms of an organized social collectivity, are the subject of laws and regulations, take their expression in actual relationships, and resonate with social practices. Land rights are embedded in broader social dynamics and serve the maintenance of social control. For example, land rights used to play an important role in the social organization of the Lozi in southern Africa, as described by Max Gluckman in his seminal work on property. Land rights combined use rights (estates of production) belonging to individual Lozi with control rights (estates of administration) held by the king, village headmen, and household heads. Rights to land consisted of a series of allocations from the king to the ultimate user, entitling every involved actor to a share in the benefits derived from land. The relation of each actor to property was a function of their sociopolitical status and depended on the fulfillment of associated obligations. In colonial India, land rights were a key element in the strategies of local power holders to maintain social control. The Indian estate holders used their control over land to expand their control over people, hence political power and wealth. To them, control over land generated economic benefits and political gains. These two examples illustrate the general observation that land rights have many functions. They influence not only the efficiency of resource use but also the patterns of political participation, processes of social continuity, and measures of social protection, among others. Land rights also include nested layers of social control in many instances, granting several units of social aggregation overlapping control rights.

The embeddedness in broader social relations implies that gender relations and land rights influence each other. In many countries, land rights belong to households. Intrahousehold relations, as shaped by gender norms and relations, shape the distribution of discretionary powers and obligations among household members. Male household heads often concentrate significant **decision-making** powers, while female household members assume a large share of labor duties. Development projects have often contributed to the gendered division of discretionary powers and labor duties in the name of modernization. For example, male farmers in Gambia have developed fruit orchards with the assistance of development projects, out-crowding females' vegetable gardens previously cultivated on the same land. Also in Gambia, the spread of commodity production in the uplands has undermined women's ability to market part of lowland **rice** production, as households consume an increasing share of rice outputs. In addition, small-scale pump-irrigated rice projects have enabled men to establish individual control over new land and draw on female labor obligations to cultivate those. Another influence of gendered notions of property is that women often gain and lose land rights through marriage, as households pass on land rights to male offspring only. Sons are considered direct successors in the family line, in contrast to female offspring, who are thought to leave the households upon marriage. These insights demonstrate that actual land rights are often quite different from those written into land laws and regulations, a central insight in the study of land rights and, more generally, property.

Land rights in advanced industrialized societies are characterized by a stark discrepancy between cultural norms on the one hand and laws and regulations on the other. Notions of private property take a prominent role at the level of social and cultural norms. Landholders are thought to enjoy private ownership of land, including extensive use and control rights, especially rights of alienation. States are accorded minimal administrative rights, such as the authority to issue title, tax land, regulate the type of use, monitor actual use, regulate transfers, and arbitrate disputes. Land rights are thought to be private and exclusive, in the sense that all rights pertaining to a particular parcel of land are bundled in the hands of one private actor. The reality is different. Land rights in the European Union are highly regulated by common European law, national laws and regulations, and local conventions. For example, codes of good agricultural practice and standards on the level of pollutants in water and air limit the range of land management practices available to farmers. British laws grant the public a right of trespassing agricultural land and decide about the **conversion** of agricultural land to other uses, including the associated financial benefits. The sale of farmland in Germany is subject to numerous written and unwritten rules, for example, requiring the seller to offer the land to fellow villagers before approaching outsiders. In the United States, 42 percent of all land is public, including federal, state, and local government holdings. On private land, the public holds the rights of eminent domain, taxation, zoning, and regulating land use. In addition to private and public lands, there is a large diversity of mixed property forms, such as common property estates, Indian estates, and public utility estates.

Notions of private, exclusive landownership have also informed the design of land legislation in post-socialist societies. **Land privatization** programs have restituted and distributed agricultural land to private actors in an unprecedented effort of land reform. Statistical data on **west-central**, **east-central**, and **southeast Europe** show a radical move from state and collective to private ownership. Nevertheless, ethnographic studies report persisting differences between actual property relations and the private, exclusive land rights promoted in land laws. Katherine Verdery has coined the term "fuzziness" to describe the ambiguity of land rights in post-socialist Europe. This fuzziness originates from multiple sources: unresolved conflicts over the ownership of single plots; overlapping claims to plots, resulting from legally pluralistic settings; contextual constraints on the exercise of the newly accorded land rights; the blurring of the boundary between private and public, related to private actors' efforts to push debts and liabilities onto the public shoulder; different meanings attributed to land, land rights being often seen as signifiers of citizenship, personhood, and social justice; different visions informing claims on land; and the absence of generally recognized rules and patterns of behavior around private property in the context of post-socialism. Fuzziness is also a key characteristic of emerging land rights in central and East Asia. The difference to post-socialist Europe is that land laws in central and East Asia have not fully adopted the notions of exclusive, private land rights. Many states grant landholders long-term usufruct rights to agricultural land only, excluding the right of permanent alienation. Land remains under the formal ownership of "the people." Intermediate layers of authority retain significant degree of control over land, allowing local and state authorities to maintain strong leverage on land use by designating land-use regulations, and allowing local communities to re-allocate use rights to agricultural land every few years. This is, in part, a legacy of socialist land rights, which granted control rights in land to multiple layers of the state.

Ambiguity and conflict have also been central themes in research on land rights in Africa. Land rights are the focus of complex material and symbolic struggles, resulting in overlapping and highly variable rights. Underlying the struggles is the overlap of customary law and European systems of property law introduced by colonial regimes. Customary law has, with some modifications, survived into the postcolonial period. Customary legal systems typically recognize several intermediate layers of authority between the ultimate sovereign (the state nowadays) and land users, resting with customary chiefs, village headmen, and family heads. The intermediate layers of authority retain residual rights on land and connect the right to land with obligations for land users, such as political allegiance, labor services, and product shares. At the same time, African land rights are rife with contestations over single resources and the specific rights associated with them. Just as in post-socialist societies, material and symbolic struggles blur the boundary between private and public, are as much about the meaning attributed to resources as they are about the resources themselves, and are informed by conflicting visions of a desirable future. States' views of rural resources as commodities contributing to foreign currency earnings are in stark contrast to local people's visions and the meanings they attribute to land.

Land rights have been the targets of repeated attempts at **land reform**. They have taken the form of redistribution, seeking a change in agrarian structure either through state-mandated land redistributions or market-assisted allocation. Redistributive land reforms have sought to achieve different goals. Some have promoted agricultural commercialization and accumulation through a landed elite, commercial farmers, or peasants. Others attempted to create shifts in the dominant rural class from capitalist landed elites to smallholders or peasants. Post-socialist redistributions have moved control over land from state and collective to private actors, being a form of land privatization. Another type of land reform does not seek to change the agrarian structure but wants to modify land rights. This type of tenurial reform has aimed at the abolition of absentee landlordism, eradication of sharecropping, replacement of sharecropping with fixed money rent, regulation of that rent, and land consolidation to eliminate fragmented holdings. Sharecropping has been a primary target of reform because of its perceived negative effects on the efficiency of land use and long-term **investments**. In Africa, land reform has often taken the form of land registration. In the 1980s, land registration programs sought to eliminate the perceived lack of secure land rights through formal titling. The outcomes of the programs in terms of their effects on agricultural performance have been largely disappointing, however. Moreover, the focus on private, exclusive land rights has weakened the land rights belonging to some actors, such as women. Overall, the outcomes of land reforms have mostly failed their goals. Legal and regulatory reforms have exerted only limited effects on actual land rights. Even where landholders have expanded their legal rights on land, they have often not been able to translate those into substantive economic benefits in the absence of accompanying measures such as credit programs.

Land rights and land use mutually influence each other, although it is difficult to come up with generalizable statements about the nature of their relationship. Programs strengthening private property rights have been shown to encourage agricultural investments and more efficient land use in some cases. The absence of enforceable land rights is often a primary driver of **deforestation**. Nationalization of land rights, in turn, has affected the management of **forests** in positive and negative ways. Current devolution programs in **forestry** and grassland management assume that assigning land rights to local communities will improve local land use. Overall, the effects of land rights on land use are highly varied and context-dependent. Vice versa, the influences of land use on the nature of land rights are equally complex. The emergence of private land rights has been hypothesized to result from increasing land scarcity due to rising population density, technical change, and expansion of trade. At the same time, there are many examples where land scarcity has not led to private land rights. Similarly, the use of **common-pool resources** such as grasslands and forests is thought to be prone to communal land rights. The world's grasslands and forests are under a large diversity of land rights, however. These insights indicate that land rights are complex social relationships, defying simple relations with land use. Land rights carry not only economic values but also social meanings, and they are embedded in broader social relations.

See also: **Access; Agro-Industry; Arena of Land Conflict; Cadastre; Community Involvement; Cultural Factors; Driving Forces; Institutions; Land Rent; Land Tenure; Mediating Factor; Transhumance.**

Further Reading

Agarwal, Bina, *A Field of One's Own: Gender and Land Rights in South Asia*, Cambridge, UK: Cambridge University Press, 1994; Bassett, Thomas, and Donald E. Crummey, eds., *Land in African Agrarian Systems*, Madison, WI: University of Wisconsin Press, 1993; Geisler, Charles, and Gail Danecker, eds., *Property and Values: Alternatives to Public and Private Ownership*, Washington, D.C.: Island Press, 2000; Gluckman, Max, *The Ideas in Barotse Jurisprudence*. Manchester, UK: Manchester University Press, 1972; Sturgeon,

Janet, and Thomas Sikor, eds., "Postsocialist Property Relations: Ethnographies of Agrarian Change in Eastern Europe and East Asia," *Conservation and Society* 2 (2004), special issue; University of Wisconsin–Madison, Land Tenure Center [Online, October 2004], Land Tenure Center Web Site, www.wisc.edu/ltc/; Verdery, Katherine, *The Vanishing Hectare: Property and Value in Postsocialist Transylvania*, Ithaca, NY: Cornell University Press, 2003.

THOMAS SIKOR

Land Stability. The scale-dependent notion of "no change" in land-cover attributes, which often contradicts prevailing scientific views or underlying assumptions, especially neo-Malthusian views of land change in **degradation narratives**. Areas of "no change" show up as diagonal elements in table formats derived from the analysis of land-cover images in post-classification "from-to" **change detection**. The Tri-Academy Panel on Population and Land Use provides examples that are actually case studies from the world's most populous countries, that is, India (Kerala, Haryana), **China** (Jitai basin region, Perl River delta), and the **United States of America** (south Florida, Chicago). Stability in land use and land cover is referred to in two respects: first, contrary to common **perceptions**, forest areas seem to be stable (or even increase) in study regions that have high population density; and second, total agricultural land in most study regions did not undergo major changes in the 1970–1995 period, even though the populations in these regions did grow significantly. Local-scale studies confirm these trends, but also stress the scale dependency of observations. For example, **agricultural intensification** and growing agricultural workforces are dominant processes associated with land change in Southeast Asia during the recent decades, but areas of **disintensification** as well as areas of production stability can be found as well. The process of land stability has involved swidden cultivators in the highlands but many more lowlanders. Summarizing results from eight sites where changes in land cover in Southeast Asian upland areas over the last fifty years were documented, Jefferson Fox and John Vogler found that, at the level of secondary forest **regrowth** after swidden cultivation, land cover remained stable across the region throughout this fifty-year period. However, the level of abstraction hid major changes that occurred within the sub-categories of **secondary vegetation** as well as in the size and number of land-cover fragments (actually, intensification of **swidden cultivation** and introduction of **cash crops**). In the lowlands, few farmers in Thailand, for example, rely solely on **agriculture** to meet their needs, and rural households are dividing their time between farm and non-farm activities instead. Thus, they construct **economic livelihoods** that are increasingly hybrid, both spatially and sectorally, which might explain the observation of land stability there. Areas of stable **forest** that are found in landscapes susceptible to **deforestation** pressures are called **forest anomalies**. Since these anomalies are located in regions where roads and/or markets primarily drive forest-cover changes, Charles Schweik and colleagues assume that they indicate the likely locations of effective forest management **institutions** that are working to mitigate the overall trend toward deforestation in these regions. Several studies in the Indian central and western **Himalayas**, and in Nepal (where the worst crisis conditions were predicted), have demonstrated that there has actually been a net regrowth of forests in several regions in the hills, following local responses of forest communities to degradation; and some extremely stable forest boundaries even maintained by local communities for hundreds of years, such as in the Kumaon region in India. Obviously, the identification of land stability can change with varying **scales** of analysis. On the other hand, it cannot be excluded that an understanding—especially by ecologists—of changing land-cover attributes due to human activities might be based on inappropriate models that emphasize equilibrium and homogeneity rather than heterogeneity and **disequilibrium dynamics**. For example, the prevailing scientific view during the colonial era was that once ecosystems reached a climax state, they maintained stability. Stability was viewed as the desired condition, while disturbance was viewed as unnatural and damaging. In this regard, indigenous land management systems that involved **slash-and-burn agriculture** in humid forest ecosystems or intensive grazing of livestock by pastoralists in drylands (both practices potentially causing periodic ecological disturbances over wide areas) were often misperceived as degrading and often outlawed.

Further Reading

Fox, Jefferson, and John Vogler, "Land-Use and Land-Cover Change in Montane Mainland Southeast Asia," *Environmental Management* 36 (2005): 394–403; Schweik, Charles M., Nagendra, Harini, and Deb Ranjan Sinha, "Using Satellite Imagery to Locate Innovative Forest Management Practices in Nepal," *Ambio* 32 (2003): 312–9; Tri-Academy Panel, "Population and Land Use in India, China, and the United States: Context, Observations, and Findings," in *Growing Populations, Changing Landscapes: Studies from India, China, and the United States*, eds. Indian National Science Academy, Chinese Academy of Sciences, and U.S. National Academy of Sciences, Washington, D.C.: National Academy Press, 2001, pp. 9–72.

HELMUT GEIST

Land Tenure.   The mode by which land is held or owned. It may vary from place to place within the same locality, or between an urban area and its surroundings (hence, the distinction between "urban" and "rural" tenure). The word "tenure" the (from the French *tenir*, meaning "to hold") has come to mean primarily holding of rights to own, occupy, use, control, and transfer land. The meaning of ownership is emphasized in the Anglo-American legalist approach to tenure studies, which deems landed rights as "**property**" vested in an individual or group and applicable separately to land or development on it. This approach places the focus on rights, not people or land, hence downplaying the role of the stakeholder, geographies, and technologies in shaping tenurial arrangements.

Several complications merit brief explication. First, land tenure is a multi-dimensional concept. It comprises the complete "bundle" of landed rights made up of many separate "sticks" that are often assigned to a multitude of landholders (quantity dimension). It can be classified into general tenure types, the "public," "communal" (or **common-pool resources**), and "private" being the primary division (quality dimension). These dimensions may be used separately or together to identify principal tenure forms: for example, ownership and tenancy are two broad classes defined in terms of quantity of interest vested in land, while the latter may be further classified into freehold and leasehold types according to duration of such interests. The spatial and hierarchical arrangements of tenure add further complications: for example, not only may several forms of tenure co-exist on the same plot, but tenure boundaries may not be coterminous with operating units. Remnants of the feudal tenure system may provide for a hierarchy of successive groups according to rank from the landowner to the farm worker. Second, it might be more appropriate to think of tenure status as a continuum (e.g., using quantity/quality of vested interests in land as a common denominator), rather than in terms of mutually exclusive classes. The continuum may run from occupancy with no legal interest (as represented by squatters) to unqualified public interest in land control (as manifest in the rights of public domain, policing, taxation, and spending vested in the U.S. government), while "sharecropping" and "hired man" may fall in between. Third, it is necessary to distinguish *de jure* from *de facto* (and formal from informal) tenure status with regard to tenure security. Although tenure security is widely accepted as a precondition for land stakeholders to invest in any improvement, occupancy alone sometimes provides security of tenure, a fact well known to squatters. Notably, different legally acceptable systems may operate in adjacent locations, so that legal tenants migrating from one location may be considered to be behaving illegally in their new locations.

We may speak of different land tenure systems (the "statutory," "customary," "Islamic," etc.) in terms of their origins, basic principles, and structures. Most actions that affect the tenure systems occur externally, where breaks with the past were first developed in practice and later formalized in statutory enactments. They often embody divergent principles: for example, many customary tenure systems regard land as sacred, and man's role as one of stewardship to protect the interests of future generations, where land is distributed by community leaders according to needs, rather than through payment. In the **United States of America**, on another hand, the overarching tenure principle has long been freedom of action without personal responsibility, where the economist's free market theory and the policymaker's practice of *laissez faire* complement the legalists' championing of freedom of contract (as seen in the "fee simple" estate). In terms of tenure structure, socialist countries vest most landed rights in the state, whereas capitalist countries restrict public ownership to a narrow range. In a larger picture, the manner by which a society distributes and assigns rights in land is an important indicator of that society, since they can be held to reflect rights in other areas of social life. Therefore, **land reform** essentially means tenure reform.

Land tenure is critical to understanding land-use/cover change processes, for it provides for an orderly way of land use, a means of conflict prevention in land transactions, and more importantly, a frame for land-use **decision-making** (if we deem it a key component in the "structure-agency" formulation with regard to structure and **institutions**). It helps establish the size of feasible farming operations, and influences urban land value and options for **access** by the urban poor. However, tenure studies remain underdeveloped in land-use/cover-change research.

See also: **Agricultural Revolution; Arena of Land Conflict; Cadastre; Human Immuno-deficiency Virus (HIV)/Acquired Immunodeficiency Syndrome (AIDS); Investments; Land Privatization; Land Rights; Mineral Extraction; Population Dynamics; Private Property; Transhumance.**

KE CHEN

Land Use Policy.   Launched in 1984, this interdisciplinary journal is currently produced in four issues per year by Elsevier Science Ltd. It publishes reports of original research and critical reviews relating to any aspect of land-use policy, broadly defined. Submissions are subjected to independent external review before acceptance for publication. Special thematic issues are published from time to time. Book reviews are also carried. Papers on urban and transport-related issues are accepted, but the majority of articles relate to rural themes. Typical topics include tropical **deforestation**, the European Union's Common Agricultural Policy, and matters of **land-use planning** in the **urban-rural fringe**. A key feature is an emphasis on implications for land-use policy. The journal is international in its readership and author origins. It seeks to provide a stimulating forum for researchers and practitioners involved in the broad field of land-use policy.

See also: **Land-Use Policies.**

Further Reading
Elsevier Science [Online, December 2004], The Land Use Policy Journal Web Site, www.elsevier.com.

ALEXANDER S. MATHER

LANDFIRE Project.   An initiative started in 2003 to provide detailed, consistent, and comprehensive landscape-scale wildfire-related vegetation and other ecological variables for the **United States of America**. LANDFIRE data sets support **fire** management planning, prioritization of fuel treatments, collaboration, community and firefighter protection, and effective resource allocation. LANDFIRE includes detailed natural vegetation characteristics maps that correspond to fuels. Mapping fire fuels requires data and information about **vegetation** types, structure, and green **biomass**. Combined with extensive field-based vegetation descriptions, **Landsat** 7 ETM+ imagery are being used as the primary source of the vegetation characteristics data, a general land-cover classification, a **forest** -type group classification, and a tree canopy density data layer. The LANDFIRE initiative is a collaborative effort among several federal agencies, including the U.S. Forest Service, U.S. Geological Survey, and the Department of the Interior.

See also: **United States Geological Survey (USGS) Program.**

Further Reading
United States Geological Survey, Landfire [Online, November 2004], The LANDFIRE Web Site, www.landfire.gov.

THOMAS R. LOVELAND

Landholding. The economic attributes of rural producers are strongly associated with the landholding size, but also to the way in which the production process is organized, and the levels of economic or **agricultural intensification**. In this light, looking at the size of landholdings provides the main distinction among rural agricultural producers that differentiates small- and large-scale farmers, or some other types. Hence, landholding size by itself is a powerful variable that allows for separating the attributes of two quite different groups of rural producers. It is well known that land-use decisions made at the farm level, and their implications on forest **conversion** to agricultural uses, tend to vary greatly in small- versus large-scale landholdings. Landholding size, therefore, can be seen as a **mediating factor** that contributes to explain **deforestation** in the tropical landscapes.

Land-use change processes and patterns are different for small- versus large-scale landholders, though their actions in some cases are interlinked. In the tropical regions, deforestation taking place in large-scale landholding is often done through clearing of larger chunks of **forests**, taking the shape of large continuous and symmetric patches. The trajectories of forest conversion are relatively simple, since forest is in most of the cases definitely converted to agricultural uses, mainly to **pasture**, and it is alternatively devoted to grains production, though some proportion may be abandoned.

Land-use change trajectories taking place in small-scale landholdings, and their imprints on the landscape, are both quite more complex. There is no single dominant land-use trajectory in small-sized lots. Forest cleared for **agriculture** may remain for a couple of years under temporal crops, or it may be converted to pasture after the first year to avoid forest succession. In other cases, forest **regrowth** is allowed once the soil fertility is exhausted with temporal crops. Some other smallholders favor permanent crops intermingled with temporal crops and forest succession areas. The diversity of situations is due to the fact that small-scale farmers often choose diversified systems of production as a way to secure their livelihoods, which leads to diversified land uses whose final imprints on the landscape will consist of multiple and dispersed small, asymmetric patches of cleared areas in combination with forest regrowth areas in different stages of succession.

Large-scale landholders' main motivation is to achieve the largest economic benefits. They are often better connected to markets, use industrial inputs, and hire labor forces. Hence, in these groups of producers, land-use change decisions, and particularly deforestation, are contingent upon the profit they will obtain from expanding arable land for commercial crops, or pasture for beef cattle production. The medium- and large-scale agriculturalists or ranchers, therefore, will deforest more in cases in which the prices of commodities are higher than their production costs. The latter is likely to happen when they benefit from prices of support, subsidized and cheap credit, preferential export taxes, favorable exchange rates for export commodities, and low prices of inputs. The medium- and large-scale establishments have in diverse contexts benefited from policies that have for different reasons tended to provide them additional institutional rents (cheap credit, access to roads, etc.).

Small-scale farmers have lower **access** to assets. These types of farmers are highly diversified because they may have a high discount rate, specializing in low-return and low-risk activities, and combining subsistence with market behaviors. While wealthier households may prefer to make **investments** in higher-return activities, such as expanding their **cash crops**, the poorest may choose to spend their economic assets in producing subsistence crops. Their difference with respect to large-scale landholders is that while the latter are able to mobilize resources to expand their scale of operations or to intensify their productive system, smallholders are constrained by their **perception** of risk, higher cost of access to financial resources, and scarce factors of production to allocate them among competing on-farm activities. Furthermore, policies have negatively biased small farmers, and limited them to expand their access to assets.

In short, the differences in the holding of land and other assets can largely influence local natural resources use by affecting landholders' resource-allocation decisions. Large-scale landholders with better access to assets are less constrained by financial resources, are better suited to interact in the markets, and adopt new technologies more readily, but this often leads to more simple land-use decisions that favor permanent forest conversion to agricultural uses. In contrast, smallholder farmers are more constrained in their access to factors of production and financial markets, and food production is at the center of their economic strategies. These factors lead smallholders to adopt more complex land-use strategies, resulting in less-homogenous landscapes in which temporal and permanent crop areas coexist intermingled with secondary forest.

See also: **Agricultural Frontier; Agrodiversity; Corporate Strategies; Decision-Making; Economic Livelihood; Food Crops; Land Abandonment; Land Concentration; Land Fragmentation; Land Privatization; Land Reform; Land Rent; Land-Use Policies; Property; Public Policy; Subsistence Agriculture.**

PABLO PACHECO

Land-Ocean Interactions in the Coastal Zone (LOICZ) Project.

A joint core project of the **International Geosphere-Biosphere Programme** and the **International Human Dimensions Programme on Global Environmental Change** since 2004. The project supports research aimed at informing scientists, policymakers, managers, and stakeholders on the relevance of **global environmental change** in the **coastal zone**.

See also: **Water-Land Interlinkages.**

Further Reading
Land-Ocean Interactions in the Coastal Zone [Online, February 2005], The LOICZ Web Site, www.loicz.org.

HELMUT GEIST

Landsat. A series of satellites designed to gather data on the earth's resources, that is, to characterize, monitor, manage, explore, and observe the earth's land surface. Landsat is an abbreviation for Landsat thematic mapper. It provides a widely used type of long-term satellite imagery—along with the **advanced very high resolution radiometer** as a second source of remote sensing information—covering the majority of the earth's surface at thirty-meter resolution. The system records the intensity of radiation reflected from the earth's atmosphere-surface system in five to seven spectral bands in the visible, near-infrared, and thermal infrared parts of the spectrum (with the information used to extract surface images). A Landsat scene is approximately 185 kilometers on a side, and each location on earth is revisited at about sixteen-to-eighteen-day intervals. The Landsat data record started in 1972, when Landsat-1 was developed and launched by the U.S. National Aeronautics and Space Administration, or NASA. The multispectral scanner on-board Landsat-1 was so valuable that a version of it was flown on each of the first five missions. Since 1982, when Landsat-4 was launched, the primary instrument has been the thematic mapper. Landsat-6 failed to reach orbit in 1993, but Landsat-5 continued to provide coverage well beyond its design lifetime. In 1999, Landsat-7 was launched to provide continuity and enhanced coverage. The Landsat program passed the Land Remote Sensing Policy Act and was then approved by Congress in 1992. The act allows a continuous collection and utilization of terrestrial **remote sensing** data from space in order to study and understand human impacts on the global environment and to better manage the earth's natural resources.

See also: **Change Detection; Earth Resources Observation Systems (EROS) Data Center; Global Terrestrial Observing System; LANDFIRE Project; NASA Land-Cover/Land-Use Change (LCLUC) Program; Normalized Difference Vegetation Index; Pixel; Sahel Land Cover; United States Geological Survey (USGS) Program.**

Further Reading
National Aeronautics and Space Administration, Landsat 7 [Online, March 2005], The NASA Web Site, landsat.gsfc.nasa.gov.

HELMUT GEIST

LandSHIFT Model. A global land-use change model currently under development at the Centre for Environmental Systems Research in Kassel, Germany. The abbreviation means "Land Simulation to Harmonize and Integrate Freshwater Availability and the Terrestrial Environment." The principal goal of the model is to translate regional demand trends for land-intensive commodities (such as field crops) into area requirements for land-use types, and to allocate these on a grid of five arc minutes resolution. Allocation assumes an idealized behavior of land users in that they seek for optimal conditions, constrained by the competition among different land-use types and regional socio-economic settings. The procedure accounts for intrinsic **feedbacks** by considering the consequences of land use and land-use changes on **land quality**, land scarcity, and water availability. LandSHIFT is expected to provide insights into the dynamic behavior of regional **land-use systems** within a globally consistent framework, driven by **scenarios** of climate change as well as socio-economic and technological development.

The set of land-use types involves **agriculture**, or agricultural land use, which is characterized by major crops and a set of management variables such as irrigation, fertilization rate, crop rotations, and/or cropping intensities. Remaining land-use types include **forestry**, urban/built-up land, and a set of **biomes** representing unused or natural land.

To allocate demands, a rule setting is applied that first translates a spatial inventory of suitability factors into a spatial measure of preference for a specific land-use type. These factors include, for example, terrain slopes, potential crop yield under specified management conditions, and **access** to market centers as a derivative of proxy data such as distance to infrastructure and distance to urban centers. Second, allocation rules account for the competitive capacity of different land-use types as well as for the **resilience** of land-use types against **conversion**. A set of spatial indicators will support the decision of how scenario assumptions about technological development and management will be spatially realized. Such indicators include measures for the abundance of suitable land, as well as maps quantifying the sensitivity of crop yields on specific management activities such as irrigation.

In terms of implementation, LandSHIFT will be coupled to two external models. First, potential crop yield will be dynamically simulated by applying the process model DayCent, which was developed by William Parton and colleagues. The adapted version of this agro-ecosystem model is able to simulate yields for major crops under specified management conditions, as well as related nitrogen, carbon, and water fluxes in the plant-soil system. DayCent allows for calculating crop water requirements as well as the impact of irrigation on yield levels, and accounts for the feedback of changing nutrient levels on crop growth. Second, the global water cycle model WaterGAP, developed by Joseph Alcamo and colleagues, is applied to compute water withdrawals and water availability on watershed levels. Water users include municipalities, industries, power plants, and agriculture. Based on daily water balances, river flows are routed through a global flow routing scheme. Both DayCent and WaterGAP are able to account for the potential impact of changes in the climate system on crop yields and water balances.

See also: Agent-Based Model; Agricultural Intensification; Century Ecosystem Model; Disintensification; Driving Forces; Land Degradation; Modeling; Proximate Causes; Urbanization; Water-Land Interlinkages.

Further Reading

Alcamo, Joseph, Doell, Petra, Henrichs, Thomas, Kaspar, Frank, Lehner, Bernhard, Roesch, Thomas, and Stefan Siebert, "Development and Testing of the WaterGAP 2 Global Model of Water Use and Availability," *Hydrological Sciences Journal* 48 (2003): 317–37; Parton, William J., Holland, Elizabeth A., Del Grosso, Steve J., Hartman,

Melanie D., Martin, Robin E., Mosier, Arvin R., Ojima, Dennis S., and David S. Schimel, "Generalized Model for NOx and N₂O Emissions from Soils," *Journal of Geophysical Research-Atmospheres* 106/D15 (2001): 17,403–19; Center for Environmental Systems Research [Online, August 2004], The Center for Environmental Systems Web Site, www.usf.uni-kassel.de/usf.

MAIK HEISTERMANN, JÖRG A. PRIESS

Land-Use History. Exploration of land change in both the distant past and over the last 300 years, and the period of the most fundamental land-cover change with many consequences for today's landscape configurations. Aside from long-term climatic shifts, changes in land cover are primarily linked to land use. Landscapes are therefore understood fully only by knowing their land-use history. This historical perspective directly informs understanding of contemporary land conditions by assessing types and rates of land change and identifying biophysical and social change processes.

The bulk of land-use history is produced by practitioners in the fields of ecology, history, and geography. Historical ecologists and physical geographers consider the interactions among the land, human activity, and natural disturbance to better understand ecological processes and changes in the structure and function of ecosystems (the magnitude and character of land change). Toward these ends, they draw on a range of tools and data sources, including pollen and macrobotanical remains, soil chemistry, climate reconstruction, radio carbon dating, cadastral information, vegetation inventories, and remotely sensed imagery.

Research in the humanities and social sciences shifts the focus of land-use history to the impact of changes in socio-economic and political organization on land use and land cover. While addressed in diverse ways, some of which rely on the same data sources as historical ecology but that also include oral histories and the written record, environmental historians and human geographers tend to rely on a narrative to weave together the role of **institutions**, **cultural factors**, and political economic characteristics in either driving land change or the reverse ways in which land conditions shape cultural norms.

In addition to addressing the questions pertinent to the practitioner's core interests, collectively, land-use history research has produced a range of important outputs. Three general contributions are emphasized here. First, land-use history empirically documents the spatial **scale**, pattern, and pace of historical land change, which helps uncover the ecological impacts of different land uses, and the role of biophysical **feedbacks** on use systems, such as rates of soil and **vegetation** recovery. The provision of historical data also contributes to understanding of biogeochemical cycles (carbon, in particular) and regional climate change.

Efforts to distinguish between natural and social causes of land change are often inconclusive. Descriptive information on historical land use, when combined with an understanding of the relationships among climate, natural perturbations (e.g., wind, **fire**, and disease), and vegetation, proves invaluable when trying to determine the relative influence of anthropogenic and biophysical forces in causing land change. In the northeastern **United States of**

America, for example, changes in **forest** composition and structure over the past 300 years are linked to both environmental variation and land use, but with climate and geology exerting a greater influence at broader scales of analysis. In another example, a recurring debate surrounds the degree to which **grassland** degradation (e.g., **soil erosion**, shrub encroachment, and falling productivity) in regions around the world is linked to climate shifts or overgrazing. In West Africa, what once were thought to be forest remnants of larger forests vastly diminished by local inhabitants are now known to be forest islands, created by land managers in what had been open grasslands. Here, land-use history has served to highlight the human role in landscape change, but it also challenges the conventional wisdom of **degradation narratives** and reinforces new ecological models that emphasize disturbance and non-equilibrium systems, which can only be understood over the long term.

Second, the reconstruction of prehistoric or indigenous landscapes and the ways in which humans have modified and transformed the land over the long term are fundamental to debunking the myth of pristine nature. Most areas on earth, from Antarctica to the Amazon basin, show a human imprint. Demonstrating the social construction of nature from the distant past to the present corrects misperceptions that human impact on the earth is recent and underscores human agency in altering biophysical environments. Such studies also demonstrate the role of humans in maintaining certain "desirable" landscapes, such as forest cover.

Recognition of the historical impact of humans on nature is relevant today. In the context of debates about sustainable development, for example, trends in the past may point to alternatives for the future. The embrace of cultural landscapes may be a necessary part of a transition to more sustainable human-environment conditions. Awareness of the outcomes of past land-use decisions, therefore, may influence choices about ongoing and future land use. Which human-environment conditions lead to more resilient or more vulnerable use systems? For example, plant and animal species (thus **biodiversity**) adapt to, and evolve with, a given disturbance regime, which in some areas is a function of long-term land use. Future alterations of past anthropogenic disturbance regimes may have implications for biodiversity. In the face of increasing dominance by human beings over land resources, which technologies and management regimes have been the most sustainable?

Land-use histories can also identify issues of path dependency, or the degree to which past land uses affect the range of use options available to present-day land managers. In the southern Yucatán peninsula, for example, past logging of hardwoods has eliminated this economic activity as a contemporary option in many locations, and long-term **slash-and-burn agriculture** seems to contribute to the rise of bracken fern (*Pteridium aquilinum*), a weed that impedes processes of **secondary vegetation** or succession and that effectively takes land out of agricultural production for an extended period of time.

Third, land-use histories inform understanding of the **driving forces** of land-use and **land-cover change** as well as efforts to model land-change dynamics. Explanations for land change are clustered, minimally, into two broad categories. The first explains land use as a product of political, economic, and social structures; and the second focuses on the immediate agents of change, usually local land managers. Land-use histories examine the relative roles of these drivers of land change as they cross distinct human-environment episodes. Within any given episode, for example, household demographics and economic explanations may be identified as the critical factors; whereas, in other periods, key land transformations may be precipitated by state policy.

Comparing regional land-use histories supports the rising consensus in nature-society studies that a web of driving forces of land change exists, the exact combination of which depends on regional and historical contexts. Land-use histories often identify the degree to which change trends are non-linear and affected by stochastic events. These events may be represented by external shocks to regional **land-use systems** (forces emanating from outside of the region and beyond the control of local inhabitants), such as federal development policy or natural perturbations, and they are forces often operating not only in distal space but in distal time. In the southern Yucatán peninsula, the most important land-use changes in the last 100 years—depleted hardwood reserves, large areas of permanently cleared forest, the establishment of **conservation** and extractive reserve areas—are all linked to particular historical episodes in which there were government policy decisions about **colonization**, federal land grants to the timber industry, agricultural development policy, and international pressure to conserve the region's tropical forests.

In **modeling** efforts, whether backcasting (to use models to "predict" the recent past, thus testing the strength of the models and their parameters) or projecting future land use, understanding the role of these external shocks is critical. Is the recent historical episode typical of land-use change in broader temporal scales, and therefore an appropriate period on which to base assumptions about future trends? By defining the **initial conditions** upon which current land-use patterns were founded, land-use history is fundamental (or should be fundamental) in setting the parameters of models and their suitability for the region in question. Models that do not include the effect of federal policy, for example, are apt to be ineffective in projecting land use beyond the short term.

See also: **BIOME 300 Project; Boserup, Ester; Cadastre; Decision-Making; Disequilibrium Dynamics; Environmental Change; Exogenous Factor; Mediating Factors; Pattern to Process; Pristine Myth; Remote Sensing.**

Further Reading

Fairhead, James, and Leach, Melissa, *Misreading the African Landscape*, Cambridge, UK: Cambridge University Press, 1996; Foster, David R., "Insights from Historical Geography to Ecology and Conservation Lessons from the New England Landscape," *Journal of Biogeography* 29 (2002): 1269–590; Klepeis, Peter, and B. L. Turner II, "Integrated Land History and Global Change Science: The Example of the Southern Yucatán Peninsular Region Project," *Land Use Policy* 18 (2001): 27–39; Lambin, Éric F., and Helmut J. Geist, "Regional Differences in Tropical Deforestation," *Environment* 45 (2003): 22–36.

PETER KLEPEIS, PAUL LARIS

Land-Use Legacy. The notion among ecologists and conservationists that land-use activities continue to influence ecosystem structure and function for decades or centuries—or even longer—after those activities have ceased. The recognition that legacies of land use can be remarkably persistent contradicts the pervasive assumption or **pristine myth** that society's **modification** of nature is a relatively recent phenomenon, and that remote or uninhabited lands are untouched or pristine wilderness areas. Indeed, there is growing recognition that most "natural" areas actually have more cultural history than assumed. For example, the **Intergovernmental Panel on Climate Change** estimates that roughly one-fifth of the total anthropogenic emissions of **greenhouse gases** during the 1990s were from changes in land use. Likewise, it has been estimated that the human impact on the terrestrial water cycle during the last fifty years has likely exceeded natural forcings of continental aquatic systems in many parts of the world. In the study of land-use/cover change, land-use legacies are best understood through a careful consideration of the environmental and **land-use history** of a given site or area, that is, that which constitutes the **initial conditions** of any land change. These conditions, together with an exploration of the drivers of change and the structure of feedbacks (or responses) to these drivers, do best account for typical pathways or trajectories of land change.

See also: **Degradation Narrative; Ecological Footprint; Ecosystem Services.**

Further Reading

Foster, David, Swanson, Fredrick, Aber, John, Burke, Ingrid, Brokaw, Nicholas, Tilman, David, and Alan Knapp, "The Importance of Land-Use Legacies to Ecology and Conservation," *BioScience* 53 (2003): 77–88; Turner, B. L. II, and Susannah McCandless, "How Humankind Came to Rival Nature: A Brief History of the Human-Environment Condition and the Lessons Learned," in *Earth System Analysis for Sustainability*, eds. Clark, William C., Crutzen, Paul, and Schellnhuber, H. John, Cambridge, MA: MIT Press, 2005, pp. 227–43.

HELMUT GEIST

Land-Use Planning.   Choosing, strategically in time and space, the uses of land in a sustainable perspective. As the land is a functional interface between the ground surface with all the physical, chemical, and biological features that influence the use of resources, land planning means the rational and optimal integration of all the dimensions of the dynamic **land-use system**. The **United Nations Environment Programme**, in collaboration with the **United Nations Food and Agriculture Organization** (FAO) and the United Nations Education, Science and Culture Organization (UNESCO), defined sustainable land use as the management of land to ensure the continuous satisfaction of human needs for present and future generations. **Sustainable land use** then automatically involves the optimization of the land-use potential and the planning of current and future uses. The implementation of land-use planning is a real challenge for land-use planners at global, regional, and national scales, but mainly for land users themselves that have in the past sometimes been excluded from the planning process. The evolution of land-planning theories, the definition of the land-planning process, and the current applications of these methodologies at different scales will be briefly described here.

Over time, land-use planning has developed into more of a science than a tool for development. Theory and methodological frameworks emerged from different scientific disciplines as landscape planning, ecosystem management, and environmental planning in the nineteenth and twentieth centuries. To better understand the reality of the land-use system, scientists built models that represented the use of resources that satisfied human needs. They evolved from models with one quantitative variable (population or climate) to more complex but qualitative land-use systems (e.g., the perspective of agro-ecosystems or **agrodiversity** that includes **indigenous knowledge**), and to quantitative **integrated models** of land-use change. Experiences of **land degradation** are progressively explained by the **vulnerability** of the population rather than by a simple misuse of the land. In the twentieth century, soil mapping widely developed and became a major method of land-resource assessment, which was based not only on field surveys but also on **remote sensing** techniques. **Decision-making** on sustainable land use requires an integrated and multidisciplinary approach of understanding land use in the long term. In its guidelines for integrated land planning, FAO proposes a step-by-step method for physical **land evaluation** that involves a consultation to formulate the relevant land uses. The **International Institute for Applied Systems Analysis**, in collaboration with FAO, developed a decision-making software tool that has been applied in Kenya. Other initiatives are currently tested, for example on the Loess Plateau in **China**. An adequate institutional context and the effective participation of all stakeholders in the negotiations is a necessity in the implementation of such participative land-use planning experiences.

The concept of planning has been used in various disciplines to describe the four steps in the implementation of plans in one organization. The first element of planning is the articulation of well-defined objectives. Land-use planning deals with multiple objectives and multi-agents with reference to specific spatial and time **scales**. The optimization of objectives requires weights and priorities in the short and long terms. Compromise goals result from **tradeoffs** between two somewhat contradictory dimensions of land use: ecological **conservation** and socio-economic subsistence. The objectives should be expressed as explicitly and quantitatively as possible, for example in terms of performance indicators. Second, the understanding of the organization and its resources is called analysis, diagnosis, or evaluation. Understanding the land-use system and its environment involves the classification of resources, production functions, populations, and influences. This analysis should define the strengths and weaknesses of all parts of the system and identify the opportunities of change. In this phase, the building of a detailed **geographical information system** (GIS) is an interesting approach. The third element of planning is the definition of possible options to achieve the objectives. In this definition, the decision support system managing a GIS proposes **scenarios** and simulations to the land user itself. These simulations consider the optimum strategies in land-use allocation and land-use changes. Development of such systems at local or regional scales is an extraordinary challenge of this century. Fourth, design and implementation of plans are achieved by a collaboration of several teams of people assembling different skills. Implementing development programs, plans, and activities should be done by governments, communities, and individuals, respectively. This fourth step in strategic planning deals with the identification and negotiation of major development priorities. This involves using a set of practical tools (land evaluation, farming systems analysis, etc.) to integrate existing information, analyzing **land-use policies** and proposals, and negotiating conflicts and competing interests.

Most of the world's countries have set up national land policies. Most developed countries have put in place physical plans or land zoning. In the United States, the first zoning ordinance was passed in New York City in 1916, and by the 1930s, most states had adopted zoning laws. From 1969 onward, a federal law required every city and county in the state to have a comprehensive land-use plan. Within the European Union (EU), many guidelines are formulated into EU directives at the supra-national level. Many developing countries have a national environmental action plan (NEAP) as well, often established with the assistance of the World Bank. Unfortunately, the strategies have been developed, but their implementation is often not yet operational. Planning at the local level (community or village) involves the identification of concrete actions within the strategic plan. Indigenous communities usually have more experience of local conditions than outsiders when managing natural resources like **pastures**, **forests**, water, and wildlife. Technical **institutions** and development organizations often have experience and knowledge of the outside world (e.g., market information and outsider demand for resources use) that is unavailable to those at the local level. Bringing these different sources of information and knowledge together opens more perspectives for sustainable development.

See also: Community Involvement; Decentralization; Integrated Assessment; Land Rehabilitation; Modeling; Participatory Land Management; Public Policy; River Basin; Urban Sprawl; Watershed.

Further Reading

Beinat, Euro, and Peter Nijkamp, "Land Use Planning," in *Encyclopedia of Global Change: Environmental Change and Human Society*, vol. 2 (J–Z), eds. Andrew S. Goudie and David J. Cuff, Oxford, UK, New York: Oxford University Press, 2002, pp. 27–33; Botequilha, Leitão A., and Jack Ahern, "Applying Landscape Ecological Concepts and Metrics in Sustainable Landscape Planning," *Landscape and Urban Planning* 59 (2002): 65–93; De Wit, Paul, and Willy Verheye, Land Use Planning for Sustainable Development, in *Encyclopedia of Life Support Systems*, developed under the auspices of UNESCO, Eolss Publishers, Oxford, UK [Online, September 2004], The EOLSS Web Site, www.eolss.net.

NATHALIE STEPHENNE

Land-Use Policies.   Land use and changes to land cover are increasingly affected by policies of various kinds. A policy usually consists of one or more defined goals and the means of achieving them: policy objectives are distinguished from policy instruments. Usually the term "policy" is used in relation to government bodies, but social groups such as environmental NGOs (non-governmental organizations) may have policies that they pursue on land they control and with which they seek to influence government

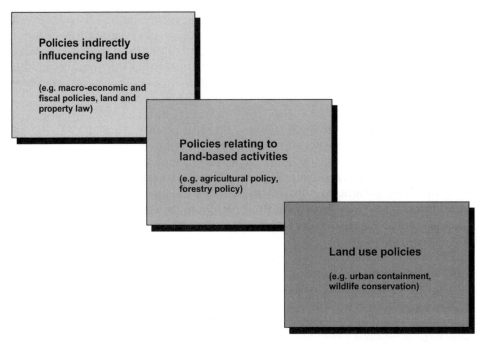

Land-use policies and policies affecting land use.

bodies. The term is a reminder that the use of land is a highly political activity and by implication indicates a belief that determination of land use should not be left to chance, to individual landowners, or to the market alone.

Policies affecting land use and land cover are diverse and wide-ranging. A distinction can be drawn between those that are focused directly on land use and that seek to influence land use per se (i.e., land-use policies *sensu stricto*), and those that influence land use but focus primarily on other objectives. Green-belt and urban-containment policies are an example of the former, while trade policies exemplify the latter. In addition, there are intermediate cases, such as policies relating to **agriculture** (see figure "Land-use policies and policies affecting land use"). Land-use policies *sensu stricto* may be spatially explicit: they may specify the land uses that are sought in specific places, whereas policies in the more general sense may have spatial implications but be less likely to be spatially explicit. It could even be claimed that a policy of not having a policy (i.e., of leaving issues of land use to be resolved by the market, or a laissez-faire policy) could have an effect on land use.

A rigid classification is probably unhelpful, because of the diversity and the numerous intermediate categories. Policies are perhaps more appropriately placed on a continuum than located in discrete categories. Almost any government policy, for example on macroeconomics, transport, or defense, is likely to have implications for land use. The indirect influences of, for example, fiscal policy on land cover or land use may be unintended, at least initially, and a lack of coherence is common. Policies have often been defined on a sectoral basis (e.g., within agriculture), and integration or coordination among sectors can be difficult to attain. A number of fields can be identified in which the influence of policy on land use has been profound. That influence can include both the purpose for which the land is used and the nature of land management.

Agricultural Policies

Much of the world's agriculture operates in the context of policies formulated by the state in which it is located, and increasingly also by policies evolving at the supra-state level. Agriculture policies illustrate several key features of land-use policies *sensu lato*. States devising such policies have (or had) concerns about **food security** and ability to feed their population. They may also have sought to control fluctuations in supplies and hence prices. In addition, there may be objectives of rural welfare—to maintain agricultural incomes and to help to stem over-rapid rural depopulation. To this end, farm prices are often supported at levels above those that the market would indicate. Support can be provided in a number of ways, operating separately or in combination. Tariff barriers may be erected to protect domestic producers from competition from cheap imports. Direct support can be provided by the setting of minimum prices, or by purchase by a state body of surplus material that cannot be cleared by the market at a price above a defined threshold. Subsidization of operations such as drainage and irrigation can occur. In addition, state sponsorship of research and development can indirectly exert a significant influence.

The conditions prevailing when state policies were first formulated usually change over time, but the policies themselves evolve more slowly and indeed may outlive the circumstances that gave rise to them. Policies are introduced in relation to a perceived problem, actual or potential. The nature of that problem changes, but once initiated the policies may persist. The initial policy instruments would (or at least should) have been selected in the light of the defined objectives, but may then become inappropriate as the objectives evolve. In other words, inertia and time lag can characterize policy instruments as well as policy objectives.

In most of the developed world there is a serious problem of balancing the maintenance of rural welfare (or at least of farm incomes) with levels of production and costs of support. If farm incomes are maintained, then the level of production may be too high in relation to demand. If they are not maintained in line with the national or regional average earnings, then a combination of political pressure and (financial) failure of farmers may result. At least this has been the traditional problem. Now the European Union is seeking a solution in a form of support that is decoupled from production. If they observe certain conditions (e.g., on environmental standards and animal welfare), farmers will receive a "single farm payment." At one level, this will weaken the state's ability to influence land use. When support was channeled through production, adjustment of the relative prices for different commodities would result in adjustment of the relative areas under different crops. Assuming that the primary objective is now rural welfare, it represents a more focused selection of policy instruments. The previous instrument, relying heavily on price support, had effects on production that were initially sought but ultimately became undesirable.

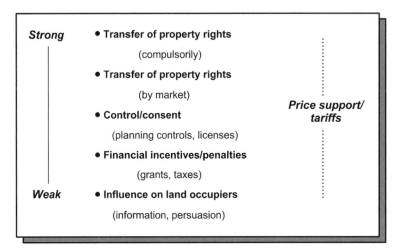

Policy instruments (the "strength" of price support is variable, depending on the level of support provided).

Paralleling agricultural policies are those relating to sectors such as **forestry**, **conservation**, and housing. All imply that an aim has been identified and that the state or other body has resolved to attempt to achieve it.

Policy Instruments

The range of potential instruments is wide: they have different characteristics, especially in terms of strength, and their selection usually involves issues of political judgment and acceptability as well as potential effectiveness. It is sometimes asserted that the number of instruments in a policy should be equal to the number of objectives, but this "law" has not always been observed in land-use policies.

Both the carrot and the stick have been widely employed (see figure titled "Policy instruments"). The latter can involve the compulsory acquisition of land and the banning of various practices or processes by legal means (e.g., removing trees or building houses without the appropriate permission). Land-use policies (*sensu stricto*) can use devices such as zoning to identify the types of land use considered appropriate in particular areas: often the implications in terms of land values are enormous. On the other hand, various incentives can be offered to encourage actions deemed desirable. For example, grants may be offered for tree-planting if forest expansion is a policy objective. They can be offered in undifferentiated form over a whole country, or targeted in particular areas.

Policy Trends

As a sweeping generalization, it can be said that through time, policies have become more numerous and complex, and policy instruments more diverse and sophisticated. More types of land use have been brought under the influence of policy, and a widening range of tools has been employed. Environmental conservation has been a particular growth area: there has been extensive use of the "protected area" or "designated area" as a basic tool, but in recent decades a wider and more imaginative range of instruments has been deployed.

The rise in significance of land-use policies parallels, in large degree, the rise of the state in the modern era, and it continues despite the alleged weakening of the state in recent decades. Some policies exist on paper but have little observable effect: to be effective, they must be backed by a capability of enforcement and implementation that states in weak or embryonic form may not possess. Nevertheless, recent decades have witnessed a trend away from traditional "top-down" approaches in which policies are simply devised and imposed by government bodies, and toward greater participation in policy formulation by the people directly affected by policies.

One of the dimensions of complexity is vertical or hierarchical. In theory, the policies pursued by local government bodies should "nest" within those of the national government. Increasingly, supra-national and global bodies are also becoming involved, through measures such as the Common Agricultural Policy and the **United Nations Convention on Biological Diversity**. Policies, in both the strict and the broad sense, have become major elements of the context in which land is used. Issues such as their evolution and the nature of **feedback** between change in land use and change of policy will increasingly attract the attention of researchers in the field of land use/cover change.

See also: **Access; Agrodiversity; Arena of Land Conflict; Community Involvement; Decentralization; Decision-Making; Green Wall Project; Land Concentration; Land Degradation; Landholding; Land-Use Planning; Participatory Land Management; Perception; Public Policy; Reforestation; Tradeoffs.**

Further Reading

European Commission, CAP Reform: A Long-Term Perspective for Sustainable Agriculture [Online, December 2004], The EC Web Site, europa.eu.int; Organisation for Economic Cooperation and Development, *Agricultural Policies in OECD Countries: At a Glance*, Paris: OECD, 2004; Platt, Rutherford H., *Land Use and Society: Geography, Law, and Public Policy*, Washington, DC: Island Press, 2004; Reid, Robin S., Tomich, Thomas P., Xu, Jianchu, Geist, Helmut, DeFries, Ruth S., Liu, Jiangho, Alves, Diogenes, Agbola, Babatunde, Lambin, Éric F., Chabbra, Abha, et al., "Linking Land-Use/Cover Change Science and Policy: Current Lessons and Future Integration," in *Land Use and Land Cover Change: Local Processes, Global Impacts*, eds. Éric F. Lambin and Helmut Geist, Berlin: Springer-Verlag, 2006; Rydin, Yvonne, *Urban and Environmental Planning in the UK*, London: Palgrave MacMillan, 2003; World Bank, Sustaining Forests: A Development Strategy [Online, December 2004], The World Bank Web Site, www.worldbank.org.

ALEXANDER S. MATHER

Land-Use System.   Systemic or semi-systemic representation and characterization of the intricate web of relationships, factors, processes, and structures that impinge on land use (change) and—to the extent that the dynamics of land-use is addressed—**land-cover change**. The simplest form of a land-use system (LUS) consists of one land-use type applied to one land unit, assuming uniform fields. Multiple land uses can be conceptualized as aggregations of such simple systems: for example, intercropping can be considered as concurrent single use sharing the same land unit, whereas rotations are their sequences. Its plural form speaks to the fact that relationships and processes that are manifest in different land-use contexts may vary widely across places; hence, we have myriad land-use systems—for example, **urbanization**, **agriculture**, **forestry**, and **mineral extraction**—coexisting over heterogeneous landscapes.

Diversity of Approaches

There is a wide diversity of LUS constructs in applications, ranging from those conceived in the context of **land evaluation** to those proposed in global change research by the **Land-Use/ Cover Change (LUCC) project**, for example (see figure title "Framework"). The former is chiefly concerned with spatial organization of such land-use elements as soil series, land units, and land utilization types; and their interrelationships in maintaining the functioning of the production system are defined mostly in economic and biophysical terms. Although it captures many operational details of land use at the field/parcel level, it often does not account for land-use change, much less their broad-scale **driving forces**. The latter attempts to identify and explain cross-scale, interactive mechanisms of change in both land use/ land cover by enlisting a different set of analytical elements (drivers, agents of change, **proximate causes**, mitigators, etc.), which are woven into a broader analytical framework to capture multiple dimensions of LUS.

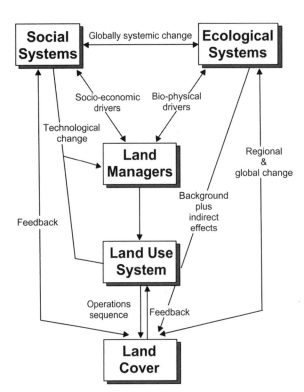

Framework for understanding land-use/cover situations.

SOURCE: Science/Research plan of the Land-Use/Cover Change (LUCC) project, 1995, p. 35.

This diversity may be attributed to differences in purpose of study, level of aggregation, dynamics considered, and functional specification, etc.; but more significantly, it is owing to the lack of theoretically informed conceptualizations of LUS that are widely accepted—those that are able to address the "what" of land-use change and the "why" of this change. In other words, such conceptualization should facilitate researchers not only to describe the patterns of land-use/land-cover changes and their impacts, but also to explain mechanisms of these changes in spatially explicit terms, in that both changes and impacts are treated as phenomena occurring on an area of the earth's surface with particular, spatially variable properties. As noted by Helen Briassoulis, many existing land-use theories refer only to either the "what" or the "why" questions, but rarely both; they attempt to describe but rarely explain. When they do address the "what," they tend not to deal directly with land-use change but with changes in its determinants, while the "why" part is often "black boxed," leaving the user to figure out the underlying mechanisms. In addition, they tend to be procrustean in choosing drivers to be considered, or fail to offer exact conceptual and operational definitions of these drivers, or not treat land-use change and their drivers at the same level of definitional detail. Moreover, many of them are not spatially explicit (in the sense that they tend to reduce land to a point in space, or land as independent variable is totally absent from analysis); and if they do, they do not cover the full array of existing and potential uses of land, their changing patterns, and drivers. Many of these inevitably filter into conceptualizations of LUS in reality. Given the lack of theoretical integration to ground LUS conceptualizations, instrumentalist use of theoretical assumptions becomes a prevailing approach. Systems analysis is something up for grabs for such use, due to substantive and practical imperatives working from both within the land-change community and without. From within, strong need exists to accord a measure of coherence to an intellectual terrain that has become increasingly fragmented by growing diversity of disciplinary inputs, crying for more integrative theoretical schemata. Now that the feasibility and desirability of seeking a general theory of land-use change remains questionable, systems analysis offers a convenient instrument for ostensible if not truly transdisciplinary integration. From without (and in a more practical sense) land-use researchers well-steeped in social science tradition have long envied (and feel inferior to) natural science researchers in earth systems science and climate change study in terms of prestige and funding; for them, systems approaches provide a partial bridge to the language barrier that has long hindered them from communicating and competing with natural science programs operating in the broad field of global change research, such as the Global Change and Terrestrial Ecosystem (GCTE) project of the **International Geosphere-Biosphere Programme**.

Elements of the Systems Approach

The very notion of LUS implies applying systems approaches (not necessarily systems theory per se) to the study of land use. Although most of the systems concepts reviewed below found their way into land-change research mainly through the "back door" of (systems) ecology, their semantic and theoretical lineages can be traced back to classical systems literature of natural sciences (notably cybernetics, informatics, dissipate structure, and synergetics). They have inspired subsequent research in hierarchy, complexity, adaptive behaviors, and nonlinear and spatial dynamics. Systems concepts recently enlisted to describe LUS as complex, dynamic systems include complexity, openness, connectivity, hierarchy, stability/resilience, non-linearity, and entropy.

With regard to complexity, lying at the interface of multiple, cross-scale social and ecological systems, LUS is complex in that it comprises biophysical, socio-economic, and ecological components (subsystems), with attributes of openness, spatiotemporal dynamics, adaptability, and multiple objectives; it represents a myriad of **feedback** mechanisms that intermediate among land uses, their driving forces, and land covers, either strengthening or attenuating their interrelationships or rendering them non-linear. To better understand the behavior of LUS as complex systems, two interrelated questions that pertain to two key dimensions of complexity need to be addressed: How is LUS functionally structured and embedded with other socio-ecological systems? (according to Kasper Kok, this is a question of continuity as well as one of "structural complexity"); and, How do changes in these embedded systems (disturbances and chaos in the shorter term, transformation in the longer term) affect the attributes of LUS and vice versa? The latter is a question of change as well as one of "functional complexity."

With regard to openness, land-use systems bear features of **metabolism**, that is, they are in constant exchange of materials, energy, and information with the outside world (e.g., through trade, tourism, pollution, and species invasion).

Connectivity refers to the fact that spatially distant locations influence each other through multiple ways: either directly (e.g., **soil erosion** leading to sedimentation) or indirectly (e.g., **land degradation** at one location triggering migration, thus leading to **deforestation** at another location). The neighborhood effect in spatial analysis is also a relevant example.

Hierarchy describes both the scale of observation as pertains to LUS, and also explains how the interplay of the two key dimensions of complexity ("structural" and "functional") drive and shape the evolution of LUS. LUS is influenced by a great number of factors operating at different scales, while the observed patterns and processes often differ with regard to the **scale** of observation. Across the hierarchically nested levels, variables and processes operating at higher levels often steer and constrain lower-level variables and processes (e.g., dam building at basin level triggering irrigation and inundation in tributaries), but higher-level phenomena might emerge from lower-level dynamics—this is the property of emergence (e.g., massive overdraft by individual households leading to aquifer degradation, thus triggering landscape change).

Concerning stability and **resilience**, the hierarchy of LUS is not static; it evolves. The evolving hierarchy may reach a point where it becomes vulnerable to change, and a combination of internal and external forces will flip LUS either into a new qualitatively different phase of stability (e.g., agricultural adaptation in the U.S. **Great Plains** after the Dust Bowl), or into decline or collapse. In this context, resilience refers to the ability of LUS to absorb the impact of changes and continue to function.

Nonlinearity refers to the scale-dependence of observations (e.g., rainfall-runoff relationships), an intrinsic property of any dynamic system. It is popularly interpreted as "the sum of the parts do not equal the whole." Non-linear relationships tend to become more linear at coarser scales of observation, as evidenced by those among PAT variables (population, affluence, and technology). As Kasper Kok notes, "There are two fundamental problems related to structural complexity when describing the land-use system. First, processes that are established at a certain scale cannot be linearly translated and used at higher scales because of the aggregation effect. Second, processes that are established at a certain level cannot be translated to other levels because of differences in the hierarchical structure" (2001, pp. 7–8).

Entropy is a measure of disorder that tends to increase while LUS evolves (e.g., accumulative land degradation and fragmentation of landscapes). Entropy created within LUS plus entropy added by natural disasters is offset by negative entropy generated through sound land management and technology progress, and so on, for LUS to maintain its normal functioning.

Needless to say, these concepts do not do full justice to what the rich systems theory has to offer. The influence of each term on the study of land use (and the way LUS is delineated) is not even, depending on the level of abstraction it means to describe and the context of real world situation it can successfully approximate. While risking crude oversimplification, two main streams of research that conceptualize LUS in (semi-)systems terms (explicitly or implicitly) may be loosely identified: one is "observe, describe, and integrate to understanding," and another is "model to understand." A third, hybrid, "act to understand" stream may be identified, mainly in the context of **land-use planning**.

Observation, Description, and Integration

Examples of LUS conceptualized in the first stream are found mainly in traditional land-use literature. This stream contributes to the better understanding of LUS mainly through taking stock of existing research findings, clarifying theoretical concepts, and suggesting new research questions and alternative explanatory frameworks, often in combination with in-depth (not necessarily rigorous) case studies. In some cases, vivid land-change stories are told, through which empirical and interpretative baselines are established to help assess the validity of systems propositions and identify random events that significantly impinge on land-use/cover changes. Its method of inquiry is mainly qualitative, allowing a combination of different reasoning modes (from inductive to dialectic). Illustrative of this stream are the following characteristic statements:

1. The role of biophysical drivers in effecting land-use change is intermediated by socio-economic drivers, mainly through influencing the response of land owners/managers to land-cover change (e.g., no farming on marginal lands).
2. Land-cover changes resulting from land-use changes may feed back on **decision-making**, causing perhaps new rounds of land-use change and a vicious cycle of land degradation (e.g., shifting cultivators in **Latin America**).
3. Special attention should be paid to **cumulative change** and cumulative impacts, those that, though not physically connected through a globally operating system, can reach a global scale and status when their occurrence in many places adds up (e.g., **wetland** drainage and grassland degradation, **biodiversity** loss, and species transfer).
4. Certain regional impacts of land-use change may take a long time to show up in that they play out as chains of events responding to slow environmental alterations, often resulting in the delayed and sudden occurrence of harms (e.g., chemical soil pollution).

Evidently, these exemplary statements variously capture ramifications of some of the above systems concepts, albeit loosely and on abstract levels. The broad scoping of LUS, though not clearly defined, is implicit in the analytical and explanatory frameworks they propose. Because these conceptualizations are cast in abstract terms, however, operationalization in rigorous research design is often difficult. In terms of theoretical position, they are eclectic at best, which come out as both an advantage and a disadvantage.

Modeling

The second stream seeks to use quantitative models as tools to describe LUS (in the narrow sense of expressing operationally the relationships wherein), and explain (statistically) or mimic parts of its systemic character. It is overwhelmingly deductive and reductive, which is both its strength and limitation. Among many types of land-use models, the **conversion of land use and its effects (CLUE) model** or modeling framework is used as an illustrative example, because it represents the recent advance in developing systems-based toolboxes for land-use analysis.

CLUE and its supplementary modeling tools represent a new genre of land-use **modeling** effort that attempts to address, all at once, several critical land-change needs identified above: integration, linking with natural science programs, spatial explicitness, and scaling (as well as decision support). It is systemically structured, at least in a mechanical sense, as evidenced by its various modules that pertain to the main systems components of LUS as it is conceptualized (and operationalized) in CLUE. It is rigor, at least in the mathematical sense but not necessarily in the logical sense. It does manage to describe operationally many aspects of LUS as complex dynamic systems; but due to hoary data problems, the systems view embodied in CLUE has to be heavily truncated in terms of factors, relationships, and processes to be examined. Most importantly, it lacks substantive explanatory power with regard to the mechanisms of changes, as with most symbolic models. This forces us to critically reconsider the modeling fervor in land-use research and its potential influence on sound LUS conceptualization. To qualify as "substantive" in this regard, the explanation of change should tell the land-use change story and reason about the processes of changes at the same level drivers of change operate; the explanation should be theoretically informed and get into the causality of the relationship under scrutiny and map out the causal structure to the finest possible scale as the availability of (not necessarily digital) data allow.

Clearly, statistical association does not pass for causal explanation in this sense, which is what the search for causal processes of land-use change is all about. Unqualified employment of quantitative procedures raises other concerns as well: for example, used alone, statistical analysis cannot capture many qualitative aspects of LUS critical to the understanding of land-use change, such as cultural, political, and institutional factors (i.e., systems of **land tenure**) that may work as both constraints and drivers. In some cases, the rich reality is often chopped to fit the peculiar specifications of a model, hence making a model self-referential and not easily falsifiable. For another instance, statistical prediction that is the main purpose of many projective models is based on relationships existing in the past—an equilibrium assumption that may not easily hold for LUS in a rapidly changing world, if the specified time scale does not match the time horizon on which the modeled LUS operates. This is both a technical and a theoretical concern, of course, which requires a corresponding theory of change that seems to be lacking

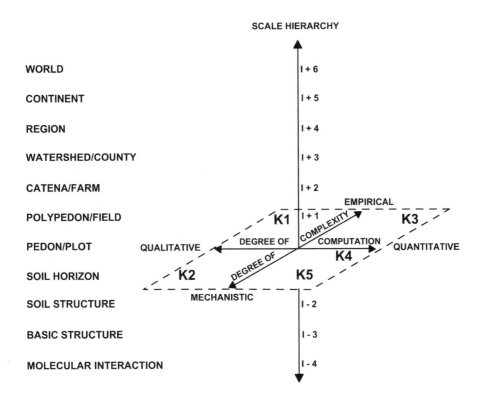

in the majority of mathematical models. Further problems arise from the general deprivation of theoretical underpinnings in operational models (and lack of interest in this by modelers): because theoretical assumptions underlie all models built, predicative value as well as functional specification of a given model have to be evaluated ultimately in light of the assumptions (implicitly embodied or explicitly articulated) and logical integrity, not by map validation, which unfortunately has become a peculiar niche business for a handful of researchers interested in technically fixing quantitative inaccuracies of models. Arguably, perfectly validated maps will never become a substitute for theoretically informed and methodologically sound analyses. In the worst cases, statistical **validation** may be used in the same way a drunkard employs a lamppost: not for illumination but for expelling the fear of darkness (in this case, the fear of lack of theoretical direction). In an overall assessment, different types of operational models conceptualize LUS differently, with some being more realistic, less abstract, and more spatially explicit than others; but their potential tends to be compromised by two common problems: the paucity of relevant data and the anemia of theories. The former is delimiting, but the latter is crippling.

The foregoing suggests that, in the context of applying systems analysis to the multidisciplinary study of land use/change, the observation by **Piers Blaikie** and **Harold Brookfield** that there is a "theoretical and methodological failure to combine social and natural science" is rightly put. Until we can find a way to move beyond the watershed dividing the two intellectual domains, no LUS construct will serve the dual goals of "**socializing the pixel**" and "**pixelizing the social**" equally well. The merits of different LUS constructs have to be evaluated ultimately in view of the value to scientific research and to management decision and policymaking. However, the criteria required in these two contexts do not always conform to each other: in the former case the criteria rest mainly on their conceptual soundness and potential in aiding hypothesis testing and theoretical development, while in the latter they are based primarily on its capabilities of helping identify **hot spots of land-cover change** and options/constraints, and on closing the inevitable gap between science and realities. The importance of scales of observation and levels of analysis is a crosscutting theme for both streams of research reviewed above, as well as for theories and models; hence they must be seriously tackled. The meaningful and useful analysis of LUS to support policymaking also hinges upon specification of land units, land-utilization types, and LUS at the right scales for effective management/policy intervention to be applied. To address these combined needs, it has been argued that LUS and its elements have to be defined differently at different spatial scales using a methodological typology that links scale specification of LUS to different level decision-making needs: the operational, the tactical, and the strategic, as illustrated in the second figure.

See also: **Agent-Based Model; Agricultural Intensification; Agrodiversity; Agroforestry; Cattle Ranching; Economic Livelihood; Exogenous Factor; Human Immunodeficiency Virus (HIV)/Acquired Immunodeficiency Syndrome (AIDS); Mediating Factor; Property; Slash-and-Burn Agriculture; Subsistence Agriculture; Swidden Cultivation; Thünen, Johann Heinrich von.**

Further Reading

Blaikie, Piers, and Harold Brookfield, *Land Degradation and Society*, London: Routledge, 1987; Briassoulis, Helen, Analysis of Land Use Change: Theoretical and Modeling Approaches [Online, August 2004], The Web Book of Regional Science Web Site, www.rri.wvu.edu/regscweb; Driessen, P. M., "Biophysical Sustainability of Land Use Systems," *ITC Journal* 3/4 (1997): 243–7; Kok, Kasper, Scaling the Land Use System: A Modeling Approach with Case Studies for Central America, PhD dissertation, Wageningen University, NL, 2001; Turner, B. L. II, and William B. Meyer, "Global Land Use and Land Cover Change: An Overview," in *Changes in Land Use and Land Cover: A Global Perspective*, eds. William B. Meyer and B. L. Turner II, Cambridge, UK: Cambridge University Press, 1994, pp. 3–10.

KE CHEN

Land-Use Transition. A possible development path or trajectory of a **land-use system** where the direction, size, and rate of change can be influenced through policy and some specific circumstances. A transition in land use is mainly but not necessarily a process of societal change in which the structural character of society (or a complex subsystem of society) and the functional character of land use (including biophysical attributes) transform, usually in a series of transitions. Broad transformations—such as those from shifting cultivation or rotational farming to sedentary, permanent, and high-intensity farming—usually result from a set of connected changes that reinforce each other but take place in several different components of the system. Multiple causality and co-evolution of different sectors (e.g., policy, meteorology, economy, hydrology, technology, and demography), caused by interacting developments, are central to the concept of transitions in land use. Thus, land-use transitions must be viewed as multiple and reversible dynamics: they do not follow a fixed pattern, and they are not deterministic. A transition in land use is not set in advance, and there is large variability in specific passages or pathways. Also, there is a strong notion of instability and indeterminacy. The theory of land-use transition has been applied in land-use change studies at different spatial and temporal **scales**. Though still in its infancy, the concept of **forest transition** is most advanced.

See also: **Boserup, Ester; Economic Restructuring; Metabolism; Public Policy.**

Further Reading
Lambin, Éric F., Geist, Helmut J., and Erika Lepers, "Dynamics of Land Use and Cover Change in Tropical and Subtropical Regions," *Annual Review of Environment and Resources* 28 (2003): 205–41.

HELMUT GEIST

Land-Use/Cover Change (LUCC) Project. A worldwide, interdisciplinary joint core project on land change of the **International Geosphere-Biosphere Programme** (IGBP) and the **International Human Dimensions Programme on Global Environmental Change**, formalized in 1995 and phased out in October 2005. The three missions of the LUCC project had been to build a compendium of information about local land-use and cover dynamics, to identify a small number of robust principles that could better knit together local insights into a predictive science, and to foster the development of common models that might then become widely available to scientists and stakeholders. In order to implement the project's science or research plan, centered around six major research themes, three interlocking strategies were encapsulated in three research foci. These had been the development of case studies that analyzed and modeled the processes of land-use change and land management in a range of generalized global situations (focus 1: land-use dynamics). After half a decade of operation, this strategy was complemented by the use of **meta-analysis** generating a general understanding, for example, of the causes, rates, and pathways of land change while preserving the descriptive richness of local studies. The second major research theme was the development of empirical, diagnostic models of **land-cover change** through direct observations and measurements of the explanatory factors (focus 2: land-cover dynamics). The third theme was the development of **integrated models** as well as of prognostic regional and global models (focus 3: integrated **modeling**). To meet its mission goals, the LUCC project reached out for a large scientific community to generate a wealth of results on its fundamental science questions. Two particular concerns had been the facilitation of interdisciplinary research work between the social and natural sciences, and to globalize research on land-change processes—such as **deforestation** and **desertification**—by contrasting results obtained from a variety of regions and geographic situations. Historically, three major periods of operation can be distinguished. From 1995 to 1998, major efforts were undertaken to establish the foundations of the project through the establishment of basic support infrastructure (i.e., scientific steering committee as core group, international project office, and regional networks); to develop and catalyze land-change science (e.g., through associated groups such as the **International Geographical Union (IGU) Commission on Land-Use and Land-Cover Change** and the **NASA Land-Cover/Land-Use Change (LCLUC) Program**); and to move toward full implementation. Following completion of the implementation strategy in 1999, major headway was made toward synthesizing land-change results and methodologies up to about 2002, while support infrastructure and dialog with funding agencies in the Americas, **Europe**, and Asia continued to be maintained. During 2003–2005, full implementation of LUCC's mission goals was achieved through final synthesis work in the form of workshops, conferences, and publications, with implications drawn for a transition toward **sustainable land use**, among other actions. Since its formalization in 1995, the LUCC project has welcomed applications by research groups, institutions, and agencies worldwide to join the community through formal status. The number of endorsed projects—such as the **Tropical Ecosystem Environment Observations by Satellite (TREES) project** and the **Large-Scale Biosphere-Atmosphere Experiment in Amazonia**—increased from two in 1997 to more than fifty, and about a dozen regional research networks were established in addition. In 2005, the scientific and institutional memory of the LUCC project was given to the **Global Land Project**, which also integrates the scientific and institutional memory of IGBP's Global Change and Terrestrial Ecosystems (GCTE) project.

See also: BIOME 300 Project; Change Detection; Hot Spots of Land-Cover Change; Land-Use System; Socializing the Pixel; Syndrome.

Further Reading

de Sherbinin, Alex, A Guide to Land-Use and Land-Cover Change (LUCC) Research: A CIESIN Thematic Guide [Online, March 2005], The CIESIN Web Site, sedac.ciesin.columbia.edu/tg; Geist, Helmut, The IGBP-IHDP Joint Core Project on Land-Use and Land-Cover Change (LUCC) [Online, March 2005], The Encyclopedia of Life Support Systems Web Site, www.eolss.net; Lambin, Éric, and Helmut Geist, eds., *Land Use and Land Cover Change: Local Processes, Global Impacts*, Berlin: Springer, 2006; Land-Use/Cover Change Project [Online, March 2005], The LUCC Web Site, www.geo.ucl.ac.be/LUCC.

HELMUT GEIST

Large-Scale Biosphere-Atmosphere Experiment in Amazonia.   An integrated, interdisciplinary study led by Brazil (and perhaps the largest research initiative or regional global change study) to explore the climatological, ecological, and hydrological functioning of the Amazon, the impact of land-use/cover change on these functions, and the interactions between **Amazonia** and the earth system. Planning for the Large-Scale Biosphere-Atmosphere Experiment in Amazonia (LBA) began in 1992, and field research programs began in 1998. Apart from having a substantial education and training component, LBA focuses on two key questions, as laid down in the science plan: How does Amazonia currently function as a regional entity? and, How will changes in land use and climate affect the biological, chemical, and physical functions of Amazonia, including the sustainability of development in the region and the influence of Amazonia on global climate? The LBA's core research focuses on the effects of tropical forest **conversion**, **regrowth**, **fire**, and selective logging on the **carbon cycle**, **nutrient cycle**, trace gas fluxes, and the prospects for **sustainable land use** in the region, with as much quantification of conditions and processes done as possible. Field research is carried out along two transects spanning gradients from wet, aseasonal **forests** to relatively dry and seasonal forest or savannas. LBA is part of a set of continental-scale terrestrial transects of the **International Geosphere-Biosphere Programme**, and it has been endorsed by the **Land-Use/Cover Change (LUCC) project**. Moving beyond the initial stages of LBA, and initiating a first major synthesis effort, the need has been identified to better integrate the human dimensions of **environmental** change, by abandoning, for example, the focus on spatially delimited transects and exploring typical pathways of land-use/cover change instead (such as **land degradation** and/or **agricultural intensification**). Several new research areas were considered essential to fostering the understanding of major processes behind large-scale changes in Amazonia. These are land-use/cover changes in relation to human **population dynamics** (mainly mobility) and **urbanization**, logistics (or infrastructure such as roads) and regional development, agrarian and **land tenure** structures, agricultural and economic systems, and environmental valuation.

See also: **Anthropological Center for Training and Research on Global Environmental Change; Arc of Deforestation; Brazilian Cerrado; Cattle Ranching; Deforestation; Forest Degradation; Secondary Vegetation; South America.**

Further Reading

Alves, Diogenes S., "An Analysis of the Geographical Patterns of Deforestation in Brazilian Amazonia in the 1991–1996 period," in *Land Use and Deforestation in the Amazon*, eds. Charles H. Wood and R. Porro, Gainesville: University Press of Florida, 2002, pp. 95–106; Keller, Michael, Silva-Dias, Maria Assunção, Nepstad, Daniel C., and Meinrat O. Andreae, "The Large-Scale Biosphere-Atmosphere Experiment in Amazonia: Analyzing Regional Land Use Change Effects," in *Ecosystems and Land Use Change*, eds. Ruth S. DeFries, Gregory P. Asner, and Richard A. Houghton (Geophysical Monograph 153), Washington, DC: American Geophysical Union, 2004, pp. 321–34; Large-Scale Biosphere-Atmosphere Experiment in Amazonia Science Plan [Online, March 2005], The LBA Web Site, lba.inpa.gov.br.

HELMUT GEIST

Latin America.   Countries in the Americas that were originally colonized by Spain and Portugal. They are located from Mexico and Central America in the Northern Hemisphere, to certain islands of the Caribbean, and throughout most of the South American continent. Land cover is strikingly variable over this vast region, ranging from the world's largest tropical rainforest areas and the arid vegetation of large deserts, to the heterogeneity of natural environments in long and high mountain chains. Land use can be characterized in terms of strong asymmetries that characterize both the physical and human geographies of the region. Not only can examples of virtually all the world's climates and resulting climatic contrasts be found, but there are deep structural imbalances in power, land, wealth, and **access** to basic human needs that divide peoples and affect land use and changes in land use. The earlier exploitive economic systems set up by the colonizing powers of Spain and Portugal in the fifteenth and sixteenth centuries were replaced by more complicated systems that perpetuated inequalities between and among the landless and landowners, the poor and rich, and ethnic and racial minorities and majorities. In addition to these historical legacies, current **land-use systems** are rapidly changing due to the growing demands and power of cities, the **colonization** of new agricultural frontiers, and increasingly intertwined socioeconomic fates precipitated by migration and **globalization**.

Due to their locations, Mexico, Central America, and the Caribbean have been most affected in recent decades by national and regional geopolitics and economics as influenced by decisions of and opportunities within the **United States of America**. This influence is especially notable in land-use changes visible near transnational borders, for example between the United States and Mexico, where industrialized or urban landscapes have replaced rural agricultural systems, and where limited supplies of water for **agriculture** and growing cities are topics of daily concern. However, habitat alterations for natural resource extraction, **urbanization**, and the effects of growing populations are also to be found in many other landscapes of the region. For example, changes in distant landscapes are now tied together through globalization and trade ties among North and South American countries and with the rest of the world.

Originally, the humid areas of Mexico, the Caribbean islands, and Central America were subtropical and tropical environments dominated by **forests** and shrublands. Northern Mexico is subtropical and affected by occasional cold air masses intruding from the North America continent from October to February. This climatic regime limits the types of tropical agriculture that can be practiced: it controls how far to the north citrics and **coffee** can be successfully raised. The **mountains** also create many elevational zones and microclimates that are home to oak and pine forests on hillsides and higher elevations, with shrublands and agricultural fields covering lower areas and valley bottoms. There are also extensive desert environments, for example along the northwestern coast of Mexico, in intermontane valleys, and on leeward sides of Caribbean mountains. These are often utilized as rangelands, with herders moving flocks of cattle, sheep, and goats. In some places, land management includes techniques such as terraces and rock walls that store erratic rain as soil moisture, permitting the growth of crops in lands otherwise marginal for agriculture.

The tropical latitudes extend from about 23°N to 23°S, and here the climate systems are regulated by the convergence of the tropical trade winds in the Intertropical Convergence Zone (ITCZ), which moves seasonally, often bringing rains and a wet season when it is overhead, and clear skies and a dry season when it moves away. These latitudes include Latin American countries from Mexico southward to Peru and northern Argentina. Especially in Central America and the Caribbean, the trade winds also carry in atmospheric disturbances from July to September that can create very damaging hurricanes and tropical storms. The humid lowlands were originally covered by extensive tropical rain forests, occasionally opened up by hurricanes or treefalls. Beginning two millennia ago, these areas were utilized by large human populations in some areas, particularly in the Mayan-dominated areas of Mexico and Central America. The Caribbean islands were also home to large indigenous populations, with resulting **conversion** of forested landscapes for agriculture. Many other people lived in the highlands of these areas, interspersing agricultural fields among forests and shrublands. The arrival of European colonists brought grazing animals and the need for more pastoral land-use systems, steel axes that facilitated tree cutting, and permanent towns and cities that prospered with trade.

Tropical **South America** has very extensive humid lowlands because of the large width of South America at those latitudes. There are tropical rain forests in the Chocó of Colombia, in the Orinoco River basin of Venezuela and Colombia, and across the immense Amazon region, including large parts of Brazil, Ecuador, Peru, and portions of other countries. There are also sizable areas of tropical savannas and herbaceous **wetlands**, especially in sites with poor drainage or with enough seasonality in rainfall that the grasses burn in the dry season; these are located around the northern, eastern, and southern limits of the tropical forests of the Amazon, characterizing, for example much of Amazonian drainages of lowland Bolivia. Coastal margins from Brazil to northernmost Peru often have mangrove forests.

The most conspicuous **land-cover changes** in tropical Latin America have been associated with the expansion of the **agricultural frontier** in the South American lowlands and equivalent areas made accessible by road or rivers in Central America and Mexico. Beginning in the 1960s and continuing to the present, this landscape alteration has often resulted from organized or spontaneous colonization, and has included the clearing of forests, increased burning for the establishment and maintenance of **pastures**, cutting of mangroves to establish shrimp-raising ponds, and the growing of crops such as **maize** and **cassava** for subsistence and others for sale. A recent phenomenon is the additional conversion of tropical savannas to agriculture, especially for soybeans, and particularly in central Brazil. The most notable sites of **deforestation** and the most rapid expansion in area of agricultural lands are in tropical lowlands that have a pronounced dry season, which facilitates tasks such as slash burning and promoting the **regrowth** of grasses. The tropical environments that are perpetually humid are generally still dominated by wet forests, except along roads and rivers. In some places, indigenous groups continue to use these areas in manners that hardly alter forest structure and cover, by extracting fish from rivers, game from the forests, and crops from small agricultural fields that approximate the size of areas disturbed naturally by treefalls.

Coastal western South America includes some of the driest places on earth, due to the limitation of atmospheric humidity and uplift caused by cold-water currents of the eastern Pacific Ocean from about 4°S to 35°S, plus a rain shadow effect created by the Andes. These drylands are in coastal Peru and northern Chile. The land cover is **mesquite** woodlands or **tropical dry**

forests in places that receive seasonal rainfall, but much of the area is too arid for most **vegetation** types to be present and instead is barren or has a light cover of seasonal or ephemeral herbs. Land use consists of the opportunistic grazing of goats or irrigated agriculture in places where water from the Andes can be tapped into. The northern part of Peru and the southern part of Ecuador are the places most affected by the ENSO (El Niño Southern Oscillation), which is triggered by shifts in pools of cool water in the Pacific Ocean, causing a low-pressure atmospheric system to form every three to eight years above these generally arid areas and creating destructive precipitation events. An interesting phenomenon is that dryland areas receive ample humidity in an El Niño year and become covered by herbs or by the seedlings of dry forest trees, "greening up" in satellite imagery.

The tropical highlands and uplands of the **Andes** Mountains (Venezuela south to Bolivia) are places of continuity in terms of land use, given that they have been utilized by people for one to ten millennia, depending on the location. These are home to extensively terraced landscapes, often with intricate systems of irrigation farming, maintained by hand tools and physical labor. Land cover includes fields and pastures, in addition to shrublands; high-elevation **grasslands**; rocky slopes; glaciers; and, in places isolated from roads and settlements, Andean forests of evergreen broadleaf trees. The biggest recent land-cover changes have been driven by deforestation along penetration roads connecting the highlands to the humid lowlands, alterations in land use due to changes in types of agriculture and sizes of local populations, and by the growth of highland cities. Other upland areas in South America were not home to large pre-Hispanic populations, but have been converted by colonists in the past several centuries. For example, the subtropical forests that once covered the coastal mountains of southeastern Brazil are now almost completely gone, cut for agriculture and pastureland.

It is likely that the tropical highlands will be among the places in the world most altered by the effects of future climate change. Ice caps have become smaller along the entire length of the Andes. There is other evidence that precipitation and cloud formation have been altered, in some places perhaps due to deforestation and changed evapotranspiration processes, but in others due to the lifting of cloud banks caused by warmer temperatures. The alteration of physical environmental conditions will have many important consequences for land use, and for nature reserves and national parks established to protect the **biodiversity** of particular locations. A partial solution being pursued throughout Latin America is to attempt to connect nature reserves together using **conservation** corridors, so that no one protected area becomes completely isolated. In addition, the uplands serve as sources of water for cities and agriculture. Changed hydrological regimes are already altering plans for supplying water to growing urban areas and for the production of hydroelectric power.

The temperate zones of South America, to the south of about 35°S latitude, have seasonally cold climates, with winter months from April to September. The westerlies bring in humid air masses from the Pacific Ocean, so the wettest environments are on the coast and in the western Andes. These are covered by wetlands and bogs or by humid forests. The exceptions are the areas that have been converted from cover by native Andean forests to plantations of exotic tree species. Large areas of Argentina are covered by humid to dry grasslands, characterized by strong winds, flat to rolling topography, and gigantic ranches for the raising of cattle and sheep.

Throughout Latin America the growth of cities has been the land-use change with the most profound implications for altering land covers and shifting socioeconomic processes. Many people have moved from rural areas to growing cities, beginning in the 1940s, but increasingly so in the 1980s and continuing to the present. Concurrent population growth in the rural areas has kept population size relatively constant, despite this out-migration. The consequences for land use include the switching of agriculture to **cash crops** and other items bought by urban populations, more raising of livestock for meat, more tree plantations for pulp and timber needs, and the continual search for more drinking water for cities. Livelihood decisions made within households on the basis of their needs, goals, and resources available (i.e., social and natural capital) affect land use on lands they control. Thus, **land tenure** and other constraints affect the rates and locations of land-cover **modifications**. The last two decades have also been years of increased democratization in Latin America, with large urban populations increasingly controlling national natural resource decisions, and increased globalization resulting in some needed products being purchased from other countries rather than bought from nearby rural populations.

The environmental shifts and dilemmas created by these global, regional, and national changes are most profound in the island nations and in other countries with limited land or resources. In addition, the places with the largest disparities in wealth, especially in terms of the distribution of lands and land tenure, also have inherent social instabilities that can cause economic crises or social injustices. These twenty-first century preoccupations also rest uncomfortably atop legacies of extractive economic systems that characterized Spain's and Portugal's colonial heritages. The borders of the Latin American countries have often been the source of conflicts in the past, and tend to be foci of rapid land-cover change today.

The dynamics of land cover reflect these broadly social processes, just as other seasonal or episodic changes in land cover are due to altering physical environmental controls of precipitation, temperature, and natural disturbances. These changes can now be tracked using **remote sensing** and analyzed using computations and models powered through **geographical information systems** and computerized simulations. Such methods have helped elucidate the current **hot spots of land-cover change**, some of which coincide spatially with Latin America's hot spots of biological diversity. These are obviously places of special global concern, and include not only the tropical forests and grasslands, but the unique upland, dryland, and wetland environments to be found in the region. Mitigation efforts, such as **carbon sequestration** by establishing tree plantations, nature reserves, extractive reserves, and environmental restoration are all taking place; all of these endeavors can alter land cover. Predicting the location and direction of land-cover change due to roads, migration, and urban expansion is greatly facilitated with these techniques. **Land-use planning** for more appropriate land use and for the sustainable extraction of aquatic and forest products now routinely uses these inherently geographical approaches. These projects are being carried out both by governmental agencies and nongovernmental organizations in many countries in Latin America.

See also: Amazonia; Arc of Deforestation; Arena of Land Conflict; Brazilian Cerrado; Cattle Ranching; Coastal Zone; Decentralization; Decision-Making; Economic Livelihood; Exotic Species; Food Crops; Investments; Land Concentration; Poverty; Remittances Landscape; Slash-and-Burn Agriculture; Southern Cone Region; Subsistence Agriculture; Tropical Humid Forest.

Further Reading

Achard, Frédéric, Eva, Hugh D., Stibig, Hans-Jürgen, Mayaux, Philippe, Gallego, Javier, Richards, Timothy, and Jean-Paul Malingreau, "Determination of Deforestation Rates of the World's Humid Tropical Forests," *Science* 297 (2002): 999–1002; Blouet, Brian W., and Olwyn M. Blouet, eds., *Latin America and the Caribbean: A Systematic and Regional Survey*, 4th ed., New York: John Wiley, 2002; Bulmer-Thomas, Victor, *The Economic History of Latin America Since Independence*, 2nd ed., Cambridge, UK: Cambridge University Press, 2003; Denevan, William M., *Cultivated Landscapes of Native Amazonia and the Andes*, Oxford, UK: Oxford University Press, 2001; Doolittle, William E., *Cultivated Landscapes of Native North America*, Oxford, UK: Oxford University Press, 2001; Economic Commission for Latin America and the Caribbean [Online, August 2004], The CEPAL Web Site, www.eclac.cl; Gwynne, Robert N., and Cristóbal Kay, eds., *Latin America Transformed: Globalization and Modernity*, 2nd ed., London: Arnold, 2004; Kellman, Martin, and Rosanne Tackaberry, *Tropical Environments: The Functioning and Management of Tropical Ecosystems*, London: Routledge, 1997; Kelly, Philip, *Checkerboards and Shatterbelts: The Geopolitics of South America*, Austin, TX: University of Texas Press, 1997; Knapp, Gregory, ed., *Latin America in the 21st Century: Challenges and Solutions*, Austin, TX: University of Texas Press, 2002; Latin American Network Information Center [Online, August 2004], The LANIC Web Site, lanic.utexas.edu; Thorp, Rosemary, *Progress, Poverty and Exclusion: An Economic History of Latin America in the 20th Century*, Baltimore, MD: Johns Hopkins University Press, 1998; Whitmore, Thomas M., and B. L. Turner II, *Cultivated Landscapes of Middle America on the Eve of Conquest*, Oxford, UK: Oxford University Press, 2001.

KENNETH R. YOUNG

Leapfrogging. A spatial process of urban development in which land closer to urban areas is skipped over while land further away is developed. The result is a discontinuous pattern of urban and rural land rather than a contiguous urban area. "Leapfrog," "scattered," and "fragmented development" are often used interchangeably to mean a pattern of discontinuous development, but leapfrog development implies such patterns with respect to an urban gradient, whereas scattered and fragmented development do not. Because the intervening rural land may be developed at a later date, these patterns can be transient. While some studies have provided evidence of extensive discontinuous development patterns in specific **urban-rural fringe** areas of the **United States of America**, it is unclear how pervasive these patterns are in general.

Explanations of leapfrog development range from exogenously driven processes of land speculation and land heterogeneity to endogenous spatial externalities. The traditional urban economics model explains leapfrogging as a temporary phenomenon resulting from optimal intertemporal **decision-making** in which developers generate higher returns by delaying the development of land that is closer to the city. Other theories appeal to the existence of dispersive forces such as congestion externalities and open-space amenities. Dawn Parker and Vicky Meretsky demonstrate that fragmented urban patterns can emerge from such effects at the urban fringe using an **agent-based model** of land use. Elena Irwin and Nancy Bockstael find empirical evidence of such externalities using data from the Washington, DC region and demonstrate that their magnitude is sufficiently large to generate systematic patterns of discontinuous residential development.

See also: **Endogenous Factor; Exogenous Factor; Land Fragmentation; Modeling; Suburbanization; Urbanization; Urban Sprawl.**

Further Reading

Irwin, Elena G., and Nancy E. Bockstael, "Interacting Agents, Spatial Externalities and the Endogenous Evolution of Land Use Patterns," *Journal of Economic Geography* 2 (2002): 31–54; Parker, Dawn C., and Vicky Meretsky, "Measuring Pattern Outcomes in an Agent-Based Model of Edge-Effect Externalities Using Spatial Metrics," *Agriculture, Ecosystems, and Environment* 101 (2004): 233–50.

ELENA G. IRWIN

M

Maathai, Wangari M. (1940–). Environmental and political activist, first woman in eastern and central Africa to earn a doctorate degree, honorary professor of biology (veterinary anatomy), assistant minister for Environment, Natural Resources and Wildlife in Kenya's ninth parliament since 2002, and winner of the 2004 Nobel Peace Prize. With the recognition by the Norwegian Nobel Committee that "peace on earth depends on our ability to secure our living environment," the "can-do person" Wangari Maathai was awarded for her holistic or integrative approach to sustainable development, which interlinked women's rights, democracy, ecological restoration, and grassroots activism in favor of sustainable development. She founded the Green Belt Movement, which for thirty years has mobilized women to plant about 30 million trees in Kenya and other African countries. More than two decades ago, she became a member and then the chair of the Environment Liaison Centre International (ELCI) in Nairobi, Kenya. The ELCI emerged from the first Earth Summit in Stockholm in 1972 and was located in Nairobi in 1974 to track international environmental processes (with the **United Nations Environment Programme** there), and to support effective techniques for networking of non-governmental organizations (NGOs) in developing countries. Locally and in Africa, ELCI campaigns relate to **forests**, **sustainable land use**, and air pollution. ELCI facilitates NGO involvement at the world summits for sustainable development and organizes the NGO community on the **United Nations Convention on Biological Diversity**. In the most recent past, Maathai was involved in campaigns against land grabbing and unsustainable forest management practices.

See also: Arena of Land Conflict; Community Involvement; Conservation; Economic Livelihood; Forestry; Green Wall Project; Institutions; Reforestation.

Further Reading
Environment Liaison Centre International [Online, February 2005], The ELCI Web Site, www.elci.org; Greenbelt Movement, Biographies [Online, February 2005], The GBM Web Site, www.greenbeltmovement.org; Maathai, Wangari [Online, February 2005], The Official Web Site of Prof. Wangari Maathai, www.wangarimaathai.or.ke.

HELMUT GEIST

Maize. After **wheat** and **rice**, the most important cereal worldwide. Maize (or corn) is one of the most widely distributed cereals grown in the tropics and subtropics and in the temperate zones, between 58° N (**Canada, Russia**) and 42° S (New Zealand, **South America**) and from below sea level (Caspian Sea plain) up to 3,600 meters high (Peruvian **Andes**).

Soil erosion in an agricultural field that has been cleared of timber and put into corn (maize) in Hamilton County, Illinois, in February 1937 (note stumps of trees and beginning of erosion gullies).
[Courtesy Library of Congress]

Maize is used as a staple food particularly in the tropics; as feed for livestock and poultry, particularly in the industrialized countries of the temperate zones; and as raw materials for many industrial products. Maize is a starch plant, and the ripe grain is prepared for human **consumption** in multiple ways. In 2000, about 19 percent of the total maize harvested in the world was used for human consumption. Most maize (about 65 percent in 2000), however, is used to feed livestock. Silage maize has become one of the most important feed crops in industrialized western countries.

Maize originated in Central America (Mexico). **Domestication** of maize started about 8,000 years ago. Cultivation of maize and its importance as a **food crop** are closely related to the rise of pre-Columbian Central and South American civilizations. In the fifteenth and sixteenth centuries, maize was introduced into **Europe** and Southeast Asia. Due to its adaptability and productivity, maize spread rapidly around the globe. It is the most domesticated crop, but its genetic origin and evolution are not well understood. Maize is a tall annual cereal grass bearing kernels on large ears. It has a single predominant stem and, different from other cereals, the formation of basal branches (tillers), which are limited and only of importance in low-density stands. Maize is a C4 plant with typically higher rates of photosynthesis compared to non-C4 plants, and requires warm temperatures and adequate soil water supply. It ideally grows in tropical and subtropical conditions and is less suitable for semiarid and equatorial climates. Minimum temperature for germinations is about 10°C, and average daily temperature during growth should be at least

20°C. The period from planting to harvesting varies considerably depending on cultivar and climatic conditions. Plants may be harvested between 70 and 200 days for very early or very late cultivars, respectively. The large number of cultivars of different maturity types allows cultivation of maize across a wide range of temperature conditions.

Maize is a highly variable species. Cultivars are divided into groups according to the structure and shape of the grain. Important cultivar groups are dent, flint, pod, pop, soft, sweet, and waxy maize, or corn.

Maize evolved into a highly productive crop within a relatively short time period. Productivity of maize was significantly increased in the early twentieth century by the discovery and development of hybrid maize, created by crossing of two or three inbred lines. Cultivation of hybrid maize requires adequate facilities for seed production and is commonly used in high-input farming with high fertilizer use. The larger area of maize (about 66 percent of the total global maize area) is, however, in developing countries characterized by low-input smallholder farming. In these areas, local, unimproved cultivars are commonly grown. Resources for production of hybrid maize seeds and intensification of maize production are generally not available.

The advantages of maize compared to other cereals with respect to its cultivation and processing and its enormous genetic variability have favored its use, geographical distribution, and production. Maize is presently grown in more countries than any other cereal, and in recent years maize production has even exceeded the production of paddy rice and wheat.

The area used for maize production has steadily increased over the last centuries due to increasing demand for food, mainly from population growth. Different from other important cereals, maize's harvested area has continued to increase up until present days. About 145 million hectares are presently cultivated with maize, which is almost 10 percent of the total arable land, and represent an increase of maize-harvested area of about 40 percent between 1961 and 2004. However, increases in maize area are mainly observed in developing countries, with more than 60 percent between 1961 and 2004. Maize-harvested area in developed countries increased only marginally, but productivity in these countries is high. About 57 percent of the global total maize production in 2004 came from developed countries.

Most maize is produced in the **United States of America**, which contributes about 42 percent (2004) to the global production. Other important producers are **China**, Brazil, Mexico, and France with 19 percent, 6 percent, 3 percent, and 2 percent of respective global maize production.

Global average maize yields have more than doubled over the last four decades and have reached almost five tons per hectare. Yields, however, differ considerably among countries, ranging between one and nine tons per hectare. Yields are more than twice as high in developed than in developing countries, which is largely due to technological, economic, and organizational reasons. In many developing countries of the tropics and subtropics, maize is grown in small farms with low inputs such as fertilizer, improved seed, water, and labor for crop weeding.

Demand for maize will further increase with population growth and economic wealth. Future production of maize will depend on biophysical and socio-economic factors. Climate change has been projected to affect the productivity of maize and its regional distribution. Increasing temperatures may drastically reduce yields in some regions, for example, in Mexico and in Mediterranean countries, or increase yields in other regions, for example, **western Europe** and North America. A northward shift of the maize-growing area has been projected for Europe and Canada. Increasing atmospheric CO_2 concentration may ameliorate some of the negative effects of climate change. However, since maize is a C4 plant, effects of increasing CO_2 on growth are less pronounced than for cereals with a C3 pathway of carbon fixation such as wheat and rice, and climate change effects may be more evident for maize than for C3 cereals. Impacts of climate change will also depend on the **vulnerability** of the agricultural systems, including their ability to adapt to changes in climatic conditions. Effectiveness of adaptation options, such as change of cultivar and/or crop, irrigation, and fertilizer will depend on the economic and social conditions of the specific agricultural system. These production conditions are less favorable in most developing countries and pose severe risks on future maize production in these countries due to climate change. The yield potential of maize is higher compared to that of wheat and rice. It will depend on the socio-economic development of the developing world to determine whether what extent this potential can be utilized in the future and the negative environmental effects associated with its intensive production can be minimized.

See also: **Agricultural Intensification; Biodiversity Novelties; Cash Crops; Exotic Species; Monoculture; Population Dynamics; Turnover.**

Further Reading

Koopmans, A., ten Have, H, and Subandi, "Zea Mays L," in *Plant Resources of South-East Asia Handbook 10: Cereals,* eds. Gerard J. H. Grubben and Soetjipto Partohardjono, Leiden, NL: Backhuys Publishers, 1996, pp. 143–49; Parry, Martin L., Rosenzweig, Cynthia, Iglesias, Ana, Livermore, Matthew, and Günther Fischer, "Effects of Climate Change on Global Food Production under SRES Emissions and Socio-Economic Scenarios," *Global Environmental Change* 14 (2004): 53–67; Ramankutty, Navin, and Jonathan A. Foley, "Estimating Historical Changes in Global Land Cover: Croplands from 1700 to 1992," *Global Biogeochemical Cycles* 13 (1999): 997–1027; United Nations Food and Agriculture Organization [Online, February 2005], The FAO Web Site, www.fao.org.

FRANK EWERT

Malaria.   A human disease caused by the vector-borne, protozoal parasite *Plasmodium spp,* affecting people in the Caribbean, **South America**, Southeast Asia, India, and sub-Saharan Africa. Malaria, together with tuberculosis and **human immunodeficiency virus (HIV)/acquired immunodeficiency syndrome (AIDS)**, is considered to be one of the most significant public health threats to development in tropical countries of the developing world. According to estimates of the World Health Organization (WHO) of the United Nations, at least a million malaria deaths occur each year, 90 percent of them in sub-Saharan Africa. Mortality is particularly high in infants and young children. Transmitted by the *Anopheles* mosquito, the plasmodium parasites affect only humans and have no other significant animal host or reservoir. In recent years, malaria has experienced a resurgence, which has been primarily associated with increased drug resistance, increased insecticide resistance, global climate change, and local land-use changes.

The links between land use and related land-cover change and malaria are predominantly associated with the impact of land-use/cover change on the habitat and survival of the *Anopheles* mosquito vector. In the case of malaria, temperature, humidity, sun exposure, and water availability can affect reproduction and breeding site availability for the mosquito vectors, though different vector and parasite species have unique temperature and environmental preferences and associated distributions. Many mosquito species breed well in stagnant water with high sunlight, such as irrigation ditches and pools. Malaria mosquitoes respond to rapid changes in their habitat and are particularly susceptible to temperature and moisture changes. As noted by Jonathan Patz and Nathan Wolfe, land-cover **modifications** resulting in changes to temperature and moisture can cause dramatic increases in vector populations and malarial transmission rates.

The name malaria derives from associations with land cover; it comes from *mal aria*, which is Italian for "bad air." This association is due to an historical association of the disease with fetid swamps and marshes. Up until the eighteenth and nineteenth centuries, malaria occurred in the **United States of America**, some parts of **Canada**, and most northern European countries. Paul Reiter argues that a dramatic decline in malaria rates in England and northwestern **Europe** during the late nineteenth century was due, in a large part, to land-use changes, including improved drainage and extensive land reclamation, increased cultivation of root crops, and increased livestock-keeping. The latter may have diverted mosquitoes from humans to animals, reducing potential transmission of the parasite within the human population, since malaria has no significant animal host. More recently, however, the contraction of malaria's extent has reversed, and the disease has extended into areas of Central America, northern South America, tropical and subtropical Asia, and parts of the Mediterranean and former Soviet Union. This reemergence has been associated with human **population dynamics** (such as changes in population distribution and density), **urbanization**, **deforestation**, drug resistance, sociopolitical instability, international travel, agricultural change, and climate change. Of these causal factors, a significant number are strongly correlated with patterns of land-use/cover change.

Land-use factors such as agricultural development, population increase, and urbanization have been associated both with declines and expansions of malarial distributions. Historical expansion of malaria in West Africa was associated with **conversion** of rain **forest** into **agriculture**; increased mosquito habitat and population growth provided breeding habitat in cleared areas with sunlight-exposed pools, as well as availability of human hosts. More recent expansion of malaria in frontier zones such as Southeast Asia and **Amazonia** is similarly associated with deforestation and conversion to agricultural cropping. By contrast, nineteenth-century disease decline in Europe has been linked to agricultural change as the resulted of improved livestock husbandry, therefore diverting mosquito feeding from humans to animals.

Decline of disease in Africa following World War II was predominantly the result of wide-scale spraying of dichloro-diphenyl trichloroethane (DDT), a pesticide lauded for its efficacy at drastically reducing mosquito densities. This campaign, however, encountered the emergence of DDT resistance in mosquitoes in conjunction with detection of persistent environmental impacts. Fifty mosquito species have now been found to be resistant to one or more insecticides, including DDT. In addition to DDT, other insecticide and larvacides are used to reduce mosquito populations and malarial rates, but associated drug resistances and environmental impacts have been noted. Use of chemicals for purposes other than malarial control has also contributed to mosquito drug resistances: the use of some pesticides and fertilizers for increased crop production has been implicated in increased mosquito resistance to DDT.

In the case of population density changes, malaria requires a minimum density of humans in order to support a continued reservoir for transmission of the parasite. Generally, malaria does not occur in high-density, established urban areas (though this has been observed in some cities of South Asia). Warm, developing cities with semi-urban slums, however, can provide reservoirs of disease surrounding urban areas, resulting in high urban infection rates. As noted by Irwin Sherman, up to 10–20 percent of incidence in some countries is attributable to urban and peri-urban malaria, due primarily to the presence of slums.

Land-development changes, including creation of dams, irrigation schemes, commercial tree cropping, deforestation, and agricultural development, have been associated with increases in mosquito vector habitat as well as increased contact between mosquitoes and humans. David Molyneux and Irwin Sherman summarize a range of examples of increased abundance of mosquito vectors following forest disturbance. Deforestation in West Africa, for example, resulted in the creation of stagnant pools ideal for larval habitat, while expansion of the species *Anopheles darlingi* is associated with deforestation in Amazonia. In Southeast Asia, adaptation of several species of mosquito to new plantations has been found. In some cases, forest conversion or **land-cover change** may decrease the habitat availability of one mosquito species while increasing available habitat for an alternate mosquito species; whether the former or the latter species is a better vector of malarial parasites will determine whether there will be an associated decrease or increase in malarial risk. Research by Kim Lindblade and colleagues found that replacement of natural swamps with agricultural cropping in Uganda led to increased temperatures and potential for increased malarial transmission. This is supported by Katrin Kuhn and colleagues, who note that mosquito densities are often associated with irrigated cropland (such as **rice** paddies) and marshland. Creation of micro-habitats such as these provides adequate sunlight and moisture appropriate for breeding in a range of mosquito species.

The types of **vegetation** fragmentation noted in the above examples, and their associated impacts on increased malarial transmission potentials, are particularly prominent in frontier areas such as the Amazon basin and in **agricultural frontiers** of Southeast Asia. Wannapa Suwonkerd and colleagues, for example, found that increased landscape diversity and forest **fragmentation** in Chiang Mai province of northern Thailand were associated with decreases in mosquito densities over twenty-three years. Irwin Sherman notes that temporary migrant workers traveling into malarial zones often become infected with malaria, particularly in association with deforestation projects in frontier forest areas.

Dam construction is another common example of land development associated with malarial incidence change. Tedros Ghebreyesus and colleagues, in a community-based study of the effect of microdam construction on malaria incidence in Ethiopia, found that the projects resulted in significantly increased transmission in local communities. Where there are unmanaged zones of vegetative growth around reservoir edges, dam development has been associated with the creation of excellent mosquito habitat. More recently, potential impacts of climate change on infectious disease rates and distributions has drawn attention to possible changes in the distribution and magnitude of global malaria risk. It is notable that climate change models predict significant increases in *P. falciparum* malaria under almost all scenarios. Jonathan Patz and colleagues provide further discussion on the potential impacts of climate change on malaria.

Reciprocally, malaria can have impacts on land use, particularly in developing countries, where disease burdens may undermine health and welfare, debilitate the active members of communities, and strain resources; these impacts affect **people at risk** in **southern Africa**.

See also: **Climate Impacts; Human Health.**

Further Reading

Ghebreyesus, Tedros A., Haile, Mitiku, Witten, Karen H., Getachew, Asefaw, Yohannes, Ambachew M., Yohannes, Mekonnen, Teklehaimanot, Hailay D., Lindsay, Steven W., and Peter Byass, "Incidence of Malaria among Children Living near Dams in Northern Ethiopia: Community Based Incidence Survey," *British Medical Journal* 319 (1999): 663–6; Kuhn, Katrin G., Campbell-Lendrum, Diarmid H., and Clive R. Davies, "A Continental Risk Map for Malaria Mosquito (*Diptera: Culicidae*) Vectors in Europe," *Journal of Medical Entomology* 39 (2002): 621–30; Lindblade, Kim A., Walker, Edward D., Onapa, Ambrose W., Katungu, Justus, and Mark L. Wilson, "Land Use Change Alters Malaria Transmission Parameters by Modifying Temperature in a Highland Area of Uganda," *Tropical Medicine and International Health* 5 (2000): 263–74; McMichael, Anthony J., *Human Frontiers, Environment and Disease: Past Patterns, Uncertain Futures*, Cambridge, UK: Cambridge University Press, 2001; Molyneux, David H., "Vector-Borne Infections and Health Related to Landscape Changes," in *Conservation Medicine: Ecology and Health in Practice*, eds. A. Alonso Aguirre, Richard S. Ostfeld, Gary M. Tabor, Carol House, and Mary C. Pearl, New York: Oxford University Press, 2002, pp. 194–206; Oaks, Stanley C., Mitchell, Violaine S., Pearson, Greg W., and Charles C. J. Carpenter, eds., *Malaria: Obstacles and Opportunities*, Washington, DC: National Academy Press, 1991; Patz, Jonathan A., and Nathan D. Wolfe, "Global Ecological Change and Human Health," in *Conservation Medicine: Ecology and Health in Practice*, eds. A. Alonso Aguirre, Richard S. Ostfeld, Gary M. Tabor, Carol House, and Mary C. Pearl, New York: Oxford University Press, 2002, pp. 167–81; Reiter, Paul, "From Shakespeare to Defoe: Malaria in England in the Little Ice Age," *Emerging Infectious Disease* 6 (2000): 1–11; Sherman, Irwin W., ed., *Malaria: Parasite Biology, Pathogenesis, and Protection*, Washington, DC: ASM Press, 1998; Suwonkerd, Wannapa, Overgaard, Hans J., Tsuda, Yoshio, Prajakwong, Somsak, and Masahiro Takagi, "Malaria Vector Densities in Transmission and Non-Transmission Areas During 23 Years and Land Use in Chiang Mai Province, Northern Thailand," *Basic and Applied Ecology* 3

(2002): 197–207; World Health Organization of the United Nations, Roll Back Malaria [Online, October 2004], The RBM Web Site, www.rbm.who.int.

LEA BERRANG FORD

Marsh, George Perkins (1802–1882). A polymath, multilingual American conservationist from Vermont who combined careers in law, farming, business, politics (U.S. congressman from 1843–1849), diplomacy (U.S. envoy to Turkey from 1849–1853 and Italy from 1861–1882), and scholarly writing. He is best known for his insights into the human imprint upon ecosystems and environmental **conservation**. David Lowenthal states that up until 1864, when Marsh wrote *Man and Nature*, most human impacts on the globe were regarded as benign or, where deleterious, trivial and self-correcting. However, Marsh demonstrated that intended as well as unintended consequences of human interventions—such as settlement or **colonization**, **deforestation**, **desertification**, and **species extinction**—deranged the normal stability of, for example, **soil erosion** and restorative processes, which otherwise altered the planet at a slow (geological) pace only. Based on extensive studies and travel, he portrayed the earth as a principally harmonious world torn apart by human greed and ignorance. Nonetheless, he never doubted that continued mastery and management of land resources was essential to civilization. Celebrated in a 1955 Princeton University symposium and again at Clark University in 1989, *Man and Nature* was republished and remains in print to this day. Although the essential message was not entirely novel—writers such as the German geographer Alexander von Humboldt, and others in France, Italy, and India, had expressed similar concerns—George Perkins Marsh was the one who synthesized in an accessible format a wide array of existing knowledge and experience (he could read in twenty languages). His approach, together with the rational utility work of **Johann Heinrich von Thünen**, has pioneered contemporary scientific studies dealing with land-use/cover changes worldwide.

George Perkins Marsh (scholar, whig congressman from Vermont, and U.S. Minister to Turkey and Italy), ca. 1850.

PHOTO: Mathew B. Brady. [Courtesy Library of Congress]

See also: **Driving Forces; Environmental Change; George Perkins Marsh Institute; Land Degradation.**

Further Reading

Lowenthal, David, "Marsh, George Perkins," in *Encyclopedia of Global Change: Environmental Change and Human Society*, vol. 2, eds. Andrew S. Goudie and David J. Cuff, Oxford, New York: Oxford University Press, pp. 60–61; Marsh, George P., *Man and Nature, or Physical Geography as Modified by Human Action*, Cambridge, MA: Harvard University Press, 1864 (reprint; ed. David Lowenthal, 1965).

HELMUT GEIST

Mediating Factor.   A causative factor of land change that shapes, modifies, or intervenes in the interplay between underlying **driving forces** and **proximate causes** and that, if related to **feedbacks**, can affect the rate of land change. Often-cited examples of mediating factors are gender, ethnic affiliation, class, and wealth status (and thus power relations), but also biophysical properties such as land suitability (e.g., degree of slopes, soil moisture, etc.). For example, the **conversion** of tropical forests into agricultural uses is often found to be mediated by the unequal relations between well-endowed large-scale farmers or corporate agricultural enterprises and smallholders eking out a living, thus creating "entrepreneurial" versus "populist" **agricultural frontiers** with rather distinct land uses. Likewise, all categories of **agrodiversity** in settled agricultural zones—that is, biophysical diversity, management diversity, agro-biodiversity, and organizational diversity—are shaped by factors that play out differently at various time and spatial **scales**. For example, crop choice and types of conservation practices often differ between poor and rich farmers, thus affecting the pattern of management diversity, and feeding back to enlarge differences in natural **land quality**. In a broader sense, mediating factors that operate through demographic drivers can shape the trajectory of land change toward degradation or restoration. Whether or not increasing population is damaging or beneficial depends upon a variety of institutional, ecological, and technological factors: population growth can cause **land degradation** in the short term, but it can also spur innovation and **agricultural intensification** as well as the adoption of conservation techniques. Mediating factors are crucial also for the response of land managers to external forces, that is, feedbacks are strongly mediated by local factors such as **access** to land, gender, education, and institutional arrangements. In particular, **institutions** need to be considered at various scales to identify those local mediating factors that—together with peoples' adaptive strategies or responses to changing market opportunities—drive land-use change. Juanita Sundberg, for example, illustrates how local participation in natural resource **conservation** is strongly mediated by a community's interaction with non-local actors such as national governments, transnational corporations, and international non-governmental organizations. Seen together with other examples, these "conservation encounters" can shape landscapes and livelihoods in rather contradictory ways. In the Mayan zone, for example, local evidence of high **deforestation** can be found close to locations where exceptionally low rates of deforestation have occurred, with intervening institutional factors making the difference. Another example relates to institutional settings that allow or impede the creation of purchasing power: especially if linked to opportunities of trade and transportation of food, purchasing power can reduce the pressure for **extensification**, as evidenced in the historical decline of New England **agriculture** and subsequent **reforestation** in the region. Mediating factors can form a **multiplier effect**. For example, the environmental effects of an external capital injection—be they foreign direct **investments** or incomes creating **remittances landscapes**—are not limited to the injection itself, but are mediated or magnified by the ways in which capital then circulates through the institutional structure of the economy.

See also: **Arena of Land Conflict; Community Involvement; Exogenous Factor; Land Concentration; Landholding; Land-Use History; Tradeoffs.**

Further Reading

Agrawal, Arun, and Gautam N. Yadama, "How Do Local Institutions Mediate Market and Population Pressures on Resources? Forest *Panchayats* in Kumaon, India," *Development and Change* 28 (1997): 435–65; Bray, David Barton, Ellis, Edward A., Armijo-Canto, Natalia, and Christopher T. Beck, "The Institutional Drivers of Sustainable Landscapes: A Case Study of the 'Mayan Zone' in Quintana Roo, Mexico," *Land Use Policy* 21 (2004): 333–46; Sundberg, Juanita, "Strategies for Authenticity and Space in the Maya Biosphere Reserve, Petén, Guatemala," in *Political Ecology: An Integrative Approach to Geography and Environment-Development Studies*, eds. Karl S. Zimmerer and Thomas J. Bassett, New York, London: The Guilford Press, 2003, pp. 50–69.

HELMUT GEIST

Medium Resolution Imaging Spectrometer. Launched in 2002 onboard the European Space Agency (ESA) ENVISAT satellite, the medium resolution imaging spectrometer (MERIS) has fifteen spectral bands between 0.39 and 1.04 μm (visible to near-infrared wavelengths). Data are acquired at a spatial resolution of 300 meters at nadir. The primary mission of MERIS is the measurement of sea color. It also has great potential to contribute to the mapping and **monitoring** of **land-cover change**. The **normalized difference vegetation index** can be computed. Good temporal resolution is achieved with a repeat-pass interval of only three days, thus ensuring frequent observations of the earth's surface.

See also: **Change Detection; GLOBCOVER; Remote Sensing.**

Further Reading

European Space Agency [Online, August 2004], The Medium Resolution Imaging Spectrometer Web Site, envisat.esa.int/instruments/meris.

KEVIN J. TANSEY

Meta-Analysis. Comparison of disparate case studies as a means of pattern recognition or providing insights about land-change dynamics at the meso- and macroscales. The purpose of meta-analytical research is to combine findings from separate but largely similar local-scale studies that allow for the application of a variety of analytical techniques such as structured literature review or formal statistical approaches, generating a general understanding, for example, of the causes, rates, and pathways of land change while preserving the descriptive richness of these studies. The potential errors produced by data biases and other methodological problems associated with the comparative method have early been raised and dealt with by Charles Ragin, Benedetto Matarazzo and Peter Nijkamp, and Thomas Cook and colleagues, among others. Given the many limitations for original research following standardized data protocols to arrive at a general understanding of land change dynamics beyond the local to subnational level, meta-analytical frameworks for analyzing case studies of **global environmental change** issues have been increasingly recognized as important analytical tools, i.e., to draw inferences on common issues with different but allied empirical backgrounds. For example, Camille Parmesan and Gary Yohe evaluated results from thousands of cases linking climate change and population ecology to better understand the assumed relevance of an **ecological footprint** of climate change upon ecosystems, while Peter Nijkamp and colleagues ran a meta-analytical comparison of empirical results related to the same notion. Likewise, Helmut Geist and Éric Lambin used a meta-analytical framework to determine whether **proximate causes** and underlying **driving forces** of tropical **deforestation** and **desertification** fall into any patterns, also identifying **mediating factors**, **feedbacks**, cross-scalar dynamics, and typical pathways.

See also: **Global Land Project; Land-Use/Cover Change (LUCC) Project; Non-Timber Forest Products; Population Dynamics; Scale.**

Further Reading
Cook, Thomas D., Cooper, H., Cordray, D., Hartmann, H., Hedges, L., Light, R., Louis, T., and F. Mosteller, *Meta-Analysis for Explanation: A Casebook*, New York: Russell Sage Foundation, 1992; Geist, Helmut J., and Éric F. Lambin, "Proximate Causes and Underlying Driving Forces of Tropical Deforestation," *BioScience* 52 (2002): 143–50; Geist, Helmut J., and Éric F. Lambin, "Dynamic Causal Patterns of Desertification," *BioScience* 54 (2004): 817–29; Matarazzo, Benedetto, and Peter Nijkamp, "Meta-Analysis for Comparative Environmental Case Studies: Methodological Issues," *International Journal of Social Economics* 24 (1997): 799–811; Nijkamp, Peter, Rossi, Emilia, and Gabriela Vindigni, "Ecological Footprints in Plural: A Meta-Analytic Comparison of Empirical Results," *Regional Studies* 38 (2004): 747–65. Parmesan, Camille, and Gary Yohe, "A Globally Coherent Fingerprint of Climate Change Impacts across Natural Systems," *Nature* 421 (2003): 37–42. Ragin, Charles, *The Comparative Method: Moving Beyond Qualitative and Quantitative Strategies*, Berkeley: University of California Press, 1987.

HELMUT GEIST

Metabolism.   A biological notion that denotes the processes that supply organisms with energy and chemical building blocks needed for growth and maintenance of their structure and functions. The notion is also used to describe energy, material, and substance flows associated with the functioning of socio-economic systems. The importance of socio-economic metabolism for land use and vice versa is increasingly recognized.

The original biological notion was coined by the biologist Jakob Moleschott (1822–1893), who described metabolism as the exchange of matter between an organism and its environment. Nowadays, biological metabolism is primarily researched by biochemists, and the notion has changed accordingly. Modern biology textbooks describe the metabolism of an organism as the sum of biological reactions that convert raw materials obtained from the environment into the macromolecules that make up the organism or are needed for its energy supply. Two main metabolic pathways are usually distinguished: catabolism denotes all biological processes by which food is broken down to simple molecules and energy is produced, while the synthesis of new molecules from simple building blocks is termed anabolism. Regulatory mechanisms that govern these complex processes are usually also included in the biological notion of metabolism. Important examples of metabolic processes are photosynthesis; the breakdown of food proteins to amino acids; energy-generating reactions such as the citric acid cycle; and the synthesis of proteins, DNA, or other important macromolecules. However, even though most ecologists acknowledge that energy flows and **nutrient cycles** in ecosystems are to some extent self-organized, biologists generally do not extend the metabolism notion to the supra-organismic level (i.e., to populations, communities, or ecosystems).

The use of the metabolism analogy to describe exchanges of energy, materials, or even single substances such as metals, carbon, nitrogen, etc., between societies and their environment can be traced back to Karl Marx (1818–1883) and Friedrich Engels (1820–1895), who borrowed the notion from Moleschott and used it in the book *Das Kapital* (*Capital*) to describe the appropriation of natural substances to human requirements in the labor process. Issues of material flows and the possible threat of related resource exhaustion played a considerable role in the 1955 conference "Man's Role in Changing the Face of the Earth" in Princeton, New Jersey.

Contemporary socio-economic metabolism research is largely divided into two strands: methods to account for inputs required (energy, materials, etc.) or outputs generated (wastes, emissions) in the production chain of products or services, an approach called life cycle analysis (LCA); and methods to analyze the energy, materials, or substance flows through geographically defined socio-economic systems (villages or cities, nation-states, supranational entities, or humanity as a whole) or parts thereof (e.g., economic sectors). The two approaches are clearly complementary and should, in principle, also be compatible; in practice, however, the two are researched by only loosely related communities. The first approach is pursued mostly by engineers and others who seek to optimize products; accordingly, analyses of substance flows (e.g., toxic or other hazardous substances) play an important role here, although accounts of the amount of land area required for a production process have also been performed. Within the second approach, most work in the last decades has gone into developing tools to account for nationwide material flows (material flow analysis, or MFA), but other spatial **scales** (e.g., cities, local communities) have also received attention. Analogous methods to account for energy flows have been proposed, and the concept can also be extended to substances such as carbon. These issues are currently researched by a community dominated by biologists, economists, and social scientists who are interested in the ability of MFA to provide accounts of national material (energy, carbon, etc.) flows in good accordance with monetary national accounts (i.e., the System of National

Accounts), thus allowing analyses of interrelations between **economic growth** and a nation's physical throughput. The metabolism approach has recently also been used in environmental history to explore changes in society-nature interaction over long periods of time.

Many aspects of socio-economic metabolism are relevant for land-use change: extraction of resources requires land area, as does their transport, processing, **consumption**, and disposal. The provision of **biomass** usually requires the largest proportion of this area. Several approaches have been used to relate systemwide analyses of socio-economic metabolism and land use. **Ecological footprint** studies basically ask how much area is needed for the provision of biomass and other resources, to host infrastructure, and to absorb carbon from fossil fuel combustion of a defined population, mostly that of a nation-state. Analyses of the **human appropriation of net primary production** demonstrate how much of the biological productivity of a defined area is harvested to support socio-economic metabolism, or foregone due to land **conversions** or land management. Historical analyses of the long-term development of land use and socio-economic metabolism have revealed that changes in socio-economic metabolism, above all those occurring during transitions from agricultural to industrial society, fundamentally alter the role of **agriculture** and **forestry** for the society's energy system and are thus a major **driving force** behind many important changes in land use.

See also: **Analysis of Land Change; Carbon Cycle; Economic Restructuring; Industrialization; Mineral Extraction; Pattern Metrics.**

Further Reading

Ayres, Robert U., and Udo E. Simonis, eds., *Industrial Metabolism: Restructuring for Sustainable Development*, Tokyo: United Nations University Press, 1994; Fischer-Kowalski, Marina, "Society's Metabolism: The Intellectual History of Materials Flow Analysis, Part I, 1860–1970," *Journal of Industrial Ecology* 2 (1998): 61–78; Haberl, Helmut, Wackernagel, Mathis, and Thomas Wrbka, eds., "Land Use and Sustainability Indicators," *Land Use Policy* 21 (2004): 193–320; Krausmann, Fridolin, Haberl, Helmut, Schulz, Niels B., Erb, Karlheinz, Darge, Ekkehard, and Veronik Gaube, "Land-Use Change and Socioeconomic Metabolism in Austria, Part I: Driving Forces of Land-Use Change 1950–1995," *Land Use Policy* 20 (2003): 1–20; Martinez-Alier, Joan, *Ecological Economics: Energy, Environment and Society*, Oxford, UK: Basil Blackwell, 1987; Purves, William K., Sadava, David, Orians, Gordon H., and H. Craig Heller, *Life: The Science of Biology*, 7th ed., Sunderland, MA: Sinauer, 2003.

HELMUT HABERL

Millennium Ecosystem Assessment.   An international work program that was designed to meet the needs of the public and decision-makers for scientific information concerning the consequences of ecosystem change for human well-being, including options for responding to those changes. The Millennium Ecosystem Assessment (MA) was launched in 2001 and will help, among others, to meet assessment needs of the **United Nations Convention on Biological Diversity**, the **United Nations Convention to Combat Desertification**, and the **Ramsar Convention on Wetlands**. The MA synthesizes information from the scientific literature and data sets, and makes use of knowledge held by the private sector, practitioners, local communities, and indigenous peoples. Designed as a multiscale assessment, the MA consists of interlinked assessments undertaken at local, **watershed**, national, regional, and global **scales**. It established mechanisms to incorporate information and knowledge from non-peer-reviewed sources including local and **indigenous knowledge**. More than 500 authors are involved in four expert working groups preparing the global assessment, and hundreds more are undertaking more than a dozen subglobal assessments. Subglobal assessments approved or linked to the MA include, among others, the tropical forest sites of the **Alternatives to Slash-and-Burn Programme** of the **Consultative Group on International Agricultural Research**. All of the MA findings undergo rigorous peer review. If the MA proves to be useful to its stakeholders, it is anticipated that an assessment process modeled on the MA will be repeated every five to ten years and that ecosystem assessments will be regularly conducted at national or sub-national scales.

See also: **Decision-Making; Desertification; Global Land Cover Map of the Year 2000; Hot Spots of Land-Cover Change; Integrated Assessment; Integrated Model to Assess the Global Environment; United Nations Environment Programme.**

Further Reading

Millennium Ecosystem Assessment, Strengthening Capacity to Manage Ecosystems Sustainably for Human Well-Being [Online, January 2005], The MA Web Site, www.millenniumassessment.org.

HELMUT GEIST

Mineral Extraction.   Economic, technological, political, and social practices through which non-renewable resources are defined, located, segregated, and removed from the immediate environment. Like **agriculture**, the extraction of minerals occurs at the interface of biophysical and socio-economic processes and is a "pressure point" through which anthropogenic activity can drive land-use change. Unlike agriculture, however, mineral extraction accounts for a very small proportion (less than 1 percent) of the terrestrial land surface, although the degree of land transformation is often extreme. The significance of mineral extraction as a driver of global land-use change, therefore, is not its cumulative areal extent but rather the location of mineral development in areas of high **biodiversity** and/or wilderness value; as well as the secondary and third-order impacts on physical ecosystems as minerals mobilized by extraction and processing move off-site (i.e., acid mine drainage, tailings disposal, or smelter emissions); and the indirect effects of mineral investment on the value of land, the structure of land ownership, and the demand of mineral extraction and processing for other environmental inputs such as wood and water.

Mineral extraction—and the role of mineral extraction in land **conversion**—surged during the nineteenth and twentieth centuries, underscoring the "prodigal and peculiar" nature of environmental transformation in the modern period. **Industrialization**, **urbanization**, and resource-intensive forms of **economic growth** expanded the demand for, and range of, minerals wrested from the earth: aluminum, nickel, oil, and natural gas, for example,

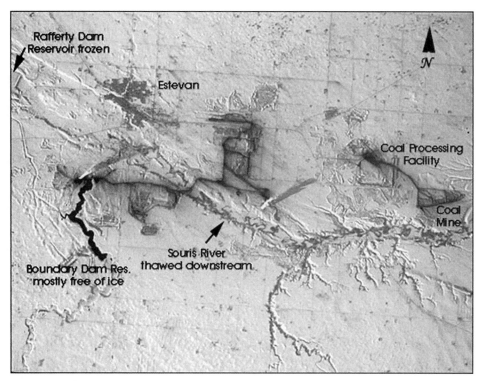

Estevan coalfield along the Souris River in Saskatchewan, Canada, where coal is mined from a trench at the surface, processed, and then used to fire two power plants, on February 17, 2001 (areas of active surface mining and processing appear black; roads used for transporting the coal are also dusted black).

PHOTO: Space Shuttle. CREDITS: Image Science and Analysis Laboratory, NASA-Johnson Space Center, 18 Mar. 2005 (Earth from Space—Image Information). http://earth.jsc.nasa.gov/sseop.efs

had no significant commercial applications 150 years ago, yet today account for a significant portion of the flow of minerals into the economy. By tapping the "extra-somatic" energy of fossil fuels, it has been possible to extract and process progressively lower-quality minerals; to develop mineral resources at great distance from centers of industry; and to produce minerals at lower unit cost, driving down their price and expanding demand still further. Demand has been met, in the main, by increased extraction of (primary) materials rather than the recycling of (secondary) materials already in the economy, as reflected in steeply upward trending resource-production curves for most minerals. The figure above provides examples of four leading non-ferrous metals for the period 1850–2000.

Mineral Resources Defined

Minerals are non-living, scarce materials found in the earth's crust that have social utility and/or economic value. They include both organic and inorganic substances and occur in solid (coal, nickel, silica), liquid (oil, brine), and gaseous states (methane, helium). Minerals are conventionally classified as fuels (oil, coal, gas), industrials (potash, salt, sulfur), metals (copper, zinc, bauxite), construction materials (sand, gravel, building stone)

Gold Coin Mine in Victor, Colorado, ca. 1900.
PHOTO: William Henry Jackson. [Courtesy Library of Congress]

and gemstones (ruby, jade), although some materials fall into more than one category. Diamonds, for example, are both industrial and gemstones; oil is a fuel mineral and an industrial feedstock; and clays and sands are used in both construction and in industrial applications. Some minerals are consumed (fuels, solvents) or dissipated in use (applications of metals like zinc and lead in paints) so that re-aggregation and recovery are economically unfeasible. Other minerals can be recycled: in the **United States of America**, for example, recycling represents 55 percent of the apparent supply of iron and steel, 30 percent of copper, 65 percent of lead, and 35 percent of aluminum. Despite an increase in the recycling of metals, resource extraction remains the primary means of meeting new metal demand, the vast majority of construction minerals, and the sole means of supplying demand for fuels and industrial minerals.

As defined here, minerals refer to a broad class of materials drawn from the earth. This definition has greater utility for research on land-use/land-cover change than the more specific definitions found in the chemical and mineralogical literature, which focus on a mineral's physical properties (atomic structure, degree of homogeneity) or mode of formation. The more inclusive definition emphasizes characteristics of minerals that are economically and environmentally significant, and includes the following:

1. Minerals are non-renewable resources and, as such, the contribution of mineral extraction to sustainable development centers on the successful

Results of deforestation during the early mining days in San Juan County, Colorado, September 1940.

PHOTO: Russell Lee. [Courtesy Library of Congress]

conversion of exhaustible natural assets into renewable forms of natural or human capital.

2. Minerals are scarce rather than ubiquitous (their occurrence in extractable quantities is limited to particular locations), and so frequently there is competition among extra-local interests for **access** to and control of mineral resources.

3. As a consequence, the direction and rate of land-use change in extractive economies is strongly influenced by relations of power between extra-local interests and local land users.

4. Whether or not a particular mineral (uranium or coal-bed methane, for example) is valued as a resource depends on geological knowledge and the economic feasibility of extraction, both of which are historically and geographically variable.

5. Thus, mineral extraction is an inherently dynamic form of land use, the "boom and bust" histories of which translate into relatively localized, yet intense forms of land-use change. For example, land use can shift to mineral extraction as new mines open, or from mineral extraction as existing operations close in response to the accumulation of geological knowledge (the development of diamond mining in northern **Canada** since the 1990s), changes in the price of minerals or the costs of extraction (the low costs of leaching technology made it profitable to develop low-grade copper and gold projects worldwide beginning in the late 1970s), a shift in the

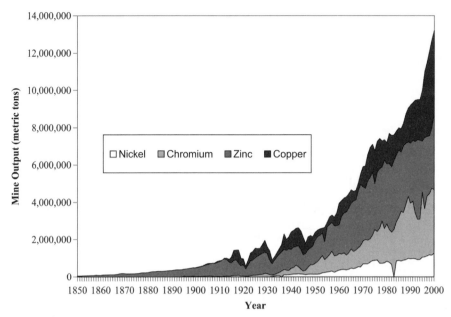

World mine output of four leading nonferrous metals, 1850–2000.

perceived risk/reward ratio of mineral **investments** relative to other invest-ment options, or a change in the perceived risk/reward ratio of one mineral region over another (the expansion of mineral investment in the Peruvian **Andes** following the adoption of economic liberalization poli-cies in Peru in the early 1990s).

Coupling Extraction and Land-Use/Cover Change

Conventionally, the concept of extraction applies to the engineering pro-cesses of removing minerals from the earth. From this perspective, the land-use changes arising from mineral extraction are best understood by referring to the technologies of extraction and via a focus on **environmental changes** at the site of extraction. Efforts to mitigate the land-use changes of extraction, for example, would center on the introduction of cleaner technology and/or best-practice environmental management techniques. This traditional view of discrete, localized impacts arising from the choice of technology or management practice has been challenged from a number of perspectives. These critiques provide the framework for a more comprehensive approach to extraction that recognizes its ecological, geographical, and institutional components.

First, research in environmental science has shown how the initial impacts of mineral extraction can be magnified or transmitted over a broad area via ecosystem processes. This work includes studies of the effect of emissions and/or solid wastes on the biological productivity of receiving ecosystems; the contribution of mineral extraction and processing to biogeochemical cycles that are regional and global in scope (mining and smelting account for 67 percent and 55 percent of the total anthropogenic release of copper and zinc, for example); and the role of mineral extraction in driving habitat fragmentation. Recent geographical shifts in the location of mining investment in the global South have intensified longstanding concerns about the impact of mining on global biodiversity and critical ecosystems. The World Resources Institute reports that 75 percent of active mines and exploration areas overlap with areas of high **conservation** value and areas of **watershed** stress, and that nearly one-third of all active mines and exploration sites are located within intact ecosystems of high conservation value.

Second, the longstanding engagement of ecological economics with thermodynamic constraints on resource production has shown that mineral extraction rests on an exchange (and geographical transfer) of organizational states: highly ordered, economically useful forms of energy are expended to add order to materials that are initially disorganized. Access to low-entropy forms of energy at low cost has underpinned the scaling up of "earth moving" capacities during the twentieth century: for example, the introduction of open pit mining for metals, or the shift from underground methods to "mountaintop removal" in Appalachian coal mining. Other work has examined how the energy exchanges of extraction produce disorder in the surrounding environment and how, as a result, extractive economies are at the center of a fundamentally uneven geographical exchange: mineral extraction requires the import and expenditure of low-entropy energy to produce an ordered material form (the useful mineral) that is subsequently exported, leaving disorder (ecosystem disruption) as a by-product. Energy analysis reveals, therefore, how the accumulation of low-entropy states in the core (the land-use changes of urbanization and industrialization that pile up sand, gravel, cement, iron, and copper, for example) and the production of high-entropy states in the periphery (the land-use changes of extraction that dismantle tropical forest cover and introduce contaminants to surface and groundwater, for example) are two sides of the same coin (although this account is not informed by energy analysis).

Third, extraction is a political and economic process through which value is created, appropriated, and transferred. The social histories of iconic mineral discoveries—the silver resources of the Comstock Lode, Nevada, in 1859, and oil at Spindletop, Texas, in 1901—vividly demonstrate how mineral production drives changes in the value of land and other critical inputs that can extend well beyond the site of extraction. Rising values for wood and water in a mine's hinterland, for example, can lead to the commodification and exchange of these resources (trees are felled for mine timbering or for the production of charcoal fuel; water is diverted for use in mining), producing extensive **land-cover change**: development of the Comstock Lode, for example, drew so extensively on lumber from the Lake Tahoe region that its mines have been called the "Tomb of the Sierras." Changes in resource ownership can also introduce new land management priorities and rules for resource access/use that affect land use and land cover. In addition, the attractiveness of mineral resources to national governments seeking to meet the requirements of multilateral lending **institutions** or deliver on economic goals means that mineral extraction is frequently the vanguard of economy-wide initiatives to commodify land and resources, create markets in resource rights, and "enclose the commons." Mineral law reform, for example, has been a central element of **economic liberalization** in many developing countries, while the prospect of investment in mineral extraction has influenced the priority and pace of land titling programs and the direction of reforms in water resource law (to facilitate water transfers away from traditional agricultural uses).

Fourth, mineral extraction may be organized in several different ways, with different implications for land-use change. Although the bulk of mineral output is produced by large, capital-intensive corporations operating a relatively small number of very large mines, some minerals (particularly gold and gemstones from placer deposits) are also extracted using labor-intensive methods. This informal mining sector is characterized by low capital intensity, relatively small volumes of output per operation, and insecurity of **land tenure**. Although the extent of informal sector mining can wax and wane in response to the same mineral price trends that govern corporate mining, the dynamics of the informal sector are quite different. Informal sector mining, for example, is often a response to **poverty** and a lack of other economic opportunities, and so it can expand even when mineral prices are relatively low. Addressing the environmental effects of mineral extraction, therefore, requires an approach that recognizes these organizational differences and that can assess their significance for addressing land-use change.

See also: Amazonia; Arena of Land Conflict; Boreal Zone; Consumption; Economic Restructuring; Ecosystem Services; Land Rights; Multiplier Effect; South Africa; South America; Taiga.

Further Reading

Brechin, Gray, *Imperial San Francisco: Urban Power, Earthly Ruin*, Berkeley, CA: University of California Press, 2001; Bridge, Gavin, "Contested Terrain: Mining and the Environment," *Annual Review of Energy and Resources* 29 (2004): 205–59; Bunker, Stephen, *Underdeveloping the Amazon: Extraction, Unequal Exchange, and the Failure of the Modern State*, Urbana, IL: University of Illinois Press, 1985; McNeill, J. R., *Something New under the Sun: An Environmental History of the Twentieth-Century World*, New York: Norton, 2000; Moore, Johnnie, and Samuel Luoma, "Hazardous Wastes from Large-Scale Metal Extraction," *Environmental Science and Technology* 24 (1990): 1278–85; Nriagu, Jerome, "History of Global Metal Pollution," *Science* 272 (1996): 223–4; World Resources Institute, *Mining and Critical Ecosystems: Managing the Risks*, Washington DC: World Resources Institute, 2004.

GAVIN BRIDGE

Miombo.   An abbreviation for miombo woodlands, which form the world's largest area of **tropical dry forest** in southeastern and central Africa, that is, 5.5 out of a total of 7.7 million square kilometers. *Miombo* is a vernacular word that has been adopted by ecologists to describe woodlands dominated by trees in the genera *Brachystegia, Julbernardia*, and *Isoberlinia*. The ecosystem extends from Tanzania and Congo in the north; through Zambia, Malawi, and eastern Angola; to Zimbabwe and Mozambique in the south. Miombo features a hot, seasonally wet climatic zone the soils of which—mainly derived from acid crystalline bedrock—are predominantly infertile. The woodlands have contracted to their present pattern of distribution in the past 1,000 years. Measured against degrees of "human disturbance," based on data from the 1970s and 1980s, it has been estimated that about 62 percent of the woodland cover has been undisturbed. However, all factors so far contributing to the **conservation** of the woodlands are seen to have drastically changed during the most recent past, and many formations are considered to be heavily modified to mostly **secondary vegetation**. The most important underlying **driving forces** of the present dynamics of miombo transformation are considered to be "macro-level phenomena," mainly national policies related to **agriculture**, **forestry**, energy, and population. Rooted in colonial interventions, these policies have grown in strength and extent since the past decades. At the proximate level, most of the **land-cover changes** in the miombo woodlands relate to agricultural economies based on export and colonial settler crops such as tea, **coffee**, **cotton**, and **tobacco**. These cash-cropping systems, practiced on commercial as well as smallholder farms, have been increasingly drawn into the world market, and they went through various stages of both cropland extension and **agricultural intensification**.

See also: **Cash Crops; Colonization; Economic Liberalization.**

Further Reading

Campbell, Bruce M., ed., *The Miombo in Transition: Woodlands and Welfare in Africa*, Bogor, Indonesia: Center for International Forestry Research, 1996; Geist, Helmut, "Soil Mining and Societal Responses: The Case of Tobacco in Eastern Miombo Highlands," in *Coping with Changing Environments: Social Dimensions of Endangered Ecosystems in the Developing World*, eds. Beate Lohnert and Helmut Geist, Aldershot, UK, Brookfield, VA: Ashgate, 1999, pp. 119–48.

HELMUT GEIST

Mode of Interaction. The interaction between the various causes or processes of land-use/cover change show different patterns that can be approached by using meta-analytical frameworks of analysis (see figure). First, one cause may completely dominate the other causes, assuming that land use in a given locality is influenced by whatever factor exerts the greatest constraint. The proportion of cases, however, in which dominant, single, or key factors operate at either the proximate or underlying level (or connecting the two levels) has been found to be low in cases of tropical **deforestation** and **desertification** (around 10 percent of the cases). Second, factors driving land change can be connected as causal chains, that is, interconnected in such a way that one or several variables drive one or several other variables. Again, the proportion of cases in which pure chain-logical causation dominates has been found to be low (less than 10 percent of the cases). Third, different factors can intervene in concomitant occurrence or concurrently, that is, independent but synchronous operation of factors leading to land change. Concomitant occurrence of causes is much more widespread than single-factor or chain-logical causation (around 25 percent of the cases). Fourth, factors and their interactions as detailed above may also play together in synergetic factor combinations, that is, several mutually interacting variables driving land-use/cover change and producing an enhanced or increased effect due to reciprocal action and **feedbacks** between causes. The pattern

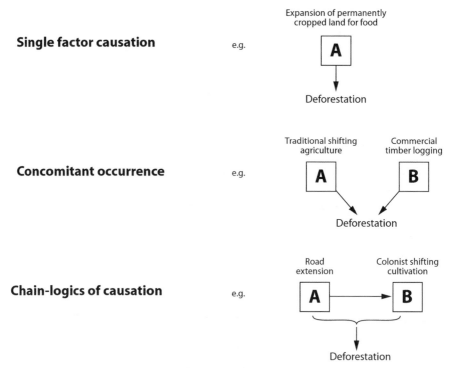

Expansion of permanently
cropped land for food

Single factor causation   e.g.

A

Deforestation

Traditional shifting   Commercial
agriculture   timber logging

Concomitant occurrence   e.g.

A   **B**

Deforestation

Road   Colonist shifting
extension   cultivation

Chain-logics of causation   e.g.

A   **B**

Deforestation

Three possible, but not widespread, modes of interaction.

of causal synergies has actually been identified as the most common type of interaction (inherent to 70–90 percent of the deforestation and desertification cases under study). Éric Lambin and colleagues propose to apply the term "land use = f (pressures, opportunities, policies, vulnerability, and social organization)," with the function f having forms that account for strong interactions among the causes of land change. In detail, these are pressures = f (population of resource users, labor availability, quantity of resources, and sensitivity of resources); opportunities = f (market prices, production costs, transportation costs, and technology); policies = f (subsidies, taxes, **property** rights, infrastructure, and governance); **vulnerability** = f (exposure to external perturbation, sensitivity, and coping capacity); and social organization = f (resource **access**, income distribution, household features, and urban-rural interactions). Though still in its infancy, the study of **forest transitions** points to such strong interaction of multiple causative factors in land-change processes, with land change as an emergent property of complex adaptive systems.

See also: Cultural Factors; Driving Forces; Economic Growth; Land-Use Policies; Land-Use Transition; Mediating Factor; Meta-Analysis; Population Dynamics; Proximate Causes; Resilience; Scale; Syndrome.

Further Reading

Lambin, Éric F., Geist, Helmut J., and Erika Lepers, "Dynamics of Land Use and Cover Change in Tropical and Subtropical Regions," *Annual Review of Environment and Resources* 28 (2003): 205–41.

HELMUT GEIST

Modeling.   Land-use/cover-change models provide simplified representations of the spatial and/or temporal dynamics of land use/cover. Since real-life experiments in land-use change are difficult to perform, computer models can be seen as a computational laboratory in which hypotheses about the processes of land-use change can be tested. Furthermore, models provide a structured way of analyzing complex interactions; scientists can make assumptions on the most important mechanisms of land-use change and then test these hypotheses through sensitivity analysis. Once a functioning and validated model of the **land-use system** is constructed, projections of future developments can be made. The user can explore system functioning through "what if" **scenarios** and explore the response of land use to these changes. Scenario simulations can provide insight in alternative futures or allow the evaluation of the effects and **tradeoffs** of different scenarios. Possible scenario simulations include the evaluation of the effect of changes in the agricultural sector (e.g., due to changes in market conditions or agricultural policies) on land-use patterns. Finally, land-use change models allow the translation of scenario conditions to visualizations of land-use patterns that can be used as a communication and learning environment for stakeholders involved in land-use **decision-making**. Projections can be used as early warning systems for the effects of future land-use changes and indicate **hot spots of land-cover change** that are priority areas for in-depth analysis and policy intervention. **Johann Heinrich von Thünen** provided one of the first spatially explicit land-use models based on economic theory. His work is based on the concept of **land rents**, which are closely related to the potential profit a farmer can make from growing a crop. As this profit is related not just to the value of a crop at the market but also to the cost to transport the products to that market, rent for any particular crop falls off with distance from the market. When farmers have a choice of crops to grow, they will, obviously, choose the most profitable one. Spatially, this will result in a series of concentric rings around the market, with crops with the highest transport costs relative to the market price grown nearest to the city.

Since the acknowledgment of the importance of land-use/cover change for **global environmental change** and following the research agenda set by the international **Land-Use/Cover Change (LUCC) project**, a large variety of modeling approaches have been developed to address different aspects of land-use change, based on different scientific disciplines.

Typology of Model Approaches

Different authors have provided reviews of land-use models using classification systems, often based on the dominant technique used in the model or the underlying disciplinary theory. For **deforestation** models, an overview is provided by Éric Lambin as well as by David Kaimowitz and Arild Angelsen, while Eric Miller and colleagues present a review of integrated urban models. Éric Lambin and colleagues review models for **agricultural intensification**, while Nancy Bockstael and Elena Irwin review a number of land-use models in terms of economic theory foundations. Chetan Agarwal and colleagues review nineteen models based on their spatial, temporal, and human-choice complexity, while Helen Briassoulis and Peter Verburg give overviews of all types of land-use models. An overview of more recent approaches is provided by the special issues edited by Tom Veldkamp and Peter Verburg.

An important distinction between different model types is the distinction between spatial and non-spatial models. This distinction is of major importance when selecting a model type for a certain application since it largely determines the type of research questions the model may answer for that application. Spatial models aim at spatially explicit representations of land-use change at some level of spatial detail, in which land-use change is indicated for individual pixels in a raster or other spatial entities such as administrative units. This group of models is, therefore, able to explore spatial variation in land-use change and account for variation in the social and biophysical environment. The group of non-spatial models focuses on modeling the rate and magnitude of land-use change without specific attention to its spatial distribution. An example of such a model is the **Sahelian land use (SALU) model**.

Apart from distinguishing models by their spatial resolution, it is possible to discuss the temporal aspects of the model by the difference between static and dynamic models. The calculation of the coefficients of a regression equation explaining the spatial distribution of land-use changes as a function of a number of hypothesized driving factors can be seen as a static model of land-use/cover change. Dynamic models often include temporal dynamics in land-use systems represented by competition among land uses, irreversibility of past changes, and fixed land-use change trajectories. Static models can be used to test our knowledge of the **proximate causes** and underlying **driving forces** of land-use change, while dynamic models are essential when projections for future land-use change should be made. Examples of dynamic models include the **conversion of land use and its effects (CLUE) model** and GEOMOD.

The next broad distinction can be made between descriptive and prescriptive models. Descriptive models aim at simulating the functioning of the land-use system and the spatially explicit simulation of near-future land-use patterns. Prescriptive models, in contrast, aim to calculate optimized land-use configurations that best match a set of goals and objectives. Descriptive models are based on the actual land-use system and dominant processes that lead to changes in this system. The model output provides insights into the functioning of the land-use system and gives projections for scenario conditions. Prescriptive models, on the other hand, mostly include the actual land-use system solely as a constraint for more optimal land-use configurations. The basic objective of most prescriptive or optimization models is that any parcel of land, given its attributes and its location, is modeled as being used in the way best matches a series of defined objectives. Prescriptive models are relevant to policymakers as a spatial visualization of the land-use pattern that is the optimal solution based on their preferred constraints and objectives. However, prescriptive models do not provide insights into the actual land-use change trajectories and the conditions needed to reach the optimized situations. Optimization models suffer from other limitations, such as the somewhat arbitrary definition of objective functions and non-optimal behavior of people, for example, due to differences in values, attitudes, and cultures. While at an aggregate level these limitations are likely to be non-significant, they are more important as one looks at fine-scale land-use change processes and takes an interest in the diversity between actors.

Another major difference between broad groups of land-use models is the role of theory. While there is no single all-compassing theory of land-use change, there are different, disciplinary theories that can describe land-use change processes. Deductive models are based on theories, and the results of model simulations are compared to actual land-use changes to test the validity of the theories. The classical example of a land-use change model based on economic theory is the Thünen model. More recent deductive models for agricultural expansion are presented by Arild Angelsen, who compares four different model specifications based on economic theory. Inductive models are based on observed processes of land-use change rather than on a theoretical model. Different types of inductive models exist, ranging from models in which decision-making by actors is specified in a range of decision rules and interactions to models in which the relation between land-use location and variability in the socio-economic and biophysical environment is captured by statistical techniques, often regression. Both inductive and deductive modeling are used to explore the processes that lead to land-use change. Whereas deductive models are able to test theories and the actual importance of a number of driving factors, inductive models help us to suggest which drivers are empirically associated with the land-use pattern.

A final distinction among model types can be made based on the simulated objects. In many spatially explicit models, the unit of analysis is based on an area of land, either a polygon representing a field, plot, or census track or a **pixel** as part of a raster-based representation. This combination of these spatial objects results in a map showing changes in land-use pattern. The disadvantage of this "land-based" approach is the poor match with the agents of land-use change. Individual farmers or plot owners are not presented, and the simulations usually do not match with the units of decision-making. The group of **agent-based models** uses individual agents as units of simulation. Several characteristics define agents: they are autonomous, they share an environment through agent communication and interaction, and they make decisions that tie behavior to the environment. Such multiagent systems give emphasis to the decision-making process of the agents and to the social organization and landscape in which these individuals are embedded. Furthermore, an agent is not necessarily an individual. An agent can represent any level of organization (a herd, a cohort, a village, etc.). Both approaches of modeling have (dis-)advantages; and the appropriate approach depends on the questions, region, and spatial extent of the model.

Generalized Structure of the Spatially Explicit Land-Use Model

In spite of the large variety of theories underlying different land-use modeling approaches and the broad range of techniques used to build the models, it is possible to identify a common structure valid for almost all spatially explicit land-use change models (except multiagent models). Such a conceptualization of the land-use system is useful for analyzing differences in land-use models: some models differ by the technique they use to quantify part of the system while others differ in the interaction of system components. The figure presents the main components found in almost all spatially

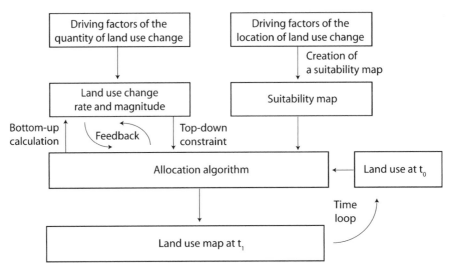

Main components of spatially explicit land-use/cover change models.

explicit land-use models. One of the important parts of the model structure relates to the distinction between the calculation of the magnitude of change and the allocation of change. Both calculations are based on a set of driving factors, some steering the magnitude of change and some others only steering the location of change. Sometimes the same driving factor can influence the quantity and location of change. Demographic drivers are often seen as important factors that can (partly) explain the rate and quantity of land-use change. However, the spatial distributions of population and migration patterns are important determinants for the location of land-use change as well.

Based on the interpretation of one or more driving factors that are supposed to be determinants of the location of land-use change, a so-called suitability or preference map is created that indicates the suitability of a location for a specific land-use type relative to the suitability of other locations. The translation of the driving factors into a suitability map can vary between the direct use of potential productivities of agricultural crops to complex empirical or rule-based systems that combine a larger set of location factors. The selection of the driving factors used in the model and its translation to a suitability map are two of the main components of a land-use model. A wide variety of approaches exist, including the following three, described below.

Rule-based systems based on expert knowledge: In this approach, a set of selected variables are assumed to best represent the decision-making concerning the location of land-use change. If it is assumed, for example, that **agriculture** is preferably located at locations where a high production can be attained, the suitability map may represent the potential or water-limited production of agricultural crops at each location. An example of this approach is the land-allocation module of the **integrated model to assess the global environment** (IMAGE).

Rule-based systems based on theory: In this approach, assumptions about decision-making are made strictly following theory. Most often, use is made of economic reasoning. In these applications, the suitability map often consists of (relative) land rent. Given the absence of data on land rent and the large number of factors determining land prices, empirical techniques are often used to estimate land rent from a number of determining factors.

Suitability maps based on empirical analysis: This is probably the most widely used approach. Because the interaction between land use and its driving factors is mostly complex and region-dependent, it is not easy to define a rule-based system for deriving a suitability map. Therefore, modelers often make use of empirical methods to explore and quantify the relation between observed (historic) land-use patterns and the spatial variation in a set of supposed driving factors. Common techniques include multiple regression, logistic or multinomial regression, and the use of neural nets. Markov models are also based on empirical analysis. They use observed transitions in land use to estimate the probability of future transitions based on the current land cover at a location.

Neighborhood interactions and **cellular automata**: Land-use patterns most often show a strong degree of spatial correlation, due to spatial clustering in driving factors and spatial interactions among locations of land use. In many models, especially those focusing on **urbanization**, special attention is given to spatial interactions that determine the suitability of a location. Cellular automata update the suitability or transition probability each time step, based on an analysis of neighboring land uses and a rule-based system that translates the neighborhood characteristics into suitabilities or transition probabilities. Examples of models making use of neighborhood interactions include those of Claudia de Almeida and colleagues, Keith Clarke and Leonard Gaydos, and Fulong Wu.

The construction of a suitability map does not directly lead to the determination of changes in land-use patterns. The actual pattern of land-use change is also determined by the requirements for the different land-use types and competition among land uses. Therefore, in most models, a rule-based system allocates the actual land-use changes based on the suitability map. These rules range from using a simple cut-off value to select the locations with the highest suitability from the suitability map, to dynamic modeling of competition between land uses based on land-use type-specific characteristics. These rule sets are incorporated in an allocation algorithm.

A wide variety of approaches are used to calculate the claims of different land uses for space. In a number of models, a bottom-up approach is chosen, in which the spatial dynamics and allocation rules determine the aggregated quantity of land-use change. However, often a top-down approach is chosen, in which a set of rules or a modeling approach is used to calculate the quantity of change based on a set of driving factors. This quantity is used as a constraint in the actual allocation procedure. In a few cases a hybrid approach is chosen in which the land-use requirements are dynamically adapted through a **feedback** between the allocation and the calculation of land-use requirements to account for land availability, for example.

Which Model Should be Chosen?

It is not possible to indicate what model approach or specific model is most suitable for land-use/cover change modeling. The choice of model is largely dependent on the research or policy questions that need to be answered. Not all model approaches are capable of answering all questions. Furthermore, the research questions might pose restrictions on the applicability and suitability of a particular question, in terms of **scale**, that is, by its spatial and temporal resolution and extent and dominant land-use change process. For example, a spatially explicit cellular automata model might be well suited to explore urban growth dynamics but be incapable of exploring the driving factors of agricultural transitions.

A number of recent models combine different theories and techniques to better account for the different processes and make best use of the specific strengths of different approaches. Such a hybrid model is often capable of simulating multiple land-use types at the same time using different techniques to specify the suitability map and allocation rules depending on the specific land-use type rather than having a similar procedure for all land-use types.

When models are used to improve and test our understanding of the driving factors of land-use change, the combined use of inductive and deductive modeling approaches might lead to additional insights. Inductive techniques might suggest possible driving factors that can be tested on causality in deductive approaches, leading to a more direct linkage between process- and pattern-based techniques in land-use analysis. In case models are used to assess the impacts of different storylines in scenario studies, other constraints will be imposed on the choice of model. Most scenarios assume changes in attitude and decision-making in line with changes in worldview. Such scenarios cannot be simulated by land-use models in which the transition rules are solely determined by (empirical) analysis of past trends. Similar considerations apply to all studies with temporal extent longer than one or two decades.

Decision Support and Uncertainty

The application of land-use models in policy decision support systems is limited. This is partly due to the relatively recent development of full-fledged modeling approaches in this field and the explorative state of most model approaches. Diversity of approaches is another drawback for users, indicating the absence of a model that might perfectly suit their needs. Furthermore, most **validation** exercises have indicated that uncertainty in land-use simulations is high. These limitations are common for models of complex integrated systems; and, although progress is made, some of these constraints are inherent to the land-use system. Therefore, visualization issues and adequate presentation of the results are most important to communicate land-use modeling results to policymakers and other stakeholders.

See also: **Century Ecosystem Model; Change Detection; Continuous Data; Discrete Data; History Database of the Global Environment; Hot Spot Identification; Integrated Assessment; Integrated Model; LandSHIFT Model; Land-Use History; Land-Use Planning; Land Use Transition; Pattern to Process; Probit/Logit Model; Socializing the Pixel.**

Further Reading

Agarwal, Chetan, Green, Glen M., Grove, J. Morgan, Evans, Tom P., and Charles M. Schweik, *"A Review and Assessment of Land Use Change Models: Dynamics of Space, Time, and Human Choice,"* Bloomington: Indiana University, Center for the Study of Institutions, Population, and Environmental Change, and USDA Forest Service, 2001; Angelsen, Arild, "Agricultural Expansion and Deforestation: Modelling the Impact of Population, Market forces and Property Rights," *Journal of Development Economics* 58 (1999): 185–218; Bockstael, Nancy E., and Elena G. Irwin, "Economics and the Land Use-Environment Link," in *The International Yearbook of Environmental and Resource Economics 1999/2000,* eds. Henk Folmer and Thomas Tietenberg, Cheltenham, UK: Edward Elgar Publishing, 2000, pp. 1–54; Bousquet, François, and Christophe Le Page, "Multi-Agent Simulations and Ecosystem Management: A Review," *Ecological Modelling* 176 (2004): 313–32; Briassoulis, Helen, Analysis of Land Use Change: Theoretical and Modeling Approaches [Online, August 2004], The Web Book of Regional Science Web Site, www.rri.wvu.edu/regscweb; Brown, Daniel G., Pijanowski, Bryan C., and Jiunn-Der Duh, "Modeling the Relationships between Land Use and Land Cover on Private Lands in the Upper Midwest, USA," *Journal of Environmental Management* 59 (2000): 247–63; Chomitz, Kenneth M., and David A. Gray, "Roads, Land Use, and Deforestation: A Spatial Model Applied to Belize," *The World Bank Economic Review* 10 (1996): 487–512; Clarke, Keith C., and Leonard J. Gaydos, "Loose-Coupling a Cellular Automaton Model and GIS: Long-Term Urban Growth Prediction for San Francisco and Washington/Baltimore," *International Journal of Geographical Information Science* 12 (1998): 699–714; de Almeida, Claudia M., Batty, Michael, Monteiro, Antonio M. V., Camara, Gilberto, Soares-Filho, Britaldo S., Cerqueira, Gustavo C., and Cerqueira L. Paennachin, "Stochastic Cellular Automata Modeling of Urban Land Use Dynamics: Empirical Development and Estimation," *Computers, Environment and Urban Systems* 27 (2003): 481–509; Kaimowitz, David, and Arild Angelsen, *Economic Models of Tropical Deforestation: A Review,* Bogor, Indonesia: Center for International Forestry Research, 1998; Lambin, Éric F., Rounsevell, Mark D. A., and Helmut J. Geist, "Are Agricultural Land-Use Models Able to Predict Changes in Land-Use Intensity?" *Agriculture, Ecosystems and Environment* 82 (2000): 321–31; Lambin, Éric F., "Modelling and Monitoring Land-Cover Change Processes in Tropical Regions," *Progress in Physical Geography* 21 (1997): 375–93; Miller, Eric J., Kriger, David S., and John D. Hunt, *TCRP Web Document 9: Integrated Urban Models for Simulation of Transit and Land-Use Policies: Final Report,* Toronto: University of Toronto Joint Program in Transportation and DELCAN Corporation, 1999; Overmars, Koen P., and Peter H. Verburg, "Analysis of Land Use Drivers at the Watershed and Household Level: Linking Two Paradigms at the Philippine Forest Fringe," *International Journal of Geographical Information Science* 19, no. 2 (2005): 125–52; Parker, Dawn C., Manson, Steven M., Janssen, Marco A., Hoffman, Matthew, and Peter Deadman, "Multi-Agent Systems for the Simulation of Land-Use and Land-Cover Change: A Review," *Annals of the Association of American Geographers* 93 (2003): 314–37; Pontius, Robert G., Cornell, Joseph D., and Charles A. S. Hall, "Modeling the Spatial Pattern of Land-Use Change with GEOMOD2: Application and Validation for Costa Rica," *Agriculture, Ecosystems and Environment* 85 (2001): 191–203; Veldkamp, A. (Tom), and Éric F. Lambin, "Editorial: Predicting Land-Use Change," *Agriculture, Ecosystems and Environment* 85 (2001): 1–6; Veldkamp, A. (Tom), and Peter H. Verburg, "Modelling Land Use Change and Environmental Impact: Introduction to the Special Issue," *Journal of Environmental Management* 72 (2004): 1–3; Verburg, Peter H., and A. (Tom) Veldkamp, "Editorial: Spatial Modeling to Explore Land Use Dynamics," *International Journal of Geographical Information Science* 19, no. 2 (2005): 99–102; Verburg, Peter H., Kok, Kasper, Pontius, Robert Gilmore, Jr., and A. (Tom) Veldkamp, "Modelling Land Use and Land Cover Change," in *Land Use and Land Cover Change: Local Processes, Global Impacts,* eds. Éric Lambin and Helmut Geist, Berlin: Springer, 2006;

Wu, Fulong, "Simulating Urban Encroachment on Rural Land with Fuzzy-Logic-Controlled Cellular Automata in a Geographical Information System," *Journal of Environmental Management* 53 (1998): 293–308.

PETER H. VERBURG

Moderate Resolution Imaging Spectroradiometer. A key instrument aboard the Terra Earth Observing System (EOS) AM, launched in late 1999 by the U.S. National Aeronautics and Space Administration, as well as aboard the Aqua EOS PM platforms, which is of particular interest from a human-environmental perspective (but also for the study of oceans and the lower atmosphere). The majority of the thirty-six spectral channels of moderate resolution imaging spectroradiometer (MODIS) have one-kilometer resolution, although a few have 250–500 meter resolution. The combination of spectral and spatial characteristics allows for an improved calculation of various **vegetation** and **land-cover change** indices, as well as for new **fire** detection and land-cover products. Through Terra MODIS and Aqua MODIS, global data sets are acquired every one to two days, with Terra passing across the equator from north to south (in the morning) and Aqua passing over the equator from south to north (in the afternoon). MODIS generates a new generation of detailed **remote sensing** data, thus complementing the **Landsat** thematic mapper and the **advanced very high resolution radiometer** as the long-term satellite imagery widely used by ecologists.

See also: **Change Detection; Earth Resources Observation Systems (EROS) Data Center; Global Terrestrial Observing System; GLOBCARBON; IKONOS; Pixel.**

Further Reading
Janetos, Anthony C., "Remote Sensing, Terrestrial Systems," in *Encyclopedia of Global Environmental Change, Volume 2: The Earth System: Biological and Ecological Dimensions of Global Environmental Change*, eds. Ted Munn, Harold Mooney, and Josep G. Canadell, Buffins Lane, Chichester, UK: John Wiley & Sons, 2002, pp. 528–30; U.S. National Aeronautics and Space Administration, MODIS Web [Online, March 2005], The MODIS Web Site, modis.gsfc.nasa.gov.

HELMUT GEIST

Modification. Subtle change that affects the character of land cover without changing its overall classification. Land cover is defined by the attributes of the earth's land surface and immediate subsurface, including biota, soil, topography, surface and groundwater, and built-up structures created by humans. Data sets used in **land-cover change** research represent the land surface by a set of spatial units, each associated with attributes that are either a single land-cover category, that is, leading to a discrete representation of land cover; or a set of values for continuous biophysical variables, that is, leading to a continuous representation of land cover. The use of **discrete data** as opposed to **continuous data** has the advantages of concision and clarity, but it has led to an overemphasis of land-cover **conversion** and a neglect of land-cover modification. However, there is growing recognition of the importance of the processes of modification of land attributes. For example, **agricultural intensification** permitted an increase in the world's food production over the last decade, and crop yield increases have outpaced global human population growth. Another example stems from **Amazonia**, where forest impoverishment caused by **fire** and selective logging affects an area at least as large as the area affected by forest-cover conversion. Likewise, the expansion of woody shrubs in the **Great Plains** of the western **United States of America** following fire suppression and overgrazing contributed to a large carbon sink. Finally, declines in tree density and species richness in the last half of the twentieth century provided evidence of **desertification** in some areas of the western Sahel, while other studies in western Sudan, a region allegedly affected by desertification, did not find any decline in the abundance of trees despite several decades of drought.

See also: **Change Detection; Grassland; Secondary Vegetation; Soil Degradation.**

Further Reading

Lambin, Éric F., Geist, Helmut J., and Erika Lepers, "Dynamics of Land Use and Cover Change in Tropical and Subtropical Regions," *Annual Review of Environment and Resources* 28 (2003): 205–41.

HELMUT GEIST

Monitoring. A planned, systematic, and continuing process of methodological collection and organization of data that aggregates results in order to evaluate the performance of the observed object(s) based on a set of indicators. Indicators are physical, chemical, biological, ecological, environmental, social, or economic measurements that represent key elements for better understanding, identifying, and anticipating trends or changes for a variety of spatial and temporal **scales**. Indicators need to be reliable, robust, interpretable, reflective (to analyze historical trends), and predictive (to enable forecasts). Monitoring tracks the actual performance or condition against what was planned or expected according to pre-determined standards. The value of any monitoring project increases with time. Long-term data are essential to evaluate inevitable short-term fluctuations. Good monitoring procedures provide the basis for learning and timely **decision-making**.

In land-change science, monitoring means the continuing collection of land observations over time against the background idea that the primary and overwhelming benefit to society from land-use change is appropriation of ecosystem goods—food, fiber, and timber—for human **consumption**. Land-use change alters ecosystems, and ecosystem responses (hydrologic, biologic, climatic) themselves become drivers of land-use change and influence socio-economic conditions. Examples are agricultural land consumption and effects on flooding (e.g., the Elbe River flood in Germany in 2002) versus drought (e.g., the Aral Sea in Uzbekistan/Kazakhstan in 1960–2000); logging and downstream river sedimentation (e.g., in the northwest region of the **United States of America** and **Canada**); and climate change, increased forest **fire** frequency, and loss of timber versus warmer growing conditions (e.g., Siberia in **Russia** in 1990–2005).

What to monitor in a global context? Human-induced transformations of ecosystems and landscapes are the largest sources of change on earth, affecting the ability of the biosphere to sustain life. Acknowledging both mankind's influence and its dependence on the environment, the first joint global earth monitoring endeavor, Global Earth Observation System of Systems (GEOSS), is designed around nine "societal benefit areas." The GEOSS ten-year implementation plan lists observational requirements, monitoring architecture, temporal targets, and performance indicators for each area in a comprehensive tabular format (observable counts given for reference): disasters, 32 observables; health, 18; energy, 31; climate, 44; water, 54; weather, 41; ecosystems, 28; **agriculture**, 32; and **biodiversity**, 12 observables.

Good Practice Procedures

Monitoring efforts range from local, to regional, to continental and global endeavors. In general, monitoring procedures should be clearly described in an implementation phase (i.e., database of strategies, implementation plan, project implementation, implementation assessment, and implementation report) and an effectiveness phase (i.e., list of desired outcomes, indicators, effectiveness assessment, and effectiveness report).

Local/regional monitoring programs should incorporate the following steps. First, plan a successful monitoring/indicators program: be specific about goals; select the most important variables or the most practical ones (and be diverse) for the specific site or region monitored; be sure to have indicators that change over a gradient of human influences; build upon existing monitoring data and accepted standards and prioritize what needs to be done next; promote agreement by the scientific and management communities on the critical questions; embrace principles of ecologically sustainable development; specify how information collected will be used in the decision-making process; and analyze cost effectiveness. Second, carry out a successful monitoring/indicators program: sampling should be regionwide, for many years, on a frequent time scale, and collaboration among agencies is important (particularly in times of limited funds) both for sampling and sharing data; utilize different monitoring groups (e.g., citizen volunteers, academics, state agencies, and federal agencies) and make data accessible; couple monitoring with experiments to determine whether the assumptions were correct; be open to new methods, techniques, and equipment; and use a long monitoring time frame (longer than a political time frame). Third, work up the data from a successful monitoring/indicators program: use a sufficient level of data resolution to assure accuracy and precision; provide synthesis to turn data into useful information; put results into a larger perspective; track cumulative impacts; gain new insights by repeated assessments and observing responses to management actions; be sure to use reliable data in order to maintain credibility; communicate regularly to the public and to decision-makers; and establish monitoring data as an integral management tool.

The framework for a national environmental monitoring and assessment program (EMAP) has been published by the U.S. Environmental Protection Agency. The EMAP approach has exemplary character for national to continental monitoring programs. It elaborates on the scientific barriers with respect to the chosen indicators and a nationally comprehensive monitoring framework and implementation efficacy with respect to capacity-building of governmental and public bodies. EMAP is shaped by its statistical design (resource sampling: sampling site selection, classification for strata), ecological indicators (indicator development, measurement, responsiveness, and variability), reference conditions, and key research issues. Special attention is given to landscape research and landscape indicators. **Remote sensing** techniques have been applied to create a national land-cover database from **Landsat** Thematic Mapper images taken from 1991 to 1993. Work has been undertaken on a revised national land-cover database for the year 2000. The landscape indicators depend on the quality of the remotely sensed landscape data from which they are built. Problems to be investigated are related to the scale of the remote sensing data—and thus of the indicators—and the number of attributes (e.g., land-cover classes).

Comparable efforts on strengthening national environmental monitoring and information systems in **Europe** are led by the European Environment Agency (EEA) with its European Environment Information and Observation Network (EIONET) and its **Coordination of Information on the Environment (CORINE) database**. Environmental assessment was undertaken in 1995 and 1998 and focused on the following twelve prominent European environmental problems: climate change, stratospheric ozone depletion, acidification, tropospheric ozone, chemicals, waste, biodiversity, inland waters, marine and coastal environment, **soil degradation**, urban environment, and technological hazards.

Responsible for global environmental monitoring activities is the **United Nations Environment Programme** (UNEP). In 1972, UNEP established the Global Environmental Monitoring System (GEMS). GEMS includes the assessment of a list of priority pollutants and monitoring of natural resources. Monitoring is recognized as a global problem with respect to institutional, spatial, and temporal scales and under consideration of the need for agricultural, **forestry**, and freshwater sustainability. In 2003, the UN Framework Convention on Climate Change (UNFCCC) published the *Good Practice Guidance for Land Use, Land-Use Change and Forestry* (GPG-LULUCF), which provides methods for estimating, measuring, monitoring, and reporting on carbon stock changes and **greenhouse gas** emissions from LULUCF activities for the **Kyoto Protocol**—that is, choice of estimation method, quality assurance procedures, documentation, assessment of inventory, and quantification of uncertainties.

Monitoring Programs

Two major, politically driven programs were launched at the beginning of this millennium, both reflecting the necessity and urgency to monitor the state of the earth's environment: the European GMES and the global Group on Earth Observation (GEO) with its GEOSS. The overall aim of GMES is to support Europe's goals regarding sustainable development and global governance, by facilitating the provision of quality data, information, and knowledge. GMES forums bring together the main stakeholders in Europe involved in the use of observations from space and *in situ* facilities for better management of the environment and enhanced security of Europe's citizens.

The draft GEOSS ten-year implementation plan states (October 15, 2004) that the purpose of GEOSS, as stated in the declaration of the first earth observation summit, is to achieve comprehensive, coordinated, and sustained earth observations for the benefit of humankind. GEOSS implementation will enable improved monitoring of the state of the earth, increased understanding of dynamic earth processes, enhanced prediction of the earth system, and further implementation of international environmental treaty obligations. GEOSS aspires to encompass all areas of the world and to cover *in situ* observations as well as airborne and space-based observations. The focus of GEOSS is on observations relevant to large parts of the world and issues that require global information to be addressed optimally.

There are international efforts at regional to global scales, partly preparing or influencing GMES and GEOSS, which are programs that often address specific monitoring gaps about earth surface dynamics to improve the availability and dissemination of baseline data, monitoring capabilities and resources, and institutional stability.

The three aims of the European Commission's CORINE program are to compile information on the state of the environment, to coordinate the compilation of data and the organization of information within the European member states or at the international level, and to ensure that information is consistent and that data are compatible. The working scale is 1:100,000; and the land-cover nomenclature is organized into three levels with the first level having five headings, the second level having fifteen headings, and the third level having forty-four headings. A further objective of CORINE is to obtain information about the way the environment is changing. The basis has been accomplished with the two Europe-wide CORINE coverages for the years 1990 and 2000.

The aim of Monitoring of Land-Use/Cover Dynamics (MOLAND) is to provide a spatial planning tool that can be used for assessing, monitoring, and **modeling** the development of urban and regional environments in Europe. The project was initiated in 1998 under the name Monitoring Urban Dynamics (MURBANDY). Particular effort is dedicated to the development of easy handling indicators, since potential users of the results include city managers, local administrators, and planners, and even up to international bodies. The spatial dimension is used to develop sets of indicators having a degree of standardization that allows for comparison within Europe. The three specific aims are to produce quantitative information on the evolution of land use and transport networks, from 1950 onward, in study areas subject to infrastructural changes (e.g., **urbanization**, construction of transport links); to develop methods for performing a harmonized analysis of historical trends, including socio-economic aspects, impact of legislation, landscape fragmentation, etc.; and to develop models for the harmonized simulation of future Europe-wide **scenarios** at local and regional scales. Since 2004, MOLAND is contributing to the evaluation and analysis of impact of extreme weather events, in the frame of research on adaptation strategies to cope with climate change. To date, the MOLAND methodology has been applied to an extensive network of cities and regions for an approximate total coverage in Europe of 70,000 km^2.

Agro-environmental indicators have been monitored in Europe since 2003 through the Land Use/Cover Area Frame Statistical Survey (LUCAS) collaboration among EEA, the Agriculture and Environment Directorate-Generals, Eurostat, and the **Joint Research Centre (JRC) of the European Commission**. LUCAS uses land-cover information, administrative data, and geo-referenced statistical surveys that are important for providing information on fluxes, stocks, and pressure indicators; and it explores data sets for the European Union. It observed that the integration of different data sets was possible and stresses the importance of data being available at periodic intervals.

The Multi-Resolution Land Characteristics (MRLC) consortium is a group of U.S. federal agencies that joined in order to develop a land-cover data set called the National Land Cover Dataset (NLCD). In 1999, a second-generation MRLC consortium was formed to coordinate the production of a comprehensive land-cover database for the United States of America called NLCD 2001, using Landsat 7 ETM+ data. The key component of this land-cover mapping effort is a database approach, which provides flexibility in developing and applying suites of independent data layers. These independent standardized data layers or themes will be useful not only within the land-cover classification but as data components for other applications. NLCD is a twenty-one-class land-cover classification scheme applied consistently over the United States. The spatial resolution of the data is thirty meters. The classification system used is modified from the Anderson land-use and land-cover classification system. No attempt was made to derive classes that were extremely difficult or "impractical" to obtain using Landsat TM data, such as the level III urban classes. In addition, some Anderson level II classes were consolidated into a single NLCD class.

The **United Nations Food and Agriculture Organization** and UNEP established the **Global Land Cover Network** (GLCN) cooperative program as an integral part of the **Global Terrestrial Observation Strategy** (GTOS). The objective of GLCN is to provide direction, focus, and guidance for harmonization of land-cover mapping and monitoring at national, regional, and global levels. The involvement of reference institutions ensures adequate monitoring standards and effectiveness of monitoring modalities. Close collaboration and linkage with ongoing regional and global geospatial information initiatives as well as environmental conventions will be pursued: for example, UNFCCC, the **Millennium Ecosystem Assessment**, and research projects that focus on measurement of regional **land-cover change** (e.g., the **Land-Use/Cover Change [LUCC] project** and the **Large Scale Biosphere-Atmosphere Experiment in Amazonia**).

Terrestrial Ecosystem Monitoring Sites (TEMS) is an international directory of sites (named T sites) and networks that carry out long-term terrestrial monitoring and research activities. The database provides information on the who, what, and where in long-term terrestrial monitoring that can be useful to both the scientific community and policymakers.

The LULUCF report of the **Intergovernmental Panel on Climate Change** addressed the implications of differing definitions of land use, land-use change, and forestry activities: definitions of these terms as used in the Climate Convention and the Kyoto Protocol, definitions of accounting and reporting, methodological definitions, and sustainable development considerations. The technical requirements to measure, monitor, and verify

Kyoto Protocol requirements will be significantly affected by these definitions. The good practice monitoring procedures as described earlier are mirrored in this report as a list including accounting objectives and inventory techniques, that is, transparency, consistency, comparability, completeness and precision, uncertainty and accuracy, verifiability, costs, and efficiency.

In September 2004, a new theme was prepared for the Integrated Global Observing Strategy (IGOS), that is, Integrated Global Observations of Land (IGOL), with the focus on human influence on the environment and vice versa. Technical deficiencies in monitoring land cover are addressed, such as spatial and temporal resolution requirements for earth observation data. Implementation strategies are suggested, for example, to obtain annual estimates of **deforestation** and land-use change. The importance of the socioeconomic component is incorporated, considering management practices, **food security**, human settlements, and infrastructure. The IGOL strategy builds on experience and requirements from the space agencies, GTOS, UN organizations, the **International Geosphere-Biosphere Programme** and the **International Human Dimension Programme on Global Environmental Change**, GEOSS, and **Global Observation of Forest and Land Cover Dynamics** (GOFC-GOLD) as the first international program since the 1990s with a comprehensive global land-cover change observation strategy. GOFC-GOLD is also responsible for coordinating harmonization and validation of global land cover and land-cover change products. GOFC-GOLD's harmonization tool is FAO's **Land Cover Classification System**, which can be utilized as a legend translation system. Global **validation** relies strongly on GLCN.

Global Monitoring Products

In 1978, global earth observation data—that is, the **advanced very high resolution radiometer** (AVHRR) with low geometric resolution and high temporal repetition frequency—became available. Global cloud-free image mosaics and higher-level products such as land-cover maps were generated. New sensors such as the **moderate resolution imaging spectrometer** (MODIS) allow the retrieval of global biophysical products, for example, net primary production, based on operational data supply and complex processing schemes. The following global advanced map products exist from earth observation: **IGBP-DIS global 1-km land cover data set** and University of Maryland AVHRR products (1.1 kilometer resolution, years 1992–1993); MODIS products, for example, of land cover (1 kilometer since 2001, future 500-meter resolution) and **vegetation** continuous fields (500 meters, since 2001); the **Global Land Cover Map of the Year 2000** (GLC 2000) (1 kilometer, **SPOT Vegetation**, year 2000); and—in preparation—**GLOBCOVER** 2005 (300 meters, **medium resolution imaging spectrometer** [MERIS], year 2005).

To improve global monitoring research with respect to temporal extent as well as information content, a combined usage of the above-listed products would be of advantage. Cross-comparisons between products however illustrate difficulties. Chandra Giri and colleagues, for example, compared GLC 2000 and the corresponding MODIS land-cover products. The authors found general agreement at the class aggregate level except for savannas/shrublands and **wetlands**, increasing disagreement when comparing detailed land-cover classes, and highly variable percent agreement between the two data sets among **biomes**. The necessity of long time series of earth surface information and the cost involved in producing global map products require an improved correspondence between products despite differing sensor systems and classification legends. GOFC-GOLD, IGOL, and GEOSS are international activities created to address (and eventually overcome) these problems.

See also: **Aerial Photography; AFRICOVER; Change Detection; Ecosystem Services; Hot Spot Identification; Land Degradation Assessment in Drylands (LADA) Project; Land Quality; Pixelizing the Social; Socializing the Pixel; TERRA-ASTER.**

Further Reading

Berounsky, Veronica, Environmental Monitoring in Rhode Island: A Synthesis, Developing Ecological Indicators: Vital Signs for Narragansett Bay and Coastal Ponds, Final Report [Online, December 2004], The University of Rhode Island Web Site, www.ci.uri.edu; DeFries, Ruth, Asner, Gregory, and Richard Houghton, eds., *Ecosystems and Land Use Change*, Washington, DC: American Geophysical Union, 2004; Giri, Chandra, Zhu, Zhiliang, and Bradley Reed, "A Comparative Analysis of the Global Land Cover 2000 and MODIS Land Cover Data Sets," *Remote Sensing of Environment* 94 (2005): 123–32; Intergovernmental Panel on Climate Change, IPCC Special Report on Land Use, Land-Use Change and Forestry [Online, December 2004], The IPCC Web Site, www.grida.no/climate/ipcc/land_use/007.htm; United Nations Framework Convention on Climate Change, International Panel on Climate Change, Good Practice Guidance for Land Use, Land-Use Change and Forestry [Online, December 2004], The IPCC Web Site, www.ipcc-nggip.iges.or.jp/public/gpglulucf/gpglulucf.htm; United States Environmental Protection Agency, EMAP Research Strategy: Environmental Monitoring and Assessment Program (EPA 620/R-02/002), Washington, DC: Environmental Protection Agency, 2002.

CHRISTIANE SCHMULLIUS

Monoculture.   A land-use practice that involves the cropping of a single plant species in an area (including trees on plantations) year after year. Monocropping removes large amounts of specific nutrients from the soil and, in order not to decrease soil fertility, large amounts of artificial inputs such as mineral fertilizer need to be integrated into any monocultural agricultural production system, often with cascading negative effects upon the provision of **ecosystem services**—on freshwater and remote marine habitats even. Monocropping also increases the **vulnerability** of plants to disease, predation, and parasitism by providing abundant hosts or food for pest organisms, thus triggering increases in pest populations and reinforcing the input of biocides such as chemical pesticides with detrimental impacts upon native species. In general, monocultures lead to a decrease in the **biodiversity** of a given area, often involving simplifications of ecosystems and landscapes. This is also true for "tree farms" or monocultural forest plantations such as pines in the **Andes**, spruce in **west-central Europe** (notably Germany), and eucalyptus in **southern Africa**. In most of the cases, monocultures in the developing world date back to the impact of colonial rule on local land management. In Africa, for example, commercialization of agricultural commodities and the nature of **cash crops** themselves have been of major importance. Where

Monocultural cropping of Burley tobacco on a farm near Lexington in Fayette County, Kentucky, September 1940 (curing barn in the background).
PHOTO: Marion Post Wolcott. [Courtesy Library of Congress]

grown in single stands, and not intercropped or relay-cropped as in traditional gardens and fields, in particular **maize**, **cotton**, groundnuts, and **tobacco** offer poor ground cover, especially in the growing season where erosivity is highest. Monocultural specialization in the form of large-scale and high-input **land-use systems**, which have existed only for the last 50–100 years, is the current answer to the **consumption** needs of the world's population. It appears, though, that these production systems perpetuate and intensify earlier agricultural impacts such as **soil degradation** or pollution from fertilizer and pesticides, **soil erosion**, and habitat **conversion**. In addition, there are several social issues related to the consolidation of smaller farms into huge, often agro-industrial operations, with many interconnected environmental changes. Continuous, intensive monocropping may be productive and profitable in the short term—with monocultural specialization still being the dominant paradigm in today's **forestry** and **agriculture**—but the many negative impacts raise serious questions about the sustainability of "artificial forests" and high-input, intensive agriculture. Jason Clay poses the hypothesis that the externalities of monocultures are in no small part responsible for shifting **agricultural frontiers** around the world. Increasingly, methods of integrated nutrient and pest management become important in the discussion of more **sustainable land use** practices, and organic farming and polycultures are just components of these methods.

See also: Agricultural Intensification; Agro-Industry; Biotechnology; Cascades; Coastal Zone; Colonization; Exotic Species; Land Concentration; Vulnerability.

Further Reading

Blaikie, Piers M., and Harold Brookfield, *Land Degradation and Society*, London, New York: Routledge, 1987; Clay, Jason, *World Agriculture and the Environment: A Commodity-by-Commodity Guide to Impacts and Practices*, Washington, DC: Island Press, 2004.

HELMUT GEIST

Mountains. Upland environment covering a third of the world's land surface due to relatively high elevation above sea level or because of the presence of hilly or mountainous terrain. The land covers of these areas are affected therefore by the unique physical environmental conditions found at high elevations or with steep slopes. Temperatures, precipitation, and air pressure can be low enough to limit the growth and presence of living organisms. Soils are often poorly developed and shallow. There are numerous different ecological zones arrayed up the slopes of mountains, in addition to dramatic local and regional shifts with different exposures to sunlight and prevailing winds. These form environmental continua and can offer opportunities for complex systems of land use and natural resource extraction. For example, people have long exploited the numerous different ecological zones of tropical mountains through the management and **agriculture** of the landscape mosaics. This usage was also the case in many middle latitude mountainous regions in the past, although nowadays these lands are more valued for their scenic beauty and water resources. Future land-use changes are likely to be most affected by climate change, and the need to provide **access** within mountainous areas for transportation and the extraction of natural resources. The monitoring of these changes by remote sensing and the prediction of potential landscape conditions are necessary activities for the sustainable use of the highlands of the world.

Ecological Zonation, Complex Mosaics

High elevations have lower partial pressures of atmospheric gases than those lands found near sea level. Thus, gases such as carbon dioxide, needed for photosynthesis of plants, can be present in limiting amounts. The low air pressure also changes the temperature balances of the lower atmosphere and land surfaces because less heat is absorbed and maintained by gases. Nights are cool or cold, while temperatures can increase dramatically during the day due to incoming solar radiation, but cool off quickly in the shade or when evening comes. Many native plants have adaptations, such as low growth forms and hairy leaves, to cope with these daily temperature fluxes. Another consequence of energetic and thermal limitations is that mountain ecosystems have less **resilience**, or capacity to recover following a disturbance, than those found at lower elevations. The environmental conditions restrict the amount and rate of **regrowth**.

The vegetation types found in mountains are typically of a relatively small stature. There are dwarf forest types found near many altitudinal treelines, in addition to on top of mountain ridges with shallow soils. Up above the

Jura Mountains of France (left) and snowcapped Bernese Alps of Western Switzerland (right), October 10, 1994 (with Lake Geneva at the bottom of the image).

PHOTO: Space Shuttle. CREDITS: Image Science and Analysis Laboratory, NASA-Johnson Space Center, 18 Mar. 2005 (Earth from Space—Image Information) http://earth.jsc.nasa. gov/sseop.efs

treeline, a variety of herbaceous plants dominate landscapes, including meadows formed by grasses or sedges, plus forbs or shrubs with prostrate or cushion forms that hug the ground. At middle elevations, temperate mountains often are covered by evergreen conifer forests in places like western North America and New Zealand, while the European Alps, the **Himalayas**, and eastern North America have mixtures of coniferous and deciduous hardwood forests. Tropical mountains typically have diverse forests of evergreen hardwoods, although there are also conifer-dominated forests in Central American highlands and in some sites in **South America**, Africa, and the Pacific Ocean islands.

Local winds flow upslope during the day due to solar heating and thermal lifting. At night, the air parcels cool on the peaks and ridges, forcing cold downslope winds to form. These mountain and valley breezes are augmented in mountainous areas with glaciers and ice caps, which further cool the air and can form katabatic winds that flow downslope with gravity. In fact, even terrains that have no current permanent ice often show evidence of earlier cold conditions. Past glaciation, for example, left behind a landscape legacy of U-shaped valleys, ice-carved peaks and ridges, and valleys filled with moraines, erratic boulders, **wetlands**, and glacial lakes in mountains around the world.

"Aspect" refers to the compass direction of a slope, which is an important control on physical environmental conditions in rough terrain. For example, north-facing slopes in the Northern Hemisphere receive less direct solar radiation than do south-facing slopes (the opposite is the case in the Southern Hemisphere). The north-facing slopes will be cooler, photosynthetic rates will be lower, soil moisture will be higher, and fires will be less common. Often the **vegetation** and land cover vary with slope aspect, as a result. Another important feature is topographic shadow, a function of the angle of sun, which varies seasonally and with latitude, plus the height and slope angle of surrounding terrain. This aspect-controlled relationship can make for important differences in land cover from one side of a valley to another. Finally, prevailing winds that come from a particular direction can also leave an imprint on the landscape. Windward slopes face into the prevailing winds. They also tend to have abundant fog and sometimes heavy precipitation, while leeward slopes are sheltered, but may be dry and warm due to the heating and drying of descending air parcels, forming rain shadows.

The result is that mountains are complex mosaics, with ecological zonation varying with elevation, but with additional patchiness originating with aspect differences, exposure to sun and winds, and the persistence of legacies of past environmental conditions. Many mountains are geologically complex too, the result of past volcanism, compression and tension faults, plate subduction, and uplift. As soils develop on top of bedrock or deposited rock material, they likely will vary in their textures, chemistries, and degrees of edaphic development, depending on the original bedrock. Place-to-place differences in mountainous terrains are remarkable as a result, with different atmospheric, soil-related, geomorphic, hydrologic, and ecologic processes at work over short vertical and horizontal distances. In addition, natural disturbances, such as fires or landslides, alter those conditions by damaging or removing vegetation and altering topography. On a microscale, cryoturbation can alter soil and vegetation properties as ice repeatedly forms and melts in the top several centimeters of topsoil. The presence of permafrost at high elevations alters soils and can create bogs and other wetlands.

Biological diversity and the composition of plant and animal communities change with elevation and with the other sources of environmental heterogeneity. Because of the harshness of high-elevation environments, only species with unique adaptations will be found. Furthermore, many mountains or mountain chains are not connected to other high-elevation environments, creating island-like conditions in terms of their remoteness and inaccessibility. This isolation leads to genetic separation and, over long time periods, to the evolution of unique (endemic) species.

Land-Use Systems, Environmental Effects

Paradoxically, the highlands of tropical latitudes are among the most important landscapes in the world for the development of sustainable agricultural practices and the survival of ethnic groups. For example, it was in the mountains of the tropical Americas that many crops were domesticated one to six millennia ago. There are thousands of land races of potatoes in the tropical **Andes** of South America, with people mixing the genes of domesticated and wild species, and crafting variants adapted to local conditions and satisfying local tastes and needs. In places where water is limited, systems of irrigation farming were developed and communal labor arrangements were institutionalized to share that water and maintain flows. Individual fields are often small and scattered, a feature that takes advantage of the multiplicity of microclimates available and lessens the risk to crop loss due to natural disturbances and calamities caused by insect pests, crop diseases, and drought.

Pastoral systems in mountains are used to graze livestock in the high-elevation **grasslands**, on fallowed agricultural fields, and in shrublands opened by burning. In some cases, seasonality of precipitation requires the movement of livestock among different landscape locations as **pastures** change in quality and productivity. Because these kinds of land cover are found in almost any mountainous areas, the **land-use systems** present are typically multiple, with some people specializing as agriculturalists, shepherds, or traders, and others acting as generalists with lands or access to lands at all available elevations. Households organize labor and efforts to acquire needed foods and products, often through community-based arrangements of labor, land, and social ties. Land use can be ancient in some places, and so land cover as a result was anciently modified, for example, in some Mexican, Andean, African, and Asian landscapes. Other sites with extensive **forests** are being deforested currently.

Socioeconomic change is driving large shifts in tropical montane land use. **Urbanization** continues to draw many people away from rural areas, as health care, jobs, and education are more readily available in cities. Rural communities need to adapt to the loss of young people to the growing cities. Roads, put in to facilitate transportation, can also provoke unsustainable extraction of timber and wildlife from newly accessible areas, such as once-isolated cloud forests. When the cities are located in lowlands, then extraction of water and other natural resources from highland areas can be especially difficult to moderate, as the users of the resources are located far from the **environmental changes** caused by extraction. Some areas in the Andes, eastern Africa, and Asia have become tourist destinations.

The snow-capped Himalaya Mountains and the impressive high-elevation Tibetan Plateau were forced up by planetary plate tectonics, which have moved India into the Asian landmass at the rate of several centimeters each year. Over 45 million years, the resulting compressive forces have raised the tallest mountains in the world. These highlands are massive enough to affect the monsoons and global biogeochemical cycles. Their forests and mountain meadows are home to numerous endemic plants and animals. The climates that affect these habitats range from continental temperate climates in the northwest to tropical environments in the eastern Himalayas.

Land-use systems depend on rainfall amounts and on elevation, with farmers of **wheat** and barley at lower altitudes and herders of yaks and other livestock in higher environs. Although extensively and anciently utilized by people, the most abrupt recent **land-cover changes** are associated with the arrival of outsiders, especially tourists and the resulting demands placed on natural resources and protected areas. Future land-use changes will likely be driven by climate change and socioeconomic fluxes as ethnic, linguistic, and political groups search for ways to maintain some traditional values, while also allowing for needed changes.

Several of the land-use/land-cover changes noted for tropical mountain and Himalayan landscapes occurred in earlier years in temperate, middle-latitude mountains. For example, seasonal **transhumance** was at one time a defining characteristic of grazing systems that connected highlands to lowlands in **Europe** and in the **United States of America**. Ironically, the urbanization of these countries in the twentieth century meant that many highland areas were abandoned by their rural residents, permitting afforestation and a decrease in **soil erosion**. Abandoned agricultural fields regrew into forests, leaving only livestock grazing and timber management as important land-use systems over extensive areas. The environmental effects of cattle and timber harvesting practices are particularly contentious issues. Today, many high-elevation areas are most useful for their supply of water resources, used for drinking water and hydroelectric generation. The scenic areas are in great demand for short-term visits for recreation. Some anciently occupied areas in the European Alps are maintained as agricultural landscapes for their cultural, historical, and tourist values. Other temperate highland areas are being settled by former city dwellers interested in owning small tracts of land, for example, in the Rocky Mountains of western North America. Places downwind from industry and cities have been affected by acidic deposition and air pollution. Examples can be found in the mountains of California and the northeastern United States and in some mountainous areas in continental Europe.

Temperate mountains differ ecologically from tropical mountains because of their seasonal shifts in temperatures. Cold winters bring precipitation in the form of snow, and the types of vegetation present are often well correlated to the length of time the snow persists. In other words, the numbers of days the high-elevation sites are free from snow correspond approximately to the lengths of the growing season. Snowpack is also important hydrologically, as that is the source of water during the warm part of the year for mountain streams and rivers.

In many middle latitude mountains that are forested, humans have altered **fire** regimes, typically through suppression and fire prevention. These activities are necessary for safety and to protect buildings, but they run the risk of building up more fuel over time, leading inevitably to larger fires in the future. Prescribed burning is now attempted in places in the western United States and Europe, as are "let burn" policies, as long as no people or infrastructure are put in harm's way as a result. There is also a past legacy and current impact of mining in many places, often accompanied by degradation of water quality.

It is in the high mountains of the world where some of the most dramatic changes in land cover and shifting treelines are predicted to occur due to future global climate shifts. This concern exists because the plants and animals found there are already near their climatic limits for survival. Increased temperatures or atmospheric carbon dioxide would be predicted to increase photosynthetic rates. However, this increased productivity is often at the expense of dominance by the original, slow-growing plant species, which are replaced by lower-elevation species and by invasives. Nature reserves and protected areas need to be redesigned to prepare for altitudinal shifts in species and ecosystems. Precipitation is predicted to increase in some areas and decrease in others, so species responses will vary with their autoecology and with the specifics of changes in particular mountains.

In some places, snowpacks and glaciers provide environmental stability to people and their livelihoods, because they can count on meltwaters to supply irrigation needs and drinking water during warmer and drier times. Most of the world's mountain glaciers have negative mass balances and are retreating upslope. The loss of snow cover and permanent ice may require large shifts in human land use and water-resource utilization. For example, a decrease of hydrological supplies and increased seasonality might necessitate the design of larger reservoirs or more thrifty irrigation systems, or even force the development of alternative land-use systems.

Monitoring

Remote sensing of changes in land cover, for example the locations of treelines and snowlines, can permit **monitoring** of environmental change over large areas. This perspective is obviously of great practical benefit for use in the mountains of the world, which are by their nature hard to observe except from above. One important limitation on land-use/land-cover classifications is the influence of topographic shadow, which in some cases obscures up to a third of remotely sensed images. The recent development of digital elevation models permits better representation of complex mountain topography. Their use, combined with pre-classification image preparation, can reduce the information lost to shadow effects. Radar imagery can be used to isolate topographic features and even measure uplift. Some of the most sophisticated landscape computer modeling is now being done in order to predict shifts in ecological processes due to changes in physical environmental controls. Other computerized models are being used to predict when and where fires will occur and the likely landscape consequences of prescribed (or planned) burning. Natural hazards mapping and **modeling** are crucial for evaluating risks due to volcanic eruptions, earthquakes, floods, and debris flows. **Integrated assessment** models allow for the connecting of global and regional climate-change predictions to the likely landscape consequences. These and other geographical information science approaches will help predict future land-cover changes in mountains.

See also: **Atmosphere-Land Interlinkages; Biodiversity; Coffee; Community Involvement; Fallow; Forest Transition; Geographical Information System; Indigenous Knowledge; Land Abandonment; Mineral Extraction; Modification; Pastoral Mobility; Reforestation; Sustainable Land Use; Watershed.**

Further Reading

Barry, Roger G., *Mountain Weather and Climate*, 2nd ed., London: Routledge, 1991; Beniston, Martin, *Environmental Change in Mountains and Uplands*, London: Arnold,

2000; Booth, Douglas E., *Searching for Paradise: Economic Development and Environmental Change in the Mountain West*, Lanham, MD: Rowman and Littlefield, 2002; Funnell, Don, and Romola Parish, *Mountain Environments and Communities*, London: Routledge, 2001; Gerrard, A. J., *Mountain Environments: An Examination of the Physical Geography of Mountains*, Cambridge, MA: MIT Press, 1990; Godde, Pamela M., Price, Martin F., and Friedrich M. Zimmermann, eds., *Tourism and Development in Mountain Regions*, Wallingford, UK: CABI Publishing, 2000; Ives, Jack D., *Himalayan Perceptions: Environmental Change and the Well-Being of Mountain Peoples*, London: Routledge, 2004; Körner, Christian, *Alpine Plant Life: Functional Plant Ecology of High Mountain Ecosystems*, Berlin: Springer, 1999; Mountain Research Initiative [Online, February 2005], The MRI Web Site, mri.scnatweb.ch; Price, Martin F., ed., *Global Change in the Mountains*, New York: Parthenon, 1999; Shoumatoff, Nicholas, and Nina Schoumatoff, *The Alps: Europe's Mountain Heart*, Ann Arbor, MI: University of Michigan Press, 2001; The Mountain Institute [Online, August 2004], The TMI Web Site, www.mountain.org; Zheng, Du, Qingsong Zhang, and Shaohong Wu, *Mountain Geoecology and Sustainable Development of the Tibetan Plateau*, Dordrecht, NL: Kluwer Academic Publishers, 2000.

KENNETH R. YOUNG

Multiplier Effect.   A basic economic concept that describes the indirect effects of an **investment** (or expenditure) in an economy as a function of the ways in which that initial investment circulates within the economy: the more it circulates, the greater the value to the economy, and the stronger the multiplier. Employment and income multipliers can be calculated and utilized to measure and predict the economic success of development strategies. Poor internal linkages and weak multipliers are often reasons for the limited success of export-oriented resource-based models of development, such as oil, gas, and metals extraction: in these settings large capital expenditures (on plant and equipment) typically produce weak developmental effects. Beyond its use as planning and evaluation tool, the value of the term lies in its emphasis on the internal structure of an economy, and the role this structure plays in magnifying the effects of an exogenous input. The term can be adapted to consider how the environmental effects of an external capital injection (foreign direct investment, remittance incomes, or central government spending, for example) are not limited to the injection itself, but are magnified (or mediated) by the ways in which it then circulates through the structure of the economy.

See also: **Exogenous Factor; Feedback; Mediating Factor; Mineral Extraction; Public Policy; Remittances Landscape.**

GAVIN BRIDGE

N

NASA Land-Cover/Land-Use Change (LCLUC) Program. Land-use/cover change research has become an important element of global change research programs on national as well as international levels, and is currently a separate element of the United States Climate Change Science Program, the scientific direction and strategy of which are coordinated by a U.S. interagency effort. Internationally, land-use/cover change research was on the agenda of the **Land-Use/Cover Change (LUCC) project** for one decade from 1995 onward, and is on the agenda of the newly developed **Global Land Project** (GLP) from 2005 onward. Following the LUCC and preceding the GLP efforts, the U.S. National Aeronautics and Space Administration (NASA) has supported several dozen projects under its Land-Cover/Land-Use Change (LCLUC) program to feed to the national and international research on land-change science. The NASA LCLUC program, which started as a combination of regional satellite-based studies, representative field-based process studies, and **modeling**, has grown into a large and multifaceted, interdisciplinary science program; and is linked to other NASA programs, such as Terrestrial Ecology, Terrestrial Hydrology, Water Cycle, Carbon Cycle, Ocean Biology, and Applications.

Land-change science is truly interdisciplinary. It requires an alliance of physical scientists (e.g., physical geographers, climatologists, ecologists, and hydrologists), remote-sensing scientists, and practitioners and social scientists (e.g., economists, demographers, human geographers, and anthropologists). Because land-change science breaks new ground in interdisciplinary research and the emergence of a new research paradigm ("integrated land-change science"), the NASA LCLUC program has made important contributions to methods on land-cover **change detection** and classification, on linking people and **pixels**, on the synergistic use of optical and radar satellite data, and in modeling the dynamic processes of land-use and **land-cover change**. The program aims to develop and use NASA **remote sensing** technologies to improve our understanding of human interactions with the environment, and thus provide a scientific foundation for understanding the sustainability, **vulnerability**, and **resilience** of human land use and terrestrial ecosystems. In doing so, a major goal of the LCLUC program is to further the understanding of the consequences of LCLUC on ecosystem goods and **ecosystem services**, the **carbon cycle**, the water cycles, and the management of natural resources.

The longer-term objectives of the LCLUC program are to develop the capability to perform repeated global inventories of land use and land cover from space to improve the scientific understanding of land-use/cover change processes and models necessary to simulate the processes taking place from local to global **scales**; to model and forecast land changes and their direct and indirect impacts, and evaluate the societal consequences of the observed and predicted changes; and to contribute to the establishment of the operational provision of land-use/cover-change data and information products, services, models, and tools for multiple users, for example, scientists, resource managers, and policymakers.

See also: Change Detection; Pixelizing the Social; Socializing the Pixel.

Further Reading

Gutman, Garik, Justice, Christopher, Sheffner, Ed, and Tom Loveland, "The NASA Land Cover and Land Use Change Program," in *Land Change Science: Observing, Monitoring and Understanding Trajectories of Change on the Earth's Surface*, eds. Garik Gutman, Anthony C. Janetos, Christopher O. Justice, Emilio F. Moran, John F. Mustard, Ronald R. Rindfuss, David Skole, B. L. Turner II, and Mark A. Cochrane, Dordrecht, NL: Kluwer Academic Publishers, 2004, pp. 17–29; U.S. National Aeronautics and Space Administration, Land-Cover/Land-Use Change Program [Online, December 2004], The NASA LCLUC Web Site, lcluc.gsfc.nasa.gov.

GARIK GUTMAN

National Institute of Public Health and the Environment. Conducts research into public health and environmental issues in the Netherlands, also operating as the Netherlands Environmental Assessment Agency (MNP in Dutch). Besides assessing implemented policy, the National Institute of Public Health and the Environment (RIVM) provides knowledge on the effectiveness and efficiency of prospective policy measures. It highlights potential bottlenecks and responds to current developments on environmental and nature issues. Both the present status of and future prospects for environment and nature policy are examined. Effects of different policy **scenarios** used in the assessments are supported by models developed by RIVM.

Effective exchange of information is needed for national and international policy development. RIVM provides information to the European Union and organizations such as the World Health Organization, the **United Nations Environment Programme**, the United Nations Economic Commission for Europe, the **Intergovernmental Panel on Climate Change**, and the World Bank. Many of these organizations have designated RIVM as a national focal point, which means that RIVM gathers and integrates information from the Netherlands, **Europe**, and/or the world and communicates this to the European Environmental Agency (EEA) and other international institutions. RIVM also hosts integrative global climate change models such as the **Integrated Model to Assess the Global Environment** (IMPEL), an international emission database (EDGAR), and the **History Database of the Global Environment**, a historical population and land-use/cover database of the past 300 years.

Today's MNP functions as the interface between science, policy, and politics, producing independent assessments on the quality of the environment for people, plants, and animals to advise national and international policymakers. MNP is one of four independent assessment agencies in the Netherlands. These agencies all have a role to play in giving substance to the World

Bank's People-Planet-Profit concept, with the Social and Cultural Planning Office of the Netherlands (SCP) dealing with "People," the Netherlands Bureau for Economic Policy Analysis (CPB), with "Profit," and MNP, along with the Netherlands Institute for Spatial Research (RPB), with the "Planet."

See also: **BIOME 300 Project.**

Further Reading

National Institute of Public Health and the Environment [Online, November 2004], The RIVM Web Site, http://www.rivm.nl; The Netherlands Environmental Assessment Agency [Online, September 2005], The MNP Web Site, http://www.mnp.nl.

KEES KLEIN GOLDEWIJK

Non-Timber Forest Products.   Forest products not derived from logging but gained from collecting, such as fruits, nuts, medicines, woodcarvings, resins, and fibers. Based on data from the Peruvian Amazon, Charles Peters and colleagues claimed that especially impoverished forest dwellers could earn more money from tropical forests by collecting non-timber forest products (NTFPs) than from logging. This captured the imagination of those concerned with **conservation**, who reasoned that, if people could sell more NTFPs, they would be less inclined to clear the **forest**.

Using data from dozens of case studies in Africa, Asia, and **Latin America**, Manuel Ruiz-Pérez and colleagues found that most cases fell into the following three groups. First, especially in Asia, farmers manage NTFPs like crops, that is, they grow them in plantations or manage them intensively in forests (which are not the least disturbed, though). The families—usually not the poorest ones—specialize in the product and, indeed, get most of their income from it. They generally have secure **land tenure** and **access** to markets, and make a pretty good living without depleting their resources. Involved in the second group are mainly African farmers who tend to be poorer and collect their products from natural forests that are not intensively managed. They rely heavily on a number of different forest products just to barely survive, and often overharvest them. Forest products provide an important safety net for these people, but the future doesn't look bright. The third group of forest products represents a smaller share of farmers' income, but they allow this income to be more diversified. These cases fall between the first and the second group in terms of the farmers' incomes and revenue and the way they manage their resources. In summary, the **meta-analysis** suggests that there are not too many cases where selling products from largely unmanaged natural forests has helped save those forests or lift people out of **poverty**. Cultivating forest products can be a good business for better-off small farmers, while collecting products from natural forests clearly helps many people to survive.

See also: **Amazonia; Arc of Deforestation; Common-Pool Resources; Desertification; Economic Livelihood; Food Security; People at Risk.**

Further Reading

Peters, Charles M., Gentry, Alwyn H., and Robert O. Mendelsohn, "Valuation of an Amazonian Rainforest," *Nature* 339 (1989): 655–6; Ruiz-Pérez, Manuel, Belcher,

Brian, Achdiawan, Ramadhani, Alexiades, Miguel, Aubertin, Catherine, Caballero, Javier, Campbell, Bruce, et al., "Markets Drive the Specialization Strategies of Forest Peoples," *Ecology and Society* 9 (2004): [online] www.ecologyandsociety.org.

HELMUT GEIST

Normalized Difference Vegetation Index.   A simple formula used in **remote sensing** studies to yield an estimate of the likelihood that **vegetation** was actively growing during a particular observation. The Normalized Difference Vegetation Index (NDVI) is computed by taking the ratio between reflected radiance values of the difference and sum from the red and near-infrared components of the electromagnetic spectrum and calculated using the formula NDVI = (near IR band – red band) / (near IR band + red band). The advantages of the NDVI are that it is easy to compute, a sensor needs to acquire data at red and near-infrared wavelengths, and it can be derived at a range of temporal and spatial resolutions from satellites such as **Landsat** Thematic Mapper, **IKONOS**, **advanced very high resolution radiometer**, **SPOT Vegetation**, and **medium resolution imaging spectrometer**. NDVI values can be correlated to parameters that can be measured in the field (e.g., leaf area index and **biomass**). They can also be compared over a time series of observations, as they are independent of illumination and viewing geometries, whereas atmospheric corrections of data are still required. The disadvantages of using NDVI values are that it is not a measurable geophysical product, and it has no clear meaning. Soil type can have a large influence on NDVI values. The NDVI has been extensively used in studies of **land-cover change**.

See also: **Change Detection.**

Further Reading

DeFries, Ruth S., and John R. G. Townshend, "NDVI-Derived Land Cover Classifications at a Global Scale," *International Journal of Remote Sensing* 15 (1994): 3567–86; Li, Zuotao, and Menas Kafatos, "Interannual Variability of Vegetation in the United States and Its Relation to El Niño/Southern Oscillation," *Remote Sensing of Environment* 71 (2000): 239–47.

KEVIN J. TANSEY

Nutrient Cycle.   Movement, transformation, and storage of nutrients within and between the spheres of the earth system (biosphere, hydrosphere, lithosphere, and atmosphere). A nutrient is any element or compound that is necessary for or contributes to an organism's **metabolism**, growth, or other functioning. Depending on the relative amount of a nutrient species needed to maintain these functions, macro- and micronutrients are distinguished. The term "nutrient cycling" is in many aspects equivalent to the notion of biogeochemical cycling. However, biogeochemical cycles include the **carbon cycle** and are generally connoted with larger **scales** in space and time, as opposed to the nutrient cycle. The notion of nutrients commonly excludes carbon and water. However, the close link between climate/weather and nutrient cycles is obvious, because temperature and water availability are the main factors driving plant growth and decomposition of organic material, thus determining both accumulation and release of carbon and nutrients. The most important macronutrient elements are nitrogen, potassium, calcium, phosphorus, sulfur, and magnesium. The characteristics of a nutrient cycle are determined by the specific transformation processes between different nutrient compounds and by the transport processes of compounds within and between the spheres of the earth system.

Nutrient cycling in all terrestrial ecosystems, no matter whether they are natural (**forests**, rangelands, **grasslands**) or anthropogenic (agro-ecosystems, **pastures**, plantations), follows similar pathways; but nutrient pools, fluxes, and transformations may vary several orders of magnitude between systems. Both natural and managed ecosystems are open systems, which means that energy or matter can enter or leave them. The question of whether the inputs and outputs are balanced is closely related to the issue of **sustainable land use**. Natural systems tend to be almost balanced, that is, carbon and nutrient gains or losses are low and the ecosystems as such may persist over a long period of time (from a human perspective). Contrastingly, many managed systems tend to be highly unbalanced, either accumulating certain nutrients, a process that frequently occurs in irrigation farming, or in over-fertilized production systems; or losing nutrients by overexploitation of nutrient stocks (i.e., nutrient mining) or by inadequate management, which may promote losses via a number of pathways including leaching, erosion, volatilization, and others.

While the majority of principles underlying the processes and flows of the nutrient cycle are well understood, the impacts on land-use/cover change often are difficult to quantify. This fuzziness on the one hand originates from the variability as inherent property of ecosystems, and on the other hand is related to our imperfect knowledge of the specific quantities of nutrients cycling through the plethora of land-use and land-cover types. For many of the world's natural and managed ecosystems, no or only vague information about the respective nutrient cycles exists, or is not accessible. However, as all ecosystems exchange matter and energy across their boundaries, they all contribute to and influence global biogeochemical cycles. Gaseous or liquid and even particulate efflux of nutrients may be transported over large distances, and enter or pass through other ecosystems. In natural systems, large imbalances of nutrient fluxes often occur after disturbances of extreme climate conditions like drought, which may promote wind erosion; weather events like rainstorms, eventually causing soil losses or wind throws; or **fire**, which may convert large fractions of biomass-bound nutrients into ashes and leave the soil unprotected. Among managed ecosystems, especially intensively fertilized agro-ecosystems and high-density livestock systems tend to leak large amounts of nutrients into the atmosphere and/or surface and groundwater. Reactive compounds of nitrogen clearly dominate the nutrient fluxes within and between ecosystems. Accumulated global emissions of food production systems sum up to forty-three teragrams of ammonia and thirty-six teragrams of nitrous oxide (NO_x), including emissions from energy production. In the Northern Hemisphere, where the major emitters of nutrients reside, all downwind and downstream systems both marine and terrestrial receive considerable inputs of nitrogen. Thus, via far-distance transport and other mechanisms, land-use activities do affect the nutrient cycles of natural and managed ecosystems.

The high relevance of nutrient cycles for **modeling** land use and **land-cover change** is obvious, all the more as the spatiotemporal dynamics of nutrient-availability (among other factors) determine land use and vice versa (i.e., every land-use type determines a specific nutrient cycle). While the general importance of land-change research is widely accepted, the integration or application of models representing nutrient cycles is still rather exceptional, especially at large spatial scales (continental to global). In the realm of nutrient cycling, a large number of different concepts and models have been developed, ranging from simple nutrient balance approaches to numerous classes of process models of varying complexity. Many of these models are specialized to simulate either certain components of the earth system like soils or plants (crops or natural **vegetation**) or coupled systems of both. Contrasting to many recent models—such as the **LandSHIFT model**, **conversion of land use and its effects (CLUE) model**, and **integrated model to assess the global environment**(IMAGE)—most nutrient cycle models are not spatially explicit but point-based, which means they simulate, for example, a single plant or multiple plants, or a homogenous field or stand.

Nutrient Balance Models

Models based on nutrient balances are not an entirely new concept. Already in 1841 the chemist Justus von Liebig realized that the nutrients taken up from the soil by agricultural crops should be replenished. A number of approaches evolved over the years, in which most nutrient balance studies followed a fairly simple routine, starting by identification of the system boundaries and the key inputs and outputs to or from a variable number of sub-components. In the most simplistic approach, the system is considered a black box, in which nutrient depletion or enrichment of the system is simply a result of the difference between total nutrient input and total nutrient output. The more detailed compartment approach considers nutrient pools or compartments, connected by nutrient pathways and transfer rates. Estimation of flows of one or more nutrients is determined either through direct measurements or literature estimates or is based on transfer functions (e.g., for volatilization, deposition, leaching, erosion losses, and others). As nitrogen is the most limiting nutrient element with respect to plant growth and crop production, followed by phosphorus and potassium, most nutrient balance studies concentrate on these key elements. Initially, nutrient balances were applied to identify nutrient shortfalls and appropriate fertilization regimes and were carried out as simple accounting approaches of nutrients in agricultural or **land-use systems** in **Europe** and North America. More recently, nutrient balances are increasingly being used for other purposes, for example, to monitor and control levels of discharge in areas of high nitrate pollution, with the goal to limit adverse effects caused by high nutrient loads. Additionally, nutrient balance studies have been used from the farm level to continental scales in Africa and elsewhere to investigate changes in soil fertility (often declines) of a broad range of different land-use types. Furthermore, nutrient balances can be applied as indicators of sustainability with respect to soil fertility. In this context, balances are related to nutrient stocks in order to estimate the fraction of stock, which is used per cropping cycle to offset potentially negative balances.

Due to the straightforward approach, nutrient balances usually produce quick and indicative results about the present status of an ecosystem. Caused by the limited availability of measurement data, the assessment of pools and fluxes usually depends on a number of assumptions, models, or transfer functions, which may limit the precision or even bias the results, for example, if applied to large spatial scales, based on imperfect data. However, this statement also holds true for the process models discussed below. Nutrient balances provide a "snapshot" of the present status, a limitation that could partly be overcome by dynamically linking the balance to a land-use/cover-change model. Nutrient balances are black-box approaches in the sense that no processes or nutrient transformations are simulated explicitly. As a consequence, if boundary conditions and processes change over time, the validity of the model decreases, and a process model should be used instead of the balance approach.

Nutrient Cycle Process Models

Process models accounting for nutrient fluxes and transformations have been developed to simulate either soils or plants or both. Nutrient cycling in soils is dominated by the processes of decomposition of organic material, weathering of rocks, exchange processes involving clays and organic substances, and diffusive-advective transport of soluble compounds. These processes are driven by temperature, moisture, pH, and the availability of nutrients and organic material. Models simulating these processes use time steps between minutes and a month and between one and eight soil compartments of different physico-chemical properties, in up to ten soil layers. A similar bandwidth of models exists for the simulation of plant growth. These are driven either by temperature (using temperature optimum curves and growing degree days) or radiation (mimicking photosynthesis) or both, and constrained by water (e.g., using transpiration coefficients or empirical stress functions) and nutrient availability (e.g., by defining admissible ranges of C/N ratios). Carbon allocation schemes mostly are being made dependent on phenological stages. But as the nutrient availability to plants is strongly controlled by soil processes, only an integrated consideration of the plant and the soil system can provide an adequate representation of nutrient cycles. This integration is, for example, implemented in models such as DayCent, DNDC, EPIC, and WOFOST. However, several reasons limit the application of such process models in land-change studies: the high requirements of site-specific data (climate, soil, vegetation; and in the case of managed systems, additional information about management practices), the complex model structure with often intransparent parameter sensitivities, and the fact that many models are tested only for specific ecosystems or regions.

Implications for Modeling Land-Use/Cover Change

Land use and land-use changes affect input and output of nutrients in soils and vegetation, which can lead to a change in soil fertility, which in turn will affect **biomass** and crop production and finally **decision-making** on future land management. Thus, improved understanding of historical and present land-use and land-cover patterns and their effects on nutrient cycles can support the development of adaptation or mitigation strategies in alternative **scenarios** of land-use change. For this purpose, nutrient cycle models could be used in combination with land-use/cover change models to simulate the spatial distribution of various land-use types over different land units and to assess potential nutrient accumulations/depletions, as many environmental problems originate from disproportionate nutrient fluxes. Additionally, land-change models simulating land-use decisions would surely benefit from the **feedback** information of how land cover and land use are affecting the environment in terms of nutrient fluxes, and thus influencing future land-use decisions. Finally, more detailed information concerning the effects on adjacent ecosystems or other components of the earth system like water and atmosphere is needed by scientific disciplines simulating estuarine and marine ecosystems or the global climate. An exemplary coupling of land-use/cover change and nutrient cycle modeling on the regional scale is realized within the Patuxent landscape model, which has been developed by Robert Costanza and colleagues. On the global scale, such coupling has partly been carried out with IMAGE, and is on the way in the LandSHIFT model.

See also: **Agricultural Intensification; Atmosphere-Land Interlinkages; Century Ecosystem Model; Coastal Zone; Cumulative Change; Ecosystem Services; Eutrophication; Land Degradation; Soil Degradation; Soil Erosion; Systemic Change; Urbanization; Water-Land Interlinkages.**

Further Reading

Costanza, Robert, Voinov, Alexej, Boumans, R., Maxwell, Tomas, Villa, Ferdinando, Wainger, L., and Helena Voinov, "Integrated Ecological Economic Modeling of the Patuxent River Watershed, Maryland," *Ecological Monographs* 72 (2002): 203–31; Galloway, Jim, "The Global Nitrogen Cycle: Changes and Consequences," *Environmental Pollution* 102 (1998): 15–24; Melillo, Jerry M., Field, Christopher B., and Bedrich Moldan, eds., *Interactions of the Major Biogeochemical Cycles: Global Change and Human Impacts*, Washington, DC: Island Press, 2003; Parton, William. J., Holland, Elizabeth A., Del Grosso, Steve J., Hartman, Melanie D., Martin, Robin E., Mosier, Arvin R., Ojima, Dennis S., and David S. Schimel, "Generalized Model for NO_x and N_2O Emissions from Soils," *Journal of Geophysical Research* 106 (2001): 17, 403–19; Tiessen, Holm, ed., *Phosphorus in the Global Environment: Transfers, Cycles and Management*, Chichester, UK: Wiley, 1995; United Nations Food and Agriculture Organization, Assessment of Soil Nutrient Balance: Approaches and Methodologies [Online, October 2004], The FAO Web Site, www.fao.org; The University of Vermont, Gund Institute for Ecological Economics, PLM-Patuxent Landscape Model [Online, October 2004], The PLM Web Site, www.uvm.edu/giee/PLM; Vitousek, Peter M., Aber, John D., Howarth, Robert W., Likens, Gene E., Matson, Pamela A., Schindler, David W., Schlesinger, William H., and David G. Tilman, "Human Alteration of the Global Nitrogen Cycle: Causes and Consequences," *Issues in Ecology* 1 (1997): 1–15.

JÖRG A. PRIESS, AMARE HAILESELASSIE, AND MAIK HEISTERMANN

P

Paleo Perspectives. Evidence for changes in land use and land cover prior to the period of **industrialization**, which comes from a variety of sources. Documentary records of various kinds reach back as far as 5,000 years in Egypt and almost 4,000 years in **China**, though in most parts of the world they span a much shorter period. Evidence of past food plants comes from macroscopic remains found at archeological sites, for example in storage jars and pits, trash middens, floor sweepings, and coprolites. By far the most widespread indications of past land-cover change come from pollen analytical evidence.

Putting all these lines of evidence together, it is clear that the earliest evidence for plant **domestication** and incipient **agriculture** in the Near East predates the opening of the **Holocene** period. The cultivation of **rice** in East Asia began no later than 8,500 years ago. Pollen analysis documents the earliest discernible impact of human activities on land cover through **fire**, forest clearance, grazing, and cultivation. In **Europe**, there are relatively few signs of human impact before the spread of Neolithic people from around 8,000 years ago onward, though it seems quite likely that Mesolithic people had a crucial impact on marginal upland ecosystems as a result of the use of fire.

Several factors impede the development of quantitative estimates of pre-industrial human impacts on land cover. Even where a long history of human impact is well recognized, the tendency in recent years has been to discount this in the urge to use pollen analysis as a tool for climate reconstruction. Where the evidence for human impact is relatively easy to identify and evaluate, deriving quantitative estimates of **land-cover change** from pollen analytical evidence is a major challenge. This is even the case in the relatively simple situation where human activity replaces temperate deciduous **forest** with open **pasture** and cultivated land. It is much more difficult in many other types of **biome** where discrimination between forested and open landscapes is more difficult. Despite these difficulties, it is important to realize that in many long-settled areas of the world, evidence for extensive human impact over many millennia is incontrovertible. In **western Europe**, for example, there are many sites where pollen analytical evidence for **deforestation** and the creation of open landscapes during the late Bronze Age, Iron Age, and medieval times confirms a massive impact on terrestrial ecosystems.

There is an urgent need to re-evaluate Holocene pollen-analytical data in light of the growing interest in the possible impact of pre-industrial land-cover change on past climate, both directly through changes in **albedo** and moisture flux and indirectly though its impact on the **carbon cycle**. There is an equally urgent need to improve the basis for quantifying landscape openness from pollen diagrams. Only through such efforts will it be possible to establish the extent to which non-forest areas that were not managed through cultivation or pastoralism at the beginning of the eighteenth century, including extensive areas over which earlier human activities were subsequently discontinued, led to major shifts in ecosystem structure and function.

See also: **Environmental Change; Human Impact on Terrestrial Ecosystems (HITE) Initiative; Pristine Myth.**

Further Reading

Alverson, Keith, Bradley, Raymond S., and Thomas F. Pedersen, eds., *Palaeoclimate, Global Change and the Future*, Berlin: Springer Verlag, 2003; Oldfield, Frank, *Environmental Change: Key Issues in and Alternative Perspectives*, Cambridge, UK: Cambridge University Press, 2005.

FRANK OLDFIELD

Participatory Geographic Information System.   The utilization of **geographical information system** (GIS) technology in the context of the needs and abilities of populations that will be involved with, and affected by, development and management decisions and projects. Participatory GIS is sometimes also termed "community-integrated GIS" or "public participation GIS." Participatory GIS distinguishes itself from traditional GIS through its being based on the assumption that stakeholders will significantly contribute to sustainable natural resource management by participating in the process of GIS application. For instance, participatory GIS can render processes of **decision-making** more transparent and accessible by producing maps that visualize development and management options. Participatory GIS emerged out of the need for new GIS methodologies that were not purely top-down and involving only GIS experts, but also the populations affected by decisions based on GIS applications. Therefore, the main objective of participatory GIS is to enable more democratic **land-use planning** processes. The challenges of participatory GIS, such as empowering weaker stakeholders and effectively integrating local knowledge, are still manifold. However, it has been successfully used, for example, in land-use planning with First Nations in **Canada**, in developing small-scale **watershed** management plans with local communities or in solving conflicts about **access** rights to natural resources in protected areas. The experience with participatory GIS so far seems to indicate that its success is influenced more by the social and political context in which GIS is used than by the technological constraints.

See also: **Arena of Land Conflict; Community Involvement; Conservation; Indigenous Knowledge; Participatory Land Management; Scenario.**

Further Reading

Abbot, Jo, Chambers, Robert, Dunn, Christine, Harris, Trevor, de Merode, Emmanuel, Porter, Gina, Townsend, Janet, and Daniel Weiner, "Participatory GIS: Opportunity or Oxymoron?" *International Institute for Environment and Development, PLA Notes 33* (1998): 27–34.

SILVIA HOSTETTLER

Participatory Land Management. Encompasses a range of different research, development, and policy approaches that promote the involvement of local people in the management of their land and/or natural resources (**forests**, wildlife, **pastures**, agricultural fields, water, etc.), referred to more generally as participatory resource management. It emerged in the 1980s as an alternative to top-down, bureaucracy-heavy, and outsider- or state-enforced **conservation**-with-development planning, which was seen as failing on pragmatic and social justice grounds. Participatory land and resource management is proposed as an effective alternative because the participation of local community members will minimize conflicts that otherwise might arise as a result of introduced management rules; because local people often have detailed knowledge about their environment that could contribute to improved environmental management; and because local communities are often best situated to protect local resources and to enforce resource/land-use rules.

Participatory approaches to land management expanded in the 1990s with most rural development and conservation projects incorporating participatory components. This led to a proliferation in the use of related terms that refer to or suggest participatory processes, including devolution, decentralization, and empowerment, as well as management approaches that claim to be bottom-up, community-based, cooperative, and collective. Participatory research methodologies are pursued to incorporate **indigenous knowledge**, involve all stakeholders, and empower local communities. Despite the popularity of these terms, participation remains a complex process that is rarely complete in practice. A growing literature interrogates claims of **community involvement** or participation in land management projects by asking: How are communities defined? Who effectively participates in community projects? When do they participate—from research and planning to implementation to the **monitoring** and review stages of a program? How do people participate? Do they contribute knowledge and skills, or only voting power and labor? In asking these questions, critics acknowledge the need for participatory land management, but recognize that participation is a negotiated and contested process, the term itself dynamic and flexible. Moreover, environmental outcomes of participatory management are not straightforward, depending not only on project design but on the local socio-ecological contexts in which participatory land management programs are inserted.

See also: **Agrodiversity; Alternatives to Slash-and-Burn (ASB) Programme; Arena of Land Conflict; Institutions; Land Rights; Land-Use Policies; Participatory Geographic Information System; Population Dynamics; Project on People, Land Management and Ecosystem Conservation; Property; World Agroforestry Centre.**

Further Reading

Chambers, Robert, Pacey, Arnold, and Lori A. Thrupp, eds., *Farmers First: Farmer Innovation and Agricultural Research*, London: Intermediate Technology Productions, 1989; Jeffery, Roger, and Bhaskar Vira, eds., *Conflict and Cooperation in Participatory Natural Resource Management*, New York: Palgrave, 2001.

MARA JILL GOLDMAN, MATTHEW D. TURNER

Past Global Changes (PAGES) Project.   A core project of the **International Geosphere-Biosphere Programme** since 1991 that supports research aimed at understanding the earth's past environment from a **paleo perspective** in order to make predictions for the future. The scope of interest of PAGES includes the physical climate system, biogeochemical cycles, ecosystem processes, **biodiversity**, and human dimensions on different time **scales** such as the Pleistocene, **Holocene**, last millennium, and the recent past. PAGES has five major research fields. Under focus 5 (past ecosystem processes and human-environment interactions), the project entails the **Human Impacts on Terrestrial Ecosystems** (HITE) initiative, Land Use and Climate Impacts on Fluvial Systems during the Period of Agriculture (LUCIFS), and Human Impact on Lake Ecosystems (LIMPACS). In March 2000, researchers from PAGES and the **Land-Use/Cover Change (LUCC) project** started a joint effort to reconstruct historical changes in permanent croplands at a global scale during the last 300 years, called the **BIOME 300 project**.

See also: **Environmental Change.**

Further Reading
Past Global Changes [Online, November 2004], The PAGES Web Site, www.pages.unibe.ch.

HELMUT GEIST

Pastoral Mobility.   Refers to human movements directly associated with the grazing management of domestic livestock. It is most often used to refer to the long-distance movements conducted by herding or pastoral peoples around the world. While the mobility of herding peoples has long been seen as an archaic feature of primitive production systems, recent advances in the ecology of highly fluctuating rangeland systems (e.g., drylands and **mountains**) have characterized pastoral mobility as a necessary adaptation to climatic risk and an effective grazing management approach for variable environments both in terms of livestock production and the environment.

The mobility of herding peoples has long captured the imagination of outside observers. Whether negative or romantic in nature, extreme **perceptions** of the "wandering nomads" have hampered any real understanding of the role that mobility plays in the social life of pastoral peoples and in the ecology of their livestock and pastures. Mobility was seen by early researchers as a key feature of the material culture of herding peoples that distinguished them from others. Various herding groups were defined with respect to their mobility. Categories such as "nomadic," "semi-nomadic," "sedentary," and "settled" were used to categorize social groups ranging from communities to whole ethnic groups. With respect to land management, mobility was seen to produce an incentive structure not conducive to good land husbandry (pastoralist societies were seen not only as outwardly invasive socially but ecologically as well)—overgrazing local pastures before moving on. Early linkages of pastoral mobility with primitiveness, aggression, and poor land husbandry have proven to be highly persistent within national governments and the international development communities. An important political factor contributing to this persistence is the difficulty that mobility presents for national governments seeking to govern, control, and provide services to pastoral populations.

Longer-term studies of pastoral communities have shown that simple characterizations of the mobility of an ethnic group, community, and even household are highly misleading. Pastoral households show significant variation within and between years in their reliance on mobility to provide fodder to their animals. This variation reflects the particular household's resources and constraints, such as the number of animals it is managing; its **access** to herding labor; its access to local and distant water and **pasture**; and its reliance on other economic pursuits such as crop **agriculture**. In addition, broader factors that change over time, including climatic conditions, political insecurity, market opportunities, viability of social networks, cropland spread, and government actions, will affect when, where, and whether herders will move their livestock. Therefore, mobility should not be seen as some inherent cultural feature but as a term that encompasses a range of grazing management options that pastoral groups adopt in response to changing social and ecological situations. As a result of household and broader factors of change, pastoral mobility has declined significantly across wide areas of the world.

Resource managers working in dryland and mountain environments generally recognize the role that livestock mobility can play in both adjusting grazing pressures to forage availabilities at a regional scale, and producing temporal grazing patterns (graze-rest cycles) at the patch level that are more consistent with rangeland sustainability. In most areas of the world, livestock mobility can be maintained only through herding and pastoral mobility. As a result, there has been an upsurge of interest in reinvigorating pastoral mobility, particularly in dryland regions. These efforts have been plagued by a number of problems. First, there remains significant conceptual confusion around the term. What features of the grazing system are managers interested in promoting? Distance covered by livestock? Spatial dispersion of livestock? Distance covered between successive pastoral encampment points? Second, there exist many household level and broader factors that now make it difficult to reinvigorate pastoral mobility. Many pastoral peoples have come to rely on crop agriculture, sedentary ranching, and wage labor to supplement their pastoral livelihoods, which are restricted due to dwindling pasturelands available to their herds, and unfavorable government regulations. The movements of livestock and people are mediated by social networks across broad areas. Disruption of these networks will make it impossible for herders to socially navigate themselves across the pastoral landscape, even if they have the economic resources to do so. Reinvigoration of pastoral mobility will require a combination of state-level policy and local project initiatives working at different spatial scales.

See also: **Arena of Land Conflict; Cattle Ranching; Cultural Factors; Degradation Narrative; Desertification; Disequilibrium Dynamics; Great Plains; Transhumance.**

Further Reading

Fratkin, Elliot, Galvin, Kathleen A., and Eric A. Roth, eds., *African Pastoralist Systems: An Integrated Approach*, Boulder, CO: Lynne Reinner Publications, 1994; Niamir-Fuller,

Maryam, ed., *Managing Mobility in African Rangelands*, London: Intermediate Technology Publications, 1999; Scoones, Ian, ed., *Living with Uncertainty: New Directions in Pastoral Development in Africa*, London: Intermediate Technology Publications, 1994.

MATTHEW D. TURNER, MARA JILL GOLDMAN

Pasture.   Landscapes that have a ground story in which grasses are the dominant **vegetation** life form. Definitions often include savannas, as well as the very sparsely grasslands of arid regions. **Grasslands** constitute one of the main land uses on earth, around 36×10^6 km², according to information from the **United Nations Food and Agriculture Organization**, including three major vegetation types. The first category is treeless grasslands, which may be temperate (the steppes of **Europe** and Asia) or tropical (Africa, **South America**, and Australia). The second type is savannas, the grasslands with a woody overstory, which often occur next to the treeless grasslands. The last category is the arid, shrubby grasslands. In Africa, South America, and Australia, these types tend to form a continuum of vegetation.

Given equivalent biophysical characteristics, grasslands are the exception rather than the rule, since long-lived woody vegetation (trees and shrub) will ultimately compete with the relatively short-lived grasses. In such contexts, the competitive advantage of woody species is significant, and grasslands tend

Cattle grazing in mountain pastures of Siskiyou County, California, July 1942.
PHOTO: Russell Lee. [Courtesy Library of Congress]

Fenced pasture on cut-over forest land in the Priest River Valley of Bonner County, Idaho, October 1939.

PHOTO: Dorothea Lange. [Courtesy Library of Congress]

to exist only in cases in which the dominance of woody species is suppressed. The factors that prevent dominance by woody plants are disturbance (fire), soil (physical and chemical properties), and temperatures (freezing). **Fires** are influential in the tropics and subtropics for suppressing woody plants. Grasslands are often located in areas with lower precipitation levels and under soil with specific characteristics (such as sandy soils or those subject to regular flooding) that inhibit woody and shrub vegetation. Finally, temperatures constitute a crucial factor that explains the distribution of grasslands at global **scales**.

Grasslands are exposed to **conversion** to croplands where temperatures and moisture availability provide conditions for agricultural expansion. In contrast, where the factors of soil or freezing temperatures are responsible for suppressing competition from woody species, grasslands remain devoted to pastoral uses. Nevertheless, in some environments such as the tropical regions, there is an increasing expansion of pasture to the detriment of

Overgrazed pasture with spreading gullies in the mountains of Bernanillo County, New Mexico, April 1940.

PHOTO: Russell Lee. [Courtesy Library of Congress]

forest. In both cases, grasslands and implanted pastures are exposed to **land degradation** as a result of overgrazing and lack of management practices. Many savannas and grasslands are under growing land-use pressure from grazing animals, cropping, and the extensive use of fire. The implanted pastures in the tropical and subtropical landscapes are exposed to the same degradation effects.

It has been documented that the most relevant **land-cover change** is tropical forest conversion to pasture for livestock production. For instance, in the Brazilian Amazon, around 65 million hectares had been deforested for agricultural uses by 2003. Pasture expansion is responsible for about 70 percent of forest conversion, while **fallow** and unutilized areas account for about one-fifth of the total deforested area, and one-tenth is likely under agricultural uses. There have been different explanations for the overwhelming expansion of pasture in tropical areas. The main explanation has stressed the influence of government policies that have provided incentives to cattle ranchers to settle down in frontier areas (such as tax holidays and subsidized credit). Although incentives have been removed over time, **cattle ranching** continues its expansion, mainly due to the development of infrastructure, and **access** to growing domestic and foreign markets that contribute to making beef cattle production a profitable activity.

In the Amazon, some grasses were introduced from Africa, principally *Brachiaria spp.* The overall stocking rate is less than one head per hectare, which is relatively low compared to other areas. Most cattle ranching in the Amazon is extensive, and there are no appropriate pasture management practices. The pastures are exposed to degradation after relatively short periods of use for cattle breeding, approximately ten to fifteen years. It is estimated that about half of the total pasture implanted might be degraded, supporting very low stocking rates. These areas show low plant cover, invasions by non-palatable native species, and often dense population with termite mounds. The main causes of pasture decline are inappropriate management practices, among them lack of maintenance and fertilization.

Intensification of cattle production in the Amazon is still incipient because of the large availability of cheap forestland and the high costs of pasture maintenance. It has been extensively argued that it could be cheaper to clearcut one hectare of forest to implant pasture than to recover the same hectare covered with degraded pasture. This trend might revert as obstacles for cattle ranching to expand over forest fringes increase (i.e., by enforcement of land-use regulations).

The solutions to pasture degradation are different, depending on both the causes and the stage of pasture degradation, which may range from a mere reduction of the pasture vegetation by the competence of some shrub species, to acute process of **soil degradation**, to reduction of vegetal **biomass**. The most commonly suggested solutions are soil fertilization, renovation of pasture, and the introduction of more intensive techniques of pasture management, using some sort of rotational systems over time and space, or interplanting pasture with some short-life crops. Furthermore, it has been suggested that pasture productivity could be enhanced by planting more diversified systems such as silvopastoral systems, allowing for some secondary succession. In some cases, however, degraded areas are abandoned and lead to forest **regrowth** to become the dominant vegetation.

Research focus on forest succession has found that forests regenerate vigorously in areas that were previously devoted to light agricultural use, and in which there was no complete removal of forest cover. Areas with moderately grazed pasture also are able to develop forest after abandonment, but biomass accumulation is lower than on light-use sites. In contrast, abandoned pastures subjected to heavy use have distinct patterns of succession because forest succession is scarce. Amazon forest can recover after large-scale pasture disturbance, and only **reforestation** is uncertain where land has been used too intensively for long periods. Some research done on soil nutrition concentration after pasture use has found that soil nutrition is independent of age and prior intensity of use in the pastures. There is some soil nutrition concentration when the pasture areas are abandoned, but then nutrient uptake by successional vegetation tends to reduce such soil nutrient stocks.

See also: **Agricultural Frontier; Agricultural Revolution; Alternatives to Slash-and-Burn (ASB) Programme; Amazonia; Brazilian Cerrado; Cumulative Change; Deforestation; Desertification; Disequilibrium Dynamics; Extensification; Great Plains; Hot Spots of Land-Cover Change; Land Abandonment; Land Rehabilitation; Pastoral Mobility; Savannization; Secondary Vegetation; Slash-and-Burn Agriculture; Tradeoffs; Transhumance; Tropical Dry Forest.**

Further Reading

Graetz, Dean, "Grasslands," in *Changes in Land Use and Land Cover: A Global Perspective*, eds. William B. Meyer and Billie L. Turner II, New York: Cambridge University Press, 1994, pp. 125–48; Uhl, Christopher, Buschbacher, Robert, and Emanuel A. S. Serrao, "Abandoned Pastures in Eastern Amazonia I: Patterns of Plant Succession," *The Journal of Ecology* 76 (1988): 663–81; Uhl, Christopher, Buschbacher, Robert, and Emanuel A. S. Serrao, "Abandoned Pastures in Eastern Amazonia II: Nutrient Stock in the Soil and Vegetation," *The Journal of Ecology* 76 (1988): 682–99.

PABLO PACHECO

Pattern Metrics.   A set of quantitative indicators of landscape configuration, generated from thematic classifications and providing insight as to the patterns impacting landscape processes. Metrics measure many types of landscape configuration, including shape complexity, diversity, interspersion, proximity, and abundance. Pattern metrics are the implementation of the basic tenet of landscape ecology that understanding landscape processes requires not only knowing what is there but also knowing how it is configured and how, therefore, energy and materials flow in the landscape.

Landscape ecology premises that the **scale** at which landscapes are observed impacts the discernible patterns used to infer process; it further purports that those patterns comprising the landscape mosaic are key in understanding ecological processes by linking form to function. That is, the configuration of landscape elements both constrains and promotes the movement of materials and energy across the landscape. The landscape mosaic comprises three components: the matrix (background or dominant class), patches (contiguous cover of a class of interest having both edge and interior), and corridors (essentially interiorless, elongated, and relatively narrow patches that link or divide other landscape components). The scale of observation (and analysis) influences not only whether a landscape element is detectable but also whether it is defined as a patch, a corridor, or part of the matrix.

Once landscape elements are delimited into these configurational categories and attributed with their thematic class, pattern metrics can be generated. Pattern metrics can be run on either vector or raster data, and while some metric equations are the same between the two, others are different. Therefore when interpreting pattern metrics results, is it important to know the processing lineage. Vector-based metrics use a circle as perfect geometry for comparing shape departures. A circle is used because it represents the greatest area-to-perimeter ratio possible, and thus provides a baseline for comparison (that is not to suggest normatively that simple geometry is necessarily better than complex geometry). Using the raster data model pushes the perfect geometry to a square for comparative purposes. There is, however, an equally important consideration in the differences between pattern metrics run on vector versus raster data sets. Vector data sets typically represent landscapes more generally and smoothly than their raster counterparts, owing to the method by which that landscape was classified. Vector classifications are usually the results of manually digitizing off of a product (map, air photo, satellite image), where the analyst through interpretive skill is able to classify complex or mixed landscape classes. For example, if studying **agriculture** and digitizing agricultural areas off an air photo, the analyst would not digitize rows of soil separately from rows of plants in areas with row crop agriculture. With raster/automated processing, most classification treatments operate at the **pixel** level, resulting both in soil and **vegetation** being classified separately and a more pixelated product (depending upon the spatial grain). Unless smoothed or filtered first, raster products translate into more fragmented landscapes than does the manual/vector approach, and this difference can be observed in metrics.

While over fifty metrics have seen widespread use, most researchers now agree that there are many fewer (four to seven) classes of metrics that measure different components of landscape configuration. Additionally, while metrics are generated at the patch level, they can be reported at the class (thematic) and landscape levels as well via weighted averaging. Pattern metrics provide indications of shape complexity, abundance, interspersion, proximity, diversity, and, most recently, orientation. Not all metrics are supported at each level, and many require a minimum number of thematic classes or number of patches for the metric to be meaningful.

Several challenges in pattern metrics analysis remain. First, many parameters must be set before running metrics, such as how to define proximity (distance threshold, and in raster data also whether diagonal neighbors "count" or not) and edge versus interior (depth of edge). A slight change in these parameters (or the classification scheme) can produce dramatically different results. Second, the comparison of metrics results across time and/or areas is difficult since there are few absolute baselines in most pattern metrics. Holding number of classes, number of patches, and area size fairly constant can minimize artificial differences. Third, while the ecological importance of landscape pattern is clear, it is less clear what the ecological meaning of particular results is and when changes in the metrics represent ecologically significant change. In partial answer to the problems of comparison and ecological meaning, landscape ecologists have developed the neutral landscape method whereby real patterns and metrics can be compared to simulated (and therefore controlled) landscape pattern metrics.

See also: **Aerial Photography; Agent-Based Model; Auto-Correlation; Metabolism; Pattern to Process; Remote Sensing.**

Further Reading

Forman, Richard T. T., *Land Mosaics: The Ecology of Landscapes and Regions*, Cambridge, UK: Cambridge University Press, 1995; Frohn, Richard C., *Remote Sensing for Landscape Ecology: New Metric Indicators for Monitoring, Modeling, and Assessment of Ecosystems*, Boca Raton, FL: Lewis Publishers, 1998; Riitters, Kurt H., O'Neill, Robert V., Hunsaker, Carolyn T., Wickham, James D., Yankee, Dennis H., Timmins, S. P., Jones, K. Bruce, and B. L. Jackson, "A Factor-Analysis of Landscape Pattern and Structure Metrics," *Landscape Ecology* 10 1 (1995): 23; Turner, Monica G., *Landscape Ecology in Theory and Practice: Pattern and Process*, New York: Springer Press, 2001.

KELLEY A. CREWS-MEYER

Pattern to Process. A phrase that refers to the analysis of a spatial pattern of land cover in order to detect and understand the mechanisms that produced the pattern. "Process to pattern" is a related phrase that refers to the analysis of how the mechanisms of a land-change process manifest as a pattern on the landscape.

The fishbone pattern of **deforestation** in **Amazonia** is perhaps the best example of a distinct pattern of land change that has been linked to a particular **agricultural frontier** process. There is a main backbone of deforestation along the Transamazon Highway, and secondary ribs of deforestation along the secondary roads that are at regular intervals emanating from and perpendicular to the highway. This pattern of deforestation is caused by a sociological process of immigration to the forest via the highway and secondary roads.

There are numerous valid ways to characterize both a process and a pattern; therefore it is a challenge to link a particular process of land change with a particular pattern of land change. For example, Amazonian deforestation could be explained as a manifestation of the sum of the decisions of many individual farmers, each of whom is trying to maximize income for his or her family. Alternatively, Amazonian deforestation could be explained as a manifestation of the decisions of a much smaller number of regional planners who are encouraging immigration in order to alleviate unemployment in Brazilian cities. Both explanations lead to the fishbone pattern. Also, there are a variety of ways to measure a pattern, such as average size of patches, perimeter-to-area ratio of patches, etc. It is not immediately obvious which **pattern metrics** are best to characterize a particular pattern.

Linking pattern to process is especially important for scientists who create simulation models of land change. A model usually attempts to simulate the process of land change and then predicts the resulting pattern of land change. If there is a clear link between pattern and process, then the model can be validated by comparing the predicted patterns against reference data of land cover.

See also: Cellular Automaton; Decision-Making; Land-Use History; Modeling; Pixelizing the Social; Socializing the Pixel; Transition Matrix; Validation.

Further Reading

Liverman, Diana, Moran, Emilio, Rindfuss, Ronald, and Paul Stern, eds., *People and Pixels: Linking Remote Sensing and Social Science*, Washington, DC: National Academy Press, 1998; Nagendra, Harini, Munroe, Darla, and Jane Southworth, "From Pattern to Process: Landscape Fragmentation and the Analysis of Land Use/Land Cover Change," *Agriculture, Ecosystems and Environment* 101 (2004): 111–65; White, Roger, and Guy Engelen, "Cellular Automata and Fractal Urban Form: A Cellular Modelling Approach to the Evolution of Urban Land Use Patterns," *Environment and Planning A* 25 (1993): 1175–79.

ROBERT GILMORE PONTIUS JR.

Peatlands. One of the most important ecosystems in the world, representing 50–70 percent of **wetlands** on earth. They are also fundamental due to their vital functions in carbon, hydrological, and geochemical cycles. Although peatlands cover only 3 percent of the land and freshwater surface of the planet, they contain 30 percent of the carbon stored in soil and 10 percent of freshwater resources. It is estimated that the total carbon pool of wetlands exceeds that of the world's **forests**. They also support a very high diversity of species, many of which are unique for this ecosystem. Peatlands exist in varying climate conditions on different continents, from sea level to high-alpine conditions, and from tropical to boreal and Arctic zones. They occur everywhere on the globe, except where the climate is too dry or too cold to support plant life. In some countries they account for up to 30 percent of the land area.

However, peatlands have generally been seen as a valuable natural resource in the form of peat itself, which has many uses, including fuel, animal bedding, and growth substrate in **agriculture** and horticulture. Increased human demands place pressure on the use of these ecosystems, significantly changing the range and importance of the system's diverse functions, services, and resources. Due to growing needs of resources from peatlands, sustainable and wise use of these ecosystems is essential in order to assure that sufficient areas of peatlands remain to carry out their vital natural resources functions.

Definitions

Peatlands are areas with or without **vegetation** with a naturally accumulated peat layer at the surface. Peat is a brownish-black soil composed of sedentarily accumulated material, consisting of at least 30 percent dead organic material that results from carbon chemical biosynthesis. Inorganic soil particles are washed or blown into peatlands and also form part of the peat. The prominent characteristic of a peatland is the near-surface water table, as this makes peat accumulation possible—thus, it occurs in areas where precipitation/groundwater exceeds evapotranspiration/drainage. There are several types of peatlands, which are classified according to vegetation type and the source of water. A mire is a peatland where peat is currently being formed. The two main mire types are fens and bogs. A bog is a peatland that is raised above the surrounding landscape. It is fed mainly from rainwater. Some bogs support forests, but they are nutrient-poor. A fen is a peatland situated in depressions or on slopes and is fed from both rainwater and groundwater that are minerotrophic, acid or alkaline, and nutrient-rich or nutrient-poor.

Inventories

Peatland estimations vary substantially among sources. Presently, there is a lack of knowledge on the effective extent and precise location of peatlands, as different criteria are used to classify peatlands, depending on the scientific discipline.

Globally, it is estimated that 4 million km^2 are covered with peatlands, or approximately 3 percent of the planet. The largest known concentrations of peat are located in the Northern Hemisphere, being widespread within the boreal fringe, in Southeast Asia, and parts of the Amazon basin. Southern Hemisphere peatlands are generally smaller, sparsely distributed, and relatively shallow. Less than 1 percent of the world's peatlands are found in **South America**.

In the Northern Hemisphere, due to higher rainfall, peatlands occur over large areas of **Canada**, Alaska, Fennoscandia (Finland, Sweden, and Norway) and **Russia**, having an average thickness of 1.5 meters. It has been estimated that northern peatlands cover 346 million hectares and hold 455 billion metric tons of carbon, slightly less than the amount contained in all living organisms or in all atmospheric carbon dioxide.

In Southeast Asia, the total area of peatlands is estimated to be about 20 million hectares, which is approximately one-tenth of the entire extent of global peatland resource and at least 60 percent of the world's tropical peatlands. Most of the peatlands in Southeast Asia are found in Indonesia, which possesses over 70 percent of total peatland area in this region. Other major peatlands are located in Thailand, Malaysia, Vietnam, Brunei Darussalam, and the Philippines.

Functions

Peatlands are a particularly significant type of wetland and perform many vital functions as centers of hydrological functioning, as carbon sinks, and in terms of **biodiversity**. Peatland hydrological functions are more or less similar to those of wetlands. However, peat enhances normal wetland functions, making peatlands particularly efficient in water storage and highly effective as filters. Peat acts as a sponge, storing water for long periods of time and releasing it when necessary. Peat and peatland vegetation are also good filters, making peatlands highly effective in removing sediment, pollutants, and pathogens.

Peatlands are organic sinks, and it is estimated that a quarter of the earth's soil carbon is stored as peat. When life on earth began, there was more than ten times more carbon dioxide in the air than at present. Carbon-based organisms absorbed and used carbon dioxide and slowly reduced atmospheric carbon dioxide. When these organisms died, much of the carbon was locked up in the underground. The carbon was stored as peat in wetlands, and over geological time peat turned first to lignite and then coal. Carbon accumulation balances between inputs (**biomass** or primary production) and outputs (microbial activity that degrades organic matter) in redoxic conditions. This process occurs mainly in active mires.

Specific conditions of habitat in peatlands such as stresses (high acidity in bogs and nutrient-poor, redoxic conditions) call for adaptations of plants or animals to these ecosystems. Then, plant species require adaptations in physiology to deal with the toxic substances that originate under anaerobic conditions; as well as anatomy, such as aerenchyma that lead oxygen from the parts above ground to the root system; and growth form, including aerial roots.

Benefits and Values

Like all wetlands, peatlands provide a variety of goods and **ecosystem services**, both directly and indirectly, in the form of water supply and groundwater recharge, flood mitigation, energy, **forestry**, and fishery products. They also have a fundamental functional significance, the carbon stored in peat representing some 25 percent of the world soil carbon pool, which would contribute to global warming and climate disruption if released. However,

some peatland uses and services are specifically predominant. In the Northern Hemisphere, peatlands are deep and extensive, which allows for their
use as fuel. Peat has a low combustion temperature and low nitrogen
oxide emissions and has been used traditionally for millennia in **Europe**,
Fennoscandia, and Russia as a fuel, more recently for electricity generation.
Industrial extraction of fuel peat has occurred in Europe since the nineteenth
century with the invention of peat-working machines. Extensive drainage of
bogs and fens for agricultural use in Europe began in the seventeenth century, although in some countries (like Sweden) cultivation of peat soils dates
back to the early Iron Age. At present, half of all peat extracted is used in
horticulture or for soil conditioning. Over 40 percent of the peat extracted
in **South Africa** is used as a basis for horticultural substrates, and almost all
Canadian peat is used for horticulture needs.

Threats to Peatlands

Since 1800, the total area of peatlands worldwide has been significantly
reduced by approximately 10–20 percent due to climate change and human
activities. All combined uses of peatland, through peat mining, drainage, and
conversion to agriculture and forestry, lead to a global rate of loss of active
peatlands estimated to be 0.1 percent per year. However, the losses have been
much greater in Europe, where 52 percent of peatlands have now been lost.
In some countries like the Netherlands, active peatlands that once covered
20 percent of the land surface are almost totally gone. Direct impacts on
peatlands include drainage, land conversion, excavation, inundation, and
visitor pressure. Peat extraction drastically reduces the water storage and fil-
tering properties of the wetlands, impacts the carbon fixation function, and
affects biodiversity negatively. Biologic or hydrologic **resilience** can occur
only when peat is not industrially extracted. Drainage for agriculture and
forestry still continue to be major causes affecting change in peatlands, especially in the tropics. **Forest degradation** through forest/peat fires, which
have increased in tropical peatlands since the 1990s, caused destruction of
these ecosystems in large areas in Southeast Asia, especially Indonesia, and
released a huge amount of carbon dioxide to the atmosphere, potentially
spurring global warming. Besides the increase of the **greenhouse gas** carbon
dioxide (CO_2), indirect impacts on peatlands are a result of air pollution,
water contamination, contraction through water removal, and infrastructure development.

Wise Use of Peatlands and Mires

The utilization of any natural resource should subscribe to the principles
of sustainability. The effective utilization of peatlands involves a compromise
among traditional practices, **conservation**, and tourism. Thus, peat extraction
and some management actions are not compatible with undisturbed sites of
high scientific and biodiversity value. Until recently, the conflict between
conservationists and the peat industry seemed to be irreconcilable: preserving
the natural heritage value and the bulk of sequestered carbon of peatlands
will remain incompatible with any form of exploitation. However, in some
cases limited peat extraction can increase biodiversity and may be sustainable in the long term. Therefore a wise use of peatlands and mires is at

present encouraged. On an international level, the **Ramsar Convention on Wetlands** encourages the protection of peatlands worldwide (recommendation 6.1 of the 1996 Ramsar meeting). However, despite being the most extensive single wetland type in the world, less than 10 percent of the global peatland area is represented in the Ramsar list: wetlands of international importance, or "Ramsar sites" with significant peatland components, amount to 4.1 million out of 120 million hectares of protected wetlands; and of the 1,399 Ramsar sites, 443 are wetlands with significant peatland components (as of March 2004).

See also: **Atmosphere-Land Interlinkages; Biome; Carbon Cycle; Deforestation; GLOBWETLANDS; Modification; Paleo Perspectives; Water-Land Interlinkages.**

Further Reading
Global Environment Centre, Peat-Portal [Online, January 2005], The Peat-Portal Web Site, www.peat-portal.net; International Mire Conservation Group [Online, January 2005], The IMCG Web Site, www.imcg.net; Lappalainen, Eino, ed., *Global Peat Resources: Peatlands in Biosphere, Country Reports, Conservation and Use,* Jyväskylä, FL: International Peat Society, 1996.

LAURENCE HUBERT-MOY

People at Risk. Though at risk, people are not mere "victims" of **global environmental change** but have the ability to affect and shape how they live with a range of risks. Humans interact with nature and natural resources, producing negative and positive impacts. These interactions can operate at a local level (e.g., household) or they may be the product of larger-scale interactions (e.g., outcomes arising from development strategies, legislation, or economic systems). Understanding how global environmental change problems are conceptualized and how decisions, at various levels, including risks coupled to environmental change, are important research foci. Knowledge of the array of things that people do to better live with risk arising from global environmental changes is a key issue receiving heightened attention by scientists, governments, and wider civic society.

The impacts and risks associated with complex global environmental changes include ecological risks, climate risks, and risks to health and the environment. This complexity and breadth of issues in the environmental arena requires multiple approaches and perspectives. Traditionally, this field has been strongly influenced by scientists working in the biophysical science domain. Here, concerns include improving the understanding of the functioning of the earth as a complex system. With the strong shift toward sustainability and sustainable development or **sustainable land use**, however, perspectives are also increasingly being drawn from the social, economic, and political sciences, including questions of how terms such as risks and hazards are derived, framed, and constructed. Security risks associated with global environmental changes and the realm of issues linked to uncertainty, sustainability, and change are thus emerging as key themes allied to global environmental change research.

The notion of risk can be used and interpreted in many ways. Several academic and practitioner communities have been trying, for example, to articulate concepts that can assist both in improved understanding of the complex earth system and also to find ways to better "live" within such a system. Various terms have consequently been identified that are now embedded in the global environmental change discourse and literature, including terms such as risk, hazards, risk society, adaptive (co)-management, and risk reduction.

Global environmental change scientists like Carl Folke, Fikret Berkes, and Crawford Holling, among others, have been grappling with such terms as ecological risks and resilience, and have opened up spaces for expanded dialog between natural and social scientists working on issues related to environmental and other risks. Sociologists have also tried to frame understandings of risk from social and cultural perspectives, including examining the role of **institutions** and the cultural interpretations and practices enabling people to live with risks associated with global environmental change. Some of these approaches and uses of risk in the international science, policy, and development arena are explored more fully below.

Before detailing notions of risks and responses to risks, one needs to have some common reference point. Despite the absence of a common language to capture a universal term such as risk, it is usual to view risks as potential threats and dangers. Hence, the term risk is often used in association with the risk of something occurring (e.g., hazard or process) in a specific social setting (e.g., region at risk), as Roger Kasperson and colleagues use it, among others. Astrid von Kotze and Ailsa Holloway (1996, p. 5) talk about "elements at risk" and refer to communities, structures, services, or activities. They define risk as "the expected losses (lives lost, people injured, damage to property, and disruption of economic activity or livelihood) caused by a particular phenomenon." Consequently, "a societal element is said to be 'at risk' or 'vulnerable' when it is exposed to known hazards and is likely to be adversely affected by the impact of those hazards if and when they occur."

One of the key issues in living with risk to global environmental change is trying to understand the ways that systems (both ecological and human) generate, respond, and try to accommodate and manage risks. In the discussion that follows, several aspects will be examined that are linked to notions of societal risk in relation to global environmental change, including a risk in society, **vulnerability**, **resilience**, and building adaptive capacity.

Risk Society

Some of the most interesting analyses of risk in society have been those proposed by Anthony Giddens and Ulrich Beck. Rooted in discussions of modernity is the notion that the differentiation of risk is often viewed as occurring outside of society. Centralized institutions, such as national governments, are usually incapable of effectively responding to environmental risks because they are intricately bound up with these institutions. The need for reflexivity and improved understanding of risk then become critical in terms of how we approach and accommodate risks. In the words of Ulrich Beck (1992, p. 81), "environmental problems are not problems from our surroundings, but—in their origins and through their consequences—are thoroughly social problems, problems of people, their history, their living conditions, their relation to the world and reality, their social, cultural and political situations." Managing or adapting to risks therefore requires new approaches, because "what were once side-effects are challenging the core of our everyday assumptions," as Alan Irwin (2001, p. 57) puts it. He continues that "in line with this, science's 'monopoly on rationality' has begun to break down in the face of the new set of risks and challenges created by science itself."

The causes and responses to risks cannot therefore be identified and understood using only biophysical science and technological approaches. One also requires inputs that are understood from a "sociological" lens. Roger Kasperson and colleagues provide examples of how such concepts can be applied in various "regions at risk," including **Amazonia**, Eastern Sundaland, Nepal, **China**, and Mexico.

Risks are therefore the results of processes of active construction within particular contexts and settings. To effectively manage such risks, people need to be able to participate in the processes of **decision-making** (or as some call it, the "social discourse") that ultimately shape their lives and overall quality of life. One research area that is enabling us to better understand risk management in relation to global environmental change is determining vulnerability to a range of hazards and risks.

Vulnerability

Renewed debates on notions of vulnerability are increasingly trying to comprehend what factors generate vulnerability; including, for example, social, cultural, environmental, and political factors or complex combinations of these. This renewed attention around vulnerability has helped clarify the concepts of risks and disaster. Vulnerability, for example, has been described by Robert Chambers as being comprised of two dimensions or components: first, an internal dimension usually linked to adaptive capacity or coping; and second, an external dimension usually linked to a hazard or trigger that produces a shock or impact on an exposure unit (e.g., person, a farm, or city). In other words, as put forward by Omar Cardona (2004, p. 37), "vulnerability represents the physical, economic, political, or social susceptibility or predisposition of a community to damage in the case of the destabilizing phenomenon of natural and anthropogenic origin."

Vulnerability, however, is not a residual or impact of a disaster, and is therefore very difficult to measure. In a social context, vulnerability is not a uniform phenomenon and varies in time and space. Some people's coping capacity (internal component of vulnerability) in the very same area, and indeed in the very same house or shelter, may be markedly different to those in the same or similar setting or environment. Likewise the external dimensions may also vary. Moreover, external stresses may be in the form of single stress events or as part of multiple stresses. For example, in **southern Africa** those at risk to **climate impacts** are also often the resource poor who lack **access** to markets and information and are also vulnerable to a variety of health risks, such as **human immunodeficiency virus (HIV)/acquired immunodeficiency syndrome (AIDS)** and **malaria**.

Increasingly, those working at a very practical level, for example, humanitarian aid workers in various parts of the world (including southern Africa) attached to, for example, the World Food Programme or Save the Children Fund have to therefore find ways of measuring "differential" vulnerability and responses to risks to a number of triggering phenomena and ongoing processes such as HIV/AIDS, war, **poverty**, and drought. These factors are further complicated and compounded because they usually occur in various guises and at various rates of change (e.g., drought "creeps"). Vulnerability is hence referred to as being "chronic" or insidious. For improved understanding of such complex systems, a variety of social and physical science approaches are being developed and tested.

Resilience

Focusing only on vulnerability often tends to narrow the focus of concern to those negative aspects and processes related to risk. Increasingly, attention is also being directed toward more positive interpretations of risks and response to risks in both ecological and social systems and opportunities in relation to threats or risks, as done by Carl Folke and colleagues, for example.

Responses of ecosystems to global environmental change, moreover, are no longer viewed as being only linear and therefore easily controlled. Rather, natural and social systems are seen to be strongly coupled, interconnected systems that behave in non-linear ways and exhibit marked **thresholds** in their dynamics. Some researchers working with the "resilience alliance," for example, have been trying to understand and motivate for a more holistic understanding of resilience as a way to respond to risks. From their collective research, resilience is described as the magnitude of disturbance that can be tolerated before an ecosystem moves into a different state with a different set of controls. In a vulnerable system, even a small change may shift the ecosystem into an undesirable state with several consequences. This work has also helped those working on social dimensions of risk to better understand the functioning and interaction of various social networks and groups as ways of managing risks.

Enhancing Adaptive Capacity

One of the overriding concerns when trying to live with risks in coupled human-environment systems is to ultimately find ways to reduce negative outcomes to risks associated with global environmental change and to harness positive changes. These concerns have revived interest in trying to improve the resilience of systems, and enhancing social adaptive management and coping capacity to global environmental change. Enhancing adaptive capacity includes understanding the learning aspect of system behavior in response to disturbance and reflects the ability of social-ecological systems to cope with change without losing options for the future, as expressed by Carl Folke. Adaptive capacity is thus linked to a variety of features in systems, including biological and sociological diversity, such as **biodiversity** and **agrodiversity**. In resource-poor farming communities in many parts of Africa, for example, access to both on-farm and off-farm products (e.g., craft production and other forms of wage labor) has often been shown to be a necessary coping mechanism to help people withstand periods of shock and disruption of **economic livelihood**.

Managing and living with risks thus may mean new and innovative systems and institutions that may have to be designed and developed to enhance our ability to adapt to **environmental change**. One such research and advocacy area that reflects these needs and concerns is disaster risk reduction. Those interested in disaster management, for example, have moved the thinking in this field from reactive, event-driven responses when a risk materializes into a reality to a longer-term, proactive risk management and risk-reduction approach. Strong efforts to reduce the impacts and risks to disasters before the risks heighten into disasters are being profiled. Examples are the International Decade for Natural Disaster Reduction (IDNDR) publications and the International Strategy for Disaster Reduction (ISDR). Such a focus, however, requires interactions with various stakeholders and agents including development planners, government agencies, scientists, conservationists, non-governmental organizations (NGOs), and others. With this type of an approach, risks, threats to livelihoods, and possible hazards are usually identified and mapped using multidisciplinary approaches (e.g., the Sustainable Livelihoods Approach; Food Insecurity and Vulnerability Information and Mapping System [FIVIMS]—**United Nations Food and Agriculture Organization** [FAO]; and Risk Map—Save the Children Fund (SCF), United Kingdom). People and areas that are exposed to risks and/or threats are identified, using as a guide their ability to respond and adapt to threats and risks. Increasingly, however, those exposed to risks are in fact also exposed to daily risks as a result of poverty and not only to once-off events. Hence there is a need to now begin to better match development and disaster risk reduction efforts.

Living with risks requires innovative thinking. For those in developing countries, for example, there is a need to better match development and disaster risk reduction efforts. Whatever context, global environmental change cannot be ignored, and efforts to better live and adapt to changes require inputs of all, including non-governmental organizations, civic society, and scientists.

See also: **Agrodiversity; Arena of Land Conflict; Cumulative Change; Desertification; Food Security; Human Health; Systemic Change.**

Further Reading

Adger, Neil, "Social Capital, Collective Action, and Adaptation to Climate Change," *Economic Geography* 79 (2003): 387–404; Beck, Ulrich, *Risk Society: Towards a New Modernity*, London: Newbury Park, New Delhi: Sage, 1992; Beck, Ulrich, *World Risk Society*, Cambridge, UK: Polity, 1999; Berkes, Fikret, and Carl Folke, *Linking Social and Ecological Systems: Management Practices and Social Mechanisms for Building Resilience*, New York: Cambridge University Press, 1998; Blaikie, Piers, Cannon, Terry, Davis, Ian, and Ben Wisner, *At Risk: Natural Hazards, People's Vulnerability and Disasters*, London, New York: Routledge, 1994; Cardona, Omar D., "The Need for Rethinking the Concepts of Vulnerability and Risk from a Holistic Perspective: A Necessary Review and Criticism for Effective Risk Management," in *Mapping Vulnerability: Disasters, Development and People*, eds. Greg Bankoff, Georg Frerks, and Dorothea Hilhorst, London: Earthscan, 2004, pp. 37–51; Chambers, Robert, "Editorial Introduction: Vulnerability, Coping and Policy," *IDS Bulletin* 20 (1989): 1–7; Egeland, Jan, Living with Risk: A Global Review of Disaster Reduction Initiatives [Online, September 2004], International Strategy for Disaster Reduction Web Site, www.unisdr.org; Folke, Carl, Colding, Johan, and Fikret Berkes, "Building Resilience for Adaptive Capacity in Social-Ecological Systems," in *Navigating Social-Ecological Systems: Building Resilience for Complexity and Change*, eds. Fikret Berkes, Johan Colding, and Carl Folke, Cambridge, UK: Cambridge University

Press, 2002; Giddens, Anthony, *The Consequences of Modernity*, Cambridge, UK: Polity, 1990; Holling, Crawford (Buzz), "Understanding the Complexity of Economic, Ecological and Social Systems," *Ecosystems* 4 (2001): 390–405; Irwin, Alan, *Sociology and the Environment: A Critical Introduction to Society, Nature and Knowledge*, Cambridge, UK: Blackwell, Polity, 2001; Kasperson, Roger E., Kasperson, Jeanne X., and Billie L. Turner II, "Risk and Criticality: Trajectories of Regional Environmental Degradation," *Ambio* 28 (1999): 562–8; O'Riordan, Timothy, and Stephen Rayner, "Risk Management for Global Environmental Change," *Global Environmental Change* 1 (1991): 91–108; Research on Resilience in Social-Ecological Systems: A Basis for Sustainability [Online, September 2004], Resilience Alliance Web Site, www.resalliance.org; Von Kotze, Astrid, and Ailsa Holloway, *Reducing Risk: Participatory Learning Activities for Disaster Mitigation in Southern Africa*, Natal, South Africa: International Federation of the Red Cross and Red Crescent Societies, Department of Adult and Community Education, University of Natal, 1996; Walker, Henry A., "Understanding and Managing Risks to Health and Environment from Global Atmospheric Change: A Synthesis," *Human and Ecological Risk Assessment* 7 (2001): 1195–1209.

COLEEN H. VOGEL

Perception.   Perception is the sum of all processes providing context and meaning to (sensorial) information from the environment, contributing to its awareness or apperception and allowing one to localize and recognize an object of perception (the "world of things" in its widest sense, which means objects as physical or social givens), to differentiate it from others, and to react to it (e.g., with language or with the body).

Environmental perception is about how an individual relates to his/her life space and makes sense of it. The life space does not only embrace the biotic and abiotic environment to provide the individual and his/her social group with all needed to survive, but also the space (including its components and the relation between these components) that has for them a social, political, cultural, or esthetic value. In the more general sense it is the space that, from an individual's perspective, has an effect on decision-making and where human actions have an impact; and where a mutual affectedness between humans and nature exists ("*Mitwelt*"). The perceived life space transcends the immediate local surroundings, encompassing to a different degree subnational, national, regional, and global dimensions. The increasing perception of a globalized environment is related to recent processes of economic, social, political, and cultural **globalization** processes and the concomitant increasing experience of a growing and accelerated interconnectedness of the world.

Environmental perception is mediated by worldviews; social, economic, and political structures; and by the immediate context of interaction between an individual and his/her environment. Factors such as knowledge (including ontologies) and skills, values, and norms; but also interpersonal influences, institutional settings, configurations of interest, technological possibilities, and constraints play an important role. It is also impossible to understand people's environmental perception without taking their history and the development, transformation, and perpetuation of their schemata of environmental knowledge and praxis, as well as their capacity to project themselves into the future, into account.

See also: Cultural Factors; Forest Transition; Land Degradation; Landholding; Land-Use Policies.

BARBARA GÖBEL

Pixel. A contraction of the phrase picture element. Pixels are the objects that compose satellite images and other raster-based digital data files. The term is most often used in association with **remote sensing** data products such as satellite imagery or digital **aerial photography**, but is sometimes used generically to refer to cells within raster data. Pixels arranged in a rectangular grid consisting of rows and columns of data form a complete image file. With passive remote sensing systems, pixel values represent a measurement of the ground surface reflectance (or brightness value) for a specific range of the electromagnetic spectrum. Different types of **vegetation** produce specific reflectances or spectral signatures. The pixel values for different bands of the electromagnetic spectrum can be combined and analyzed to produce classified land-cover maps using a variety of methods. Different satellite sensors are designed to detect different types of environmental features, with some sensors designed for terrestrial applications while others are designed for ocean or atmospheric applications. The spatial area covered by a single pixel is a product of the spatial resolution of the sensor. For example, the pixels in the panchromatic band of **IKONOS** data cover a 1×1 meter area on the ground. In contrast, pixels in some bands of **moderate resolution imaging spectroradiometer** imagery are 500×500 meters in size on the ground. Land-cover classification is complicated by the fact that at coarse spatial resolutions one pixel may include many different types of land cover on the ground.

See also: Change Detection; Pattern Metrics; Pixelizing the Social; Thermal Band Analysis.

Further Reading

Jensen, J. R., *Remote Sensing of the Environment: An Earth Resource Perspective*, Upper Saddle River, NJ: Prentice Hall, 2000; Liverman, Diana, Moran, Emilio F., Rindfuss, Ronald R., and Paul C. Stern, *People and Pixels: Linking Remote Sensing and Social Science*, Washington, DC: National Academy Press, 1998.

THOMAS P. EVANS

Pixelizing the Social. To "socialize the pixel" and to "pixelize the social" is to discern information embedded within spatial data on land-use/cover change that is directly relevant to the core themes of the land-change science agenda, by making **remote sensing** (RS) and **geographical information systems** (GIS) relevant to the social, political, and economic issues driving land-use/cover change. While a **pixel** is a unit of observation associated with satellite data on land cover, the term is used more broadly here to encompass all spatial analysis on land-use/cover change, even if the data are from other sources. Remote sensing, both data and image processing, and analysis through GIS have played crucial roles in the development of land-use/cover change models in both data creation and analysis.

During the course of the past ten years, significant progress has been made in acquiring spatial land-cover data sets from remotely sensed data, conceptualizing the basic geographic and environmental processes that are associated with land-use change. Numerous spatially explicit and heterogeneous land-use change models now exist, spurred by the vast amount of spatial land-use/cover data that are now available. The research agenda of the **Land-Use/Cover Change (LUCC) project** (and its associated groups) has demanded an interdisciplinary approach to **modeling**, with major contributions made by landscape ecologists, geographers, anthropologists, political scientists, economists, and demographers.

While the goal of both **socializing the pixel** and pixelizing the social is to better understand the spatial pattern of land-use/cover change, there are broad generalizations of the differences between the two, as envisioned here. To socialize the pixel is to develop land-change models that move from the RS imagery beyond its use in the natural sciences and toward the concerns of the social sciences per se. These models usually have as the unit of observation the satellite data pixel and include explanatory variables that can be "seen" from the remotely sensed data and calculated using GIS, such as distance measures; other spatial biophysical variables (e.g., soil, slope, elevation); and occasionally socio-economic "drivers," such as population or other sociodemographic characteristics, usually measured at some aggregate level, such as village or state.

To "pixelize the social" is to apply social science theories to spatial data in order to develop an underlying structural model that seeks to explain the human behavior that generates these patterns of land-use/cover change. These models usually take as the unit of observation the land under the control of an individual decision-maker and directly link the decision-maker to specific pixels. Of course, as research has progressed, and as more "socialization" has been done in the "socialize the pixel" realm and more "pixelization" has been done in the "pixelize the social" area, the boundaries between the two, rightly so, have become blurred.

The impacts of RS data and GIS to date have been strongest among the environmental and policy sciences because, respectively, space-based imagery observes the physical attributes of the biosphere, and various stakeholders and decision-makers require spatially explicit assessments of changes in the biosphere. While some areas of social science, specifically geography and regional science, as well as some specializations within economics such as urban economics, have always had a spatial focus; the majority of the human sciences have been slower to incorporate spatial issues as basic elements of research. However, land-use change is an inherently spatial process, so ignoring the spatial dimension in analysis is analogous to analyzing a dynamic process without knowing the chronological order of events. In order to explain and predict the spatial processes that result in land-use change, theoretical and empirical social science models must be developed to address where, when, and why these processes happen. In order to do this, improved spatial data, increased theoretical understandings of human behavior in space, and new methods to use these data and test these hypotheses are needed, that is, to pixelize the social.

During the past ten years there has been an increase in interest in spatial analysis within the broader social science community, as theoretical advances have been made in understanding human interactions across space, as spatial data sets of data of interest to social scientists have become more available. In addition to this increase in spatial data (remotely sensed data, such as satellite data, as well as other geo-referenced data), the advances in GIS and **global positioning systems** (GPS) availability and ease of use has increased the interest in spatial issues among economists. GIS software has a myriad of uses, including data visualization, organization, and integration across sources, as well as variable creation. Often spatial data for a study will come from a wide variety of sources as well as different **scales**, such as satellite data on land use, digital soil maps, tax assessment data on property values and structural characteristics, census data of different kinds, road networks, and school district boundaries. An important attribute of GIS is that it allows the linking of these data via their location in space to permit the assignment of the value of these data to the appropriate unit of observation in an analysis. GPS technology, which gives the coordinates of any location, allows a researcher to create a spatial database from "scratch" by linking household survey data with individual land-use management units. Finally, as spatial data and spatial data software applications have become more available, so too have the statistical tools to use these data appropriately. However, while there have been advances in each of these realms, further advances must be made in data generation and use, theoretical understanding of how individuals make choices concerning land use across space, and the methods used to develop models and test the hypotheses generated by both the data and theory, to truly pixelize the social.

Much of location-based spatial social science theory is derived from the monocentric city model. Briefly, the monocentric city model is an equilibrium model of urban spatial structure, where the distribution of land uses on a featureless plain around a central business district is the result of an equilibrium between a declining land price gradient and increasing transportation costs. This model results in concentric rings around the city made up of different land uses, with agricultural and **forest** lands as the residual land use beyond the developed land uses. For the most part, the only heterogeneity in the landscape is the distance to the central business district or some amenity or disamenity that is implicitly tied to this distance. In many situations, space matters in terms of complex spatial processes and must then be modeled as such at the appropriate scale of analysis. While some modeling approaches, such as **agent-based models**, can incorporate different spatial relationships and complexities, further advances are needed in the understanding of human behavior across space.

While many land-use change research projects use RS data for information on land use and land-use change, the issues discussed here do not involve the science of developing these data sets, but rather the creation of the necessary associated social science data to develop models of land-use change and how these data are linked with RS data. In many countries, different types of government census data sets exist that are at some level spatially explicit, but due to confidentiality issues, these data are aggregated to some level so that individuals cannot be identified. It is possible, through the use of GPS and household survey work, to create individual spatially explicit data that link individuals with specific parcels of land, but this is time-consuming and expensive. While any given piece of land is immobile in space, both the land manager and the boundaries of a parcel under an individual land manager can change over time, for example, as a farm is subdivided into residential lots, requiring longitudinal data to understand the land-use change process. An individual parcel can be under different land management units, such as owner and renter relationships, and individuals can control more than one piece of land. In addition, the unit of analysis is often at a different scale than other ancillary data, so decisions must be made on how to best combine diverse data sets to include in a land-use change model. Therefore, expertise in survey design, administration, and issues in modeling with such data are required.

When using spatial data, two related issues must be considered: how to use the data "creatively" and how to use the data "correctly." The former refers to developing ways of visualizing data to assist with the analysis as well as creating variables from spatial data that can be used in a model; the latter refers to issues of using these data in models. While many studies use data that are based on location, such as the impact of a new road on nearby forest clearing, a truly spatial study should contain a creative spatially explicit model of behavior or the interaction over space of the natural and human environments, or the explicit analysis of the spatial pattern of outcomes; or else some combinations of these and further development of these kinds of models are needed. The second modeling issue deals with the methodological concerns that follow from explicit consideration of spatial effects in such models, such as spatial **auto-correlation**. Finally, while theoretical conceptualization of land-use change processes lead to testable hypotheses, often the actual empirical specification of the model to be tested must take into consideration assorted limitations in data availability and quality, such as the use of proxies for variables of interest.

See also: **Georeferencing; Land Rent; Land-Use System; Pattern to Process; Property Thünen, Johann Heinrich von.**

Further Reading

Fox, Jefferson, Mishra, Vinod, Rindfuss, Ronald R., and Stephen J. Walsh, eds., *People and the Environment: Approaches for Linking Household and Community Surveys to Remote Sensing and GIS*, Amsterdam: Kluwer Academic Publishers, 2003; Geoghegan, Jacqueline, Pritchard, Lowell Jr., Ogneva-Himmelberger, Yelena, Chowdhury, Rinku Roy, Sanderson, Steven, and B. L. Turner II, " 'Socializing the Pixel' and 'Pixelizing the Social' in Land-Use and Land-Cover Change," in *People and Pixels: Linking Remote Sensing and Social Science*, eds., Diana Liverman, Emilio Moran, Ronald Rindfuss, and Paul Stern, Washington, DC: National Academy of Science Press, 1998, pp. 51–69; Nelson, Gerald C., and Jacqueline Geoghegan, "Modeling Deforestation and Land Use Change: Sparse Data Environments," *Agricultural Economics* 27 (2002):201–16.

JACQUELINE M. GEOGHEGAN

Population Dynamics. Demographic factors—including population size, growth, and density; fertility, mortality, and migration; and the age and sex composition of households—are known to be important factors influencing land use and cover change. Some early papers on **deforestation** and **desertification** focused on population factors as **driving forces** of land-cover change to the near exclusion of other variables. This may have been due in part to a proclivity among some researchers to identify population change as the single most important driver of all **environmental changes**. But it is likely that there were also practical reasons, such as the relative abundance of population data from censuses and surveys when compared to data on other drivers such as markets, government policies, agricultural practices, technology, roads, and settlements. Whatever the reasons, contemporary land-use/cover change research generally seeks a more balanced view of population relative to other drivers of land use and cover change.

In a **meta-analysis** of deforestation case studies, Helmut Geist and Éric Lambin found little support for the conclusion that population or other single factor causes such as shifting cultivation could explain the majority of tropical deforestation. Instead, they found that deforestation was driven by identifiable regional patterns of causal factor synergies, of which the most prominent were economic factors, **institutions**, and national policies. These, in turn, drove a range of **proximate causes**, including agricultural expansion, wood extraction, and infrastructure extension. Demographic factors such as natural increase or in-migration were explicitly mentioned in 61 percent of all case studies examined. Most of the explanatory power of population variables tends to be derived from interlinkages with other underlying forces. Many cases did not specify beyond broad notions of population pressure and growth, but those that did tended to identify in-migration more frequently than natural increase. Migration, especially in forest frontier areas, can lead to much more rapid increases in population than would be possible through natural increase (births minus deaths) alone.

Land degradation and desertification have also been attributed to population growth, especially in marginal environments that are characterized by constraints such as steep hillsides, poor soils, and drought-prone climates. Some have posited that increasing population densities in rural areas inevitably lead to land degradation and declining crop yields. Although there is evidence in parts of the world to support the notion of a downward spiral of population, **poverty**, and the environment, population size or density is rarely the sole contributing factor. Rather, as with deforestation, population variables tend to be part of a matrix of factors that include, in the case of land degradation, failed institutions, climate conditions, inherently poor soils, and lack of incentives for proper soil management (e.g., tenure insecurity and low market prices). Population displacements into marginal lands due to wars and civil conflicts, as has happened frequently in Africa and Central America, can also contribute significantly to land degradation.

There is much to be learned from regions where the downward spiral of **soil degradation** and poverty has not inevitably resulted from increasing population density. Case studies from the Machakos district in Kenya and in Chivi communal area in Zimbabwe suggest that improved soil productivity and more secure livelihoods can and do occur in a context of increased population density, and even benefit from it. The factors that can be important for averting land degradation include appropriate technologies for soil management adopted and propagated by local farmers (rather than imposed from outside); **access** to markets, road infrastructure, and development of local market towns for food processing; cash-cropping as opposed to purely **subsistence agriculture**; and development of local management capacity and skills through education and agricultural extension. Rather than blanket solutions, Ian Scoones urges more fine-tuned and people-centered development interventions in which the historical context and local specificity of needs are acknowledged. Such an approach recognizes local environmental knowledge, and sees the agricultural researcher as a facilitator rather than an expert prescribing solutions.

The **scale** of analysis is very important in terms of understanding the influence that population has on **land-cover changes**. If a researcher chooses to examine land-cover changes at the scale of the municipality simply because that is the level at which population data are reported, there is a potential to fall into an ecological fallacy. As an example of an ecological fallacy, say that population growth rates were found to be highly correlated with deforestation rates at the county level in a particular forest region. The researcher might conclude that population growth was a significant driver of deforestation. Yet, it may be that the population increased dramatically in urban areas contained within those counties, and that in fact the rural population, located where most of the deforestation occurred, remained relatively constant. This would be a form of spurious correlation; the real "culprit" might be something quite different, such as government policy or price mechanisms. Increasingly, spatially disaggregated population maps with the grid cell as the unit of analysis are being used to analyze land-cover changes. Examples include the Gridded Population of the World of the **Center for International Earth Science Information Network** and Oak Ridge National Laboratory's LandScan. These need to be interpreted with caution, since the underlying data still come from census administrative units.

Population size, density, or growth rate are not the only demographic factors of relevance to land-use and land-cover change. For example, household size and composition in **agricultural frontier** areas have been found to significantly affect land-use patterns, in particular the choice of more labor-intensive annual crops versus less-labor-intensive tree crops or livestock raising. Size and composition, in turn, are often related to the length of tenure. Security of **land tenure** has been found to affect fertility **decision-making** at the household level, with more secure tenure, *ceteris paribus*, resulting in lower fertility rates. This corroborates the theory that children are seen by rural households as a form of risk insurance and social security in highly uncertain environments.

Government policies can directly influence population dynamics in ways that lead to land-cover changes. Both Indonesia and Brazil have practiced various forms of resettlement to forest frontier areas. Some governments have deliberately encouraged migration of the rural landless as an alternative to less politically palatable **land reforms**.

See also: **Agricultural Intensification; Agrodiversity; Biodiversity; Boserup, Ester; Cash Crops; Center for the Study of Institutions, Population and Environmental Change;**

Consumption; Degradation Narrative; Economic Livelihood; Indigenous Knowledge; Malaria; Participatory Land Management; People at Risk; Population-Environment Research Network; South America; Urbanization.

Further Reading

Geist, Helmut J., and Éric F. Lambin, "Proximate Causes and Underlying Driving Forces of Tropical Deforestation," *BioScience* 52 (2002): 143–50; Scoones, Ian, "The Dynamics of Soil Fertility Change: Historical Perspectives on Environmental Transformation from Zimbabwe," *The Geographical Journal* 163 (1997): 161–9; Tiffen, Mary, Mortimore, Michael, and Francis Gichuki, *Population Growth and Environmental Recovery: Policy Lessons from Kenya* (IIED Gatekeeper Series no. 45), London: International Institute for Environment and Development, 1994.

ALEXANDER DE SHERBININ

Population-Environment Research Network.   A project of the International Union for the Scientific Study of Population and the **International Human Dimensions Programme on Global Environmental Change** that seeks to advance academic research on **population dynamics** and the environment by promoting online scientific exchange among researchers from social and natural science disciplines worldwide. The Population-Environment Research Network (PERN) receives technical support from the Socioeconomic Data and Applications Center at the **Center for International Earth Science Information Network** of Columbia University, funded by the U.S. National Aeronautics and Space Administration.

Further Reading

The Population-Environment Research Network [Online, December 2004], The PERN Web Site, www.populationenvironmentresearch.org.

HELMUT GEIST

Poverty.   Income below which a minimum nutritionally adequate diet plus essential non-food requirements are no longer affordable. Poverty is seen as both a potential cause and effect of land-use and **land-cover change**, although in each instance these assertions are contentious. Poverty, it is argued, can lead individuals, households, communities, and even states to make short-sighted land-management decisions. Degrading land resources, in turn, has an impact on the human livelihoods that depend on them, and these livelihoods are assumed to be those of the poor in many instances.

Defining Poverty

Poverty is a term that is open to multiple interpretations. It can be located at a number of spatial-social **scales**, from the individual, to the household, community, ethnic group, nation, and region. Poverty may be conceived of in relative terms (e.g., comparing households within a community, regions within a state, or states with one another). It may also be measured against a specific benchmark, such as an international poverty line. Finally, poverty may be assessed in a variety of ways, for example, in terms of monetary wealth or income, certain types of assets, or entitlements. The way in which poverty is conceptualized has an influence on who is defined as poor and how their interactions with the landscape are perceived.

Poverty as a Cause of Land Change

At the individual and household scale, it is argued that the poor are more likely to engage in environmentally deleterious behavior for at least two reasons. First, it is suggested that the poor (defined as households lacking resources) fail to maintain the natural resource base because they lack the means to invest in environmental **conservation** and remediation. As such, it is often argued that economic growth and environmental stewardship go hand in hand, as **economic growth** provides the resources necessary for sustainable environmental management. Second, it is asserted that the poor are more preoccupied with their survival in the present and therefore fail to appreciate higher-order amenities (such as environmental goods) and make natural resource management decisions that are short-sighted.

At the national scale, the theoretical framework of the environmental Kuznets curve has been used to suggest that there is a relationship between per capita income and a number of environmental measures. The model posits that countries in the early stages of development (and low per capita income) have a nominal impact on the environment. In the intermediate stages of development and income, it is suggested that countries often experience high rates of environmental degradation associated with rapid **industrialization**. Finally, the most developed and highest-income countries are supposed to have low levels of natural resource degradation given increasing **investments** in, and concern for, environmental amenities.

A number of studies have been undertaken (in all of the world's major regions) that at least complicate, and often contest, the narrative of poverty-induced environmental degradation. First, sustainable land management is not necessarily a capital-intensive process requiring monetary investment. In developing countries, many cultural ecologists have argued that traditional modes of land management are not necessarily inefficient or destructive. Furthermore, while industrial agricultural practices may lead to the higher production of certain crops, they often do so at an economic and environmental cost. In other words, the process of wealth creation may result in as much environmental destruction as it potentially remediates.

Second, some evidence regarding the famine behavior of extremely poor households in Africa suggests that they are often more concerned about the longer-term viability of their livelihoods than they are about shorter-term consumption (contradicting the notion that the poor are short-sighted). Furthermore, the concern for environmental amenities that is supposed to accompany wealth presumes that households and communities actually experience the environmental impacts of their production decisions. This increasingly is not the case in the most developed countries as, in an ever more globalized economy, sites of production and **consumption** are separated in space. There are burgeoning environmental justice and political ecology literatures describing how socially marginalized communities often bear the environmental costs of producing crops and goods for wealthier consumers in the other areas of the country or world. Wealthier households have little incentive to deal with the environmental costs of their consumption if they are unaware or spatially removed from such degradation.

Third, the trend hypothesized by the environmental Kuznets curve (and supported by a number of cross-national studies) often has as much to do with the export of dirty industries as it does with tighter environmental regulations and cleaner production. As such, analysis undertaken on a country-by-country basis may support the trend posited by the environmental Kuznets curve, whereas analysis undertaken at other scales may yield quite different results. This seeming contradiction, and the effects of varying analytical scales, is sometimes referred to as the modifiable areal unit problem.

It is not that poverty does not contribute to land-use and land-cover change in some instances (it clearly does), but that important definitional and scale issues, as well as approaches to technology and environmental accounting, influence characterizations of the poverty-land interface.

Land-Cover Change as a Cause of Poverty

Land-cover change has implications for land-dependent livelihoods. Declining productive capacity of the land may impoverish land-dependent livelihoods, just as increasing productive capacity may lead to enrichment. A variety of livelihood activities may be influenced by land-cover change. For example, decreased **forest** cover may lead to fewer opportunities for wild food collection, hunting, and some forest-dependent artisanal activities. Declining surface **biomass** may also alter the local hydrologic cycle, with implications for fisheries, flood plain **rice**, and groundwater availability (if there is a drop in the water table). Changing soil conditions have obvious implications for crop farming, but may also influence off-season grazing opportunities.

There is some debate regarding the degree to which poor (as opposed to rich) households, communities, and states are more susceptible to the adverse consequences of land-cover change. This debate about **vulnerability** revolves around at least two points. First is the degree to which the poor are more implicated in land-based livelihoods. Obviously, one's definition of poverty (and the social-spatial scale of this conception) influences such characterizations. In many rural communities, for example, the relatively wealthy actually control a bigger piece of the resource pie than the poor, which means that they may have more to lose from declines in surface biomass. In contrast, it may more convincingly be argued that the world's poorest countries frequently have more natural resource-based economies, and thus stand more to lose from the declining productive potential of the land. The second way in which the poor may be more sensitive to land-cover change concerns the ability of such households, communities, and states to cope with adverse **environmental change**. Here, the concern is that the poor have fewer reserves to fall back on in times of crisis.

Dynamic Adaptation

Very often, there is a circular process of dynamic adaptation. In many instances, this process has been depicted as a vicious cycle wherein poor households overtax the environment in order to survive, and this degraded environment further impoverishes the household. It is also quite possible, and well documented in several instances, that poor households will diversify into other activities or simply leave the landscape. In either case, the human burden on a particular landscape often declines.

See also: Decision-Making; Degradation Narrative; Desertification; Driving Forces; Economic Livelihood; Human Immunodeficiency Virus (HIV)/Acquired Immunodeficiency Syndrome (AIDS); Land Degradation; Landholding; Mediating Factor; People at Risk; Population Dynamics; Resilience; Tradeoffs; Tragedy of Enclosure.

Further Reading

Broad, Robin, "The Poor and the Environment: Friend or Foes?" *World Development* 22 (1994): 811–2; Gray, Leslie C., and William G. Moseley, "A Geographical Perspective on Poverty-Environment Interactions," *Geographical Journal* 171 (2005): 29–43; Moseley, William G., "African Evidence on the Relation of Poverty, Time Preference and the Environment," *Ecological Economics* 38 (2001): 317–26; Reardon, Thomas, and Scott A. Vosti, "Links Between Rural Poverty and the Environment in Developing Countries: Asset Categories and Investment Poverty," *World Development* 23 (1995): 1495–1506; Scherr, Sara, "A Downward Spiral? Research Evidence on the Relationship between Poverty and Natural Resource Degradation," *Food Policy* 25 (2000): 479–98; Swinton, Scott M., Escobar, German, and Thomas Reardon, "Poverty and Environment in Latin America: Concepts, Evidence and Policy Implications," *World Development* 31 (2003): 1865–72; Watts, Michael, "Poverty Gap," in *The Dictionary of Human Geography*, 4th ed., (reprinted), eds. Ron J. Johnston, Derek Gregory, Geraldine Pratt, and Michael Watts, Oxford, UK, Malden, MA: Blackwell Publishers Ltd., 2001, p. 627.

WILLIAM G. MOSELEY

Pristine Myth.   The pervasive assumption that society's **modification** of nature is a relatively recent phenomenon—that remote or uninhabited lands are untouched or pristine. Humans have always altered the environments in which they live. Most areas on earth, even the hearts of **Amazonia** and the Australian outback, show a human imprint. Research that debunks the pristine myth demonstrates the history of human agency in modifying the land and contributes to basic understanding of nature-society relationships: it highlights the potential for nature to recover from anthropogenic disturbance, it shows that both ecosystems and human cultures exhibit dynamic change rather than remaining static, and it suggests that the path to more **sustainable land use** may require embracing humanized landscapes rather than the twentieth-century goal of "wilderness" preservation.

The pristine myth is often linked with what has been called *la leyenda verde* (the green legend) or the myth of the "ecological Indian," which asserts that indigenous groups embrace an environmental ethic that contrasts with their purportedly more environmentally destructive European colonizers. Commonly associated with both the Americas and Australia, scholars and lay people alike—perhaps due to a "nostalgic longing for the past and a simpler life"—often celebrate the idea that indigenous peoples are "in and of nature," living in balance with it, as noted by Shepard Krech. Complementing this view is the notion that the Americas, for example, were sparsely populated, and therefore the capacity for Amerindians to modify the land was low.

Instead of living in harmony with nature and treating it in a benign way, however, a wide range of evidence from a **paleo perspective** shows that indigenous groups in regions throughout the world modified the environment in profound ways. The degree of impact reflects the size and duration of occupation and use of an area. First, in the case of the Americas, there were far more people living in the New World than is widely known, with reasonable estimates lying somewhere between 43 and 65 million. Second, terraces and raised fields, the creation of anthropogenic soils (e.g., *terra preta* in the Amazon basin), canals, dams, and other components of irrigation farming, and the use of **fire** to produce and maintain productive landscapes, are among the many examples of sophisticated and intensive production systems found throughout the Americas before the Columbian encounter. Finally, these production systems are linked to many of the same kinds of environmental problems witnessed today: large-scale **deforestation**; **soil erosion** and nutrient loss; **wetland** sedimentation; weed and pest invasion; and water stress.

Debunking the pristine myth is more than a historical exercise. That the landscapes in precolonial regions were socially constructed holds implications for how humans relate to nature today. Debate over whether pristine nature exists reflects conflict between those who argue that humans are a part of wild nature, and those who embrace a wilderness-civilization dichotomy. Those who celebrate humanized landscapes see lessons for how to foster sustainable development.

First, the perpetuation of the pristine myth coincides with the idea of *terra nullius* (land belonging to no one), the two ideas mutually reinforcing each other and justifying the **colonization** of thinly settled lands in the Americas and Australia. In Mexico, for example, throughout colonial history and continuing into the twentieth century, the two concepts have been used to justify the colonization and control of frontier locations. Despite the presence of indigenous populations, Porfirio Díaz (the president of Mexico from 1876–1880 and 1884–1911) declared the bulk of the country's tropical forests to be *terrenos baldíos* (unoccupied or vacant lands), which he quickly subdivided and awarded to colonists or international logging companies. Despite evidence of past human impact as well as the ongoing occupation and use of forests, for over 100 years nature **conservation** has focused primarily on using reserves to exclude people and to protect pockets of "wilderness." Uncovering flaws in the assumptions of *terra nullius* and pristine nature exposes the injustice of taking away lands from indigenous groups in the past and present.

Second, research on the pristine myth not only identifies the agency of humans to alter the earth, but also the **resilience** of nature and its capacity to recover from significant anthropogenic shocks. Caution is needed, therefore, in referring to "environmental destruction." The most widely cited examples of environmental recovery are from the New World. The first is in the Maya lowlands of Mesoamerica, which were almost completely deforested during the Classic Maya period, but returned to mature forest after the civilization's collapse around AD 800–900. Second, with a 90–95 percent depopulation of the Americas after Europeans and their Old World diseases arrived, environmental conditions "improved" (e.g., forest cover expanded) as cultural features (e.g., raised fields and terraces) were abandoned. While the scale and circumstances of the depopulation in both cases are unique, there are multiple examples in more recent periods of environmental recovery on smaller scales, and without associated extreme population shifts.

Third, by showing how societies worldwide have modified and adapted to nature, detailed regional environmental histories not only reflect the dynamism of ecosystems, but also the evolution of cultures. That neither is static draws into question attempts to preserve, fixed in time, a snapshot of a particular historical period. By showing that aboriginal groups treated nature in far from a benign way, present-day environmentalists are forced to question calls for a return to a "simpler" or more traditional way of interacting with nature. While the pace and the scale of human impact over the last 200 years is unprecedented, cultures throughout history have exploited nature for their own self-interest.

Finally, with the vast majority of nature worldwide either inhabited or being actively used by people, research that questions the pristine myth implicitly endorses the notion that society today needs to embrace humanized landscapes. The modern meaning of wilderness, which encompasses sublime landscapes where nature rules and humans are absent, may be obstructing rather than advancing sustainable development. If humanized landscapes are the norm, then a solution to environmental problems that creates barriers between people and nature is problematic. The solution to environmental problems is not to put a box around select pockets of "wilderness" while continuing with business as usual elsewhere. In both the **United States of America** and Chile, for example, the bulk of wilderness areas under protection are remote and with low population densities, not in those landscapes that are most vulnerable to human modification. In short, the focus on wilderness preservation may distract from the need to change nature-society relationships in society writ large. There is evidence, however, that the influence of the pristine myth may be declining in the Americas. In the 1980s and 1990s, there was a dramatic expansion across **Latin America** not only in protected areas but also in a celebration of humanized landscapes as part of nature conservation.

But not all are so quick to celebrate the social construction of nature. In stark contrast to the sustainable development movement, many environmentalists see the embrace of humanized landscapes as justifying continued human impact on the land. Many point out that while landscapes such as the forests of the Maya lowlands may have recovered from past use, the world today is more populous, more interconnected, and has a much greater capacity to alter the land through ever more efficient technologies; therefore, it is unlikely that the kinds of shifts in human-environment conditions that took place in the past will occur in the foreseeable future.

Other critics argue that the debate over the character of the pre-Columbian American landscape has often focused on the polar assertions that the continent was either a pristine landscape or a humanized one. This polarization detracts from a reasoned assessment of the impact of the First Americans on the land. As such, it hampers efforts to establish ecological restoration and conservation plans.

In the case of North America, Thomas Vale argues that the pre-Columbian landscape was neither entirely pristine nor intensely humanized. Rather, the landscape was a mosaic of pristine wilderness and differentially humanized patches. There were intensely humanized farmed areas along the East Coast; more moderately modified plains; the West Coast was altered to varying degrees, primarily by anthropogenic fire; and there were large tracts of wilderness (or nearly so) in the Southwest and at high elevations.

Thomas Vale identifies intensity, spatial, and temporal ambiguity in the pristine myth debate. First, the intensity of human impact in a settlement and its associated agricultural fields was profound, while a forest where gathering and hunting occurred was much less so. Second, human settlements in the western United States were highly uneven in their distribution across the landscape, often concentrated along streams. Third, some human impacts, such as hunting, may have had a relatively short-lasting impact on the land when compared with **agriculture**. Thomas Vale argues, therefore, that the American landscape was a mosaic, ranging from intensely humanized to largely pristine wilderness.

See also: **Agricultural Frontier; Degradation Narrative; Driving Forces; Great Plains; Land-Cover Change; Land-Use History; Proximate Causes.**

Further Reading

Cronon, William, ed., *Uncommon Ground: Toward Reinventing Nature*, New York: W. W. Norton & Co., 1995; Denevan, William M., "The Pristine Myth: The Landscapes of the Americans in 1492," *Annals of the Association of American Geographers* 82 (1992): 369–85; Head, Lesley, *Second Nature: The History and Implications of Australia as Aboriginal Landscape*, Syracuse, NY: Syracuse University Press, 2000; Oelschlaeger, Max, *The Idea of Wilderness: From Prehistory to the Age of Ecology*, New Haven, CT: Yale University Press, 1991; Sluyter, Andrew, *Colonialism and Landscape: Postcolonial Theory and Applications*, Landham, MD: Rowman & Littlefield Publishers, 2002; Vale, Thomas, ed., *Fire, Native Peoples and the Natural Landscape*, Washington, DC: Island Press, 2002; Whitmore, Thomas M., and B. L. Turner II, "Landscapes of Cultivation in Mesoamerica on the Eve of Conquest," *Annals of the Association of American Geographers* 82 (1992): 402–25.

PETER KLEPEIS, PAUL LARIS

Private Property. Refers to a form of **property** in which a single actor enjoys extensive use and control rights on a resource (ownership). All rights associated with a resource are bundled in the hands of an actor. Private property has received much attention in theory and policy on land use, motivating **land privatization** programs and **land reforms** in many countries. The prominence of the concept derives from its constitutive role in political liberalism and capitalist economies at the level of theory and practice. Private property is thought to promote political participation and efficient resource use. Private property on land is far from common, however. **Land rights** around the world remain subject to extensive control by states and other collective entities. For example, land rights pertaining to **common-pool resources** often take the form of common property regimes.

Private property and land use mutually influence each other. Historically, the intensification of **agriculture** has been associated with the emergence of private property. For example, agricultural intensification in eighteenth-century

Private property owners who have found a little coal on their land in roadside West Virginia, mining and selling it along the highway in September 1938.
PHOTO: Marion Post Wolcott. [Courtesy Library of Congress]

Europe went along with the enclosure of common lands. In the contemporary world, programs of land privatization and land registration are major drivers of land-use change. There is empirical evidence from Africa and Thailand that land registration has contributed to **agricultural intensification** and **investments** in **soil conservation** in some regions. Forestland allocation has been connected with the rapid expansion of smallholder tree plantations in Vietnam, re-greening previously barren hills. At the same time, policy promoting private property has also caused **land degradation**. Privatization has been linked with overgrazing and range degradation in **China** and Central Asia. Land privatization in central and eastern Europe may have allowed agricultural intensification in some regions, but it has also contributed to environmental degradation.

Land use influences the nature of property. **Ester Boserup** has hypothesized that changes in agricultural practices driven by rising population densities lead to the emergence of private property in land. As land becomes scarce, individual land users expand the bundles of rights to land, eventually achieving private property. At a more empirical level, research has shown that settlers in frontier regions often gain private property in land by being the first to clear a particular parcel of land. In Africa, tree planting strengthens private claims on land, even though they may not amount to complete private property. In Vietnam, terracing and other investments in land strengthen individual land rights, which may be relatively restricted otherwise.

See also: Cadastre; Institutions; Land-Use Policies; Public Policies.

Further Reading

Boserup, Ester, *The Conditions of Agricultural Growth*, New York: Aldine, 1965; Macpherson, C. B., ed., *Property: Mainstream and Critical Positions*, Toronto: University of Toronto Press, 1978.

THOMAS SIKOR, JOHANNES STAHL

Probit/Logit Model. Probit and logit models are members of a wider class of non-linear probability models that associate a set of explanatory variables with the probability of observing a dependent variable that is discrete. It is common to model land-cover types, for example, forest and non-forest (cleared land), as possible outcomes that are distinct from one another.

A linear probability model is used when the dependent variable takes on a continuous value as a function of the independent variables and their associated parameters. For example, one could predict the level of agricultural output in tons as a function of land, labor, and capital inputs. A linear model is less useful at predicting discrete breaks in the data. For example, **forest** may be cleared for agricultural uses when its opportunity cost exceeds the benefit in keeping the land in forest. One could use a linear dependent variable formulation of the probability of land **conversion** (i.e., anything beyond 0.50 could be cleared; below 0.50 left in forest), but there is no guarantee that predicted values will be well-behaved; that is, fall in the probability range from 0 to 1.

Thus, non-linear transformations to represent probability of a particular event or outcome are the norm. Logit models are derived from a logistic function, or the probability of a particular outcome relative to all other alternatives. Probit models are derived from the normal distribution, with the outcome taking a particular value (i.e., 1 = forest; 0 = cleared) depending on some **threshold** determined by the linear combination of independent variables and their associated parameters. The simplest formulations of both models represent a binary framework (1,0), but multinomial expansions exist for both models; that is, multiple categories of a single dependent variable.

One challenge for land-use/cover change research is the extreme likelihood of spatial and/or temporal non-stationarity; land-change processes are often spatially or temporally dependent. These problems are of course compounded by other violations of ideal conditions (heteroskedasticity, nonormality, etc.). However, corrections for both temporal and/or spatial dependence to nonlinear probability models are cumbersome. Logistic models by nature do not have an error structure; that is, explicit unmodeled or autonomous variation. Thus, it is generally impossible to correct for spatial or temporal **auto-correlation**. Probit models, because they are derived from the normal distribution, have a defined error structure, but correspondingly, they are much more sensitive to misspecification, which could lead to bias and inefficiency of the estimated parameters.

See also: **Change Detection; Continuous Data; Discrete Data; Modeling; Hot Spot Identification.**

Further Reading

Fleming, Mark M., "Techniques for Estimating Spatially Dependent Discrete Choice Models," in *Advances in Spatial Econometrics*, eds. Luc Anselin, Raymond Florax, and Serge Rey, Heidelberg, GE: Springer 2004, pp. 145–68; Long, J. Scott, *Regression Models for Categorical and Limited Dependent Variables* (Advanced Quantitative Techniques in the Social Sciences, vol. 7), Thousand Oaks, CA: Sage Publications, 1997; Mertens, Benoît, and Éric F. Lambin, "Land-Cover Change Trajectories in Southern Cameroon," *Annals of the Association of American Geographers* 90 (2000): 467–94.

DARLA K. MUNROE

Project on People, Land Management and Ecosystem Conservation.

A collaborative effort among scientists from across the developing world to develop sustainable and participatory approaches to **conservation**, especially of biodiversity, within small farmers' agricultural systems and in participation with the farmers. The Project on People, Land Management and Ecosystem Conservation (PLEC) has been developed since 1993 by the United Nations University (UNU), and received funding from 1998 to 2002 from the Global Environmental Facility (GEF), managed by the World Bank. The project is executed by UNU through a network of five locally based clusters that have been established in West Africa (Ghana, Guinea), East Africa (Kenya, Tanzania, Uganda), the Asia-Pacific (**China**, Thailand, Papua New Guinea), and America (Brazil, Peru, Mexico, Jamaica). About 80 percent of all participants are from these countries. Scientists from Australia, **Japan**, the **United States of America**, and the United Kingdom are also involved, with **Harold Brookfield** being a driving force behind the network. PLEC uniquely provides both for South-to-South cooperation and South-to-North twinning arrangements. The specific objectives are to establish historical and baseline comparative information on **agrodiversity** and **biodiversity** at the landscape level; to develop participatory and sustainable models of biodiversity management based on farmers' technologies and knowledge within agricultural systems at the community and small-area levels; to recommend approaches and policies for sustainable agrodiversity management to key government decision-makers, farmers, and field practitioners; and to establish national and regional networks for capacity strengthening within participating institutions, and to carry forward the aims of PLEC. The core of PLEC's work is in its demonstration site villages. Here, PLEC becomes the farmers' own enterprise, and scientists are the facilitators, not the instructors. The scientists identify and demonstrate farmers' practices that are environmentally, socially, and financially sustainable, and which sustain biodiversity. They help farmers in achieving their own conservationist goals. Collaborating farmers manage varied biophysical conditions, growing a range of crops and using biodiversity with discretion. The PLEC approach differs from mainstream agricultural research at experiment stations under controlled conditions. By integrating locally developed knowledge of soils, climate, and other physical factors with scientific assessments of their quality in relation to crop production, a set of sustainable agricultural technologies can be devised so that agricultural diversity is maintained. The participatory process will eventually enhance farmers' and local communities' ability to adapt to **environmental change** as well as to social and economic change. Since 2002, PLEC has focused on mainstreaming lessons it has learned into national and international policies and training **institutions**.

See also: **Alternatives to Slash-and-Burn (ASB) Programme; Community Involvement; Consultative Group on International Agricultural Research; Indigenous Knowledge; Land-Use Policies; Participatory Land Management; World Agroforestry Centre.**

Further Reading

Project on People, Land Management and Ecosystem Conservation [Online, November 2004], The PLEC Web Site, www.unu.edu/env/plec.

HELMUT GEIST

Property. Refers to social relationships between persons with respect to goods of material or symbolic value. Analysis of property involves attention to various kinds of social actors recognized to take part in property relationships; property objects, being material and cultural goods considered as valuable; and types of relationships, often expressed in terms of rights and obligations, including their temporal dimensions. For example, an analysis of property on a particular tree examines the products associated with the tree (leaves, fruits, timber, etc.), the actors making claims on the tree (male loggers, female gatherers, herders, etc.), and the types of rights and duties making up the relationships among actors with regard to the tree products (use for **consumption**, collection for sale, etc.). Property is a fundamental concept for the study of **land-use systems** because property affects land-use practices, and because land-use practices influence property relationships. **Private property**, in particular, has been at the core of a longstanding debate about the relationship between **land rights** and **land change as a forcing function in global environmental change**. But property also has other functions beyond production, such as the protection of social stability and political participation.

The rights constituting property relationships are commonly divided into use rights and control rights. Use rights include rights of cultivation, passage, grazing, and collection, among others. Control rights refer to the rights of excluding other claims, transfer the object, make major alterations to it, and sell the object. Edella Schlager and Elinor Ostrom suggest a simple classification scheme into five types of natural resource rights. The right of **access** entitles its holder to walk on to a resource system. The right of withdrawal allows its holder to enjoy resource products. The right of management includes the right to regulate use patterns within a local group of resource users and the right to transform the resource by making improvements and investments. The right of exclusion is about determining who will have an access right, and how that right may be transferred. The right of alienation refers to selling or leasing of the rights of management and exclusion. One can speak of ownership, or private property, if one actor holds all five types of rights concurrently. Ownership is far from universal, as many resource users do not have the rights of alienation. Their "bundles of rights" are limited to the rights of access, use, management, and, perhaps, exclusion.

The distinction between use and control rights informs a customarily used classification of property regimes into four broad types. One speaks of private property when a single actor enjoys extensive use and control rights on a resource. Under state property, it is the state that exercises significant use and control rights, although it may grant limited use rights to other actors. Common property refers to a situation in which some collective entity, such as a village community, holds control rights and grants use rights to its members. Open access, in turn, occurs when many actors assert use rights in the absence of control rights. This classification is a highly simplified representation of actual property relations, however. Property relations are in reality much more varied, being characterized by complex combinations of use and control rights among multiple actors.

Property exists simultaneously at various layers of social organization. Notions of property are fundamental to cultural norms and social values. Property is a key object in laws and regulations enacted by states, religious entities, international treaties, and other forms of social organization. Property exists at the level of actual social relationships, being embedded in wider social relationships, such as kinship, religious, or political affiliations. Finally, property relationships become manifest in social practices, that is, in the interactions of various types of actors with concrete property objects and rights. By way of concrete practices, the actors reproduce and modify actual social relationships, laws and regulations, and cultural norms in material and symbolic struggles. Property relations at the different layers are interconnected, although property at one layer is analytically distinct from property at another layer. In addition, common are discrepancies between the forms property assumes at the different layers. The effects of legal property reforms on land use, therefore, depend on the degree to which they are translated into actual social relationships and practices.

The relations between property and land use are mediated by broader access relations. The concept of access, in its sociological interpretation, describes the ability of actors to derive benefits from a resource. It is broader than the property concept, as property focuses on rights-based access. In contrast, access to a resource includes attention to other structural and relational mechanisms influencing the benefits derived from rights. These include access to technology, capital, markets, labor, and knowledge. For example, many actors beyond wood collectors themselves enjoy access to benefits derived from charcoal. They include transporters moving charcoal to the city and retailers selling it to urban customers. The relations between property and land use, therefore, are mediated by numerous other factors, as research on the effects of land rights and registration has demonstrated.

See also: Cadastre; Common-Pool Resources; Institutions; Mediating Factor; Transhumance.

Further Reading

Benda-Beckmann, Franz von, and Keebet von Benda-Beckmann, "A Functional Analysis of Property Rights, with Special Reference to Indonesia," in *Property Rights and Economic Development: Land and Natural Resources in Southeast Asia and Oceania*, eds., Toon van Meijl and Franz von Benda-Beckmann, London, New York: Kegan Paul

International, 1999, pp. 15–65; Ribot, Jesse C., and Nancy L. Peluso, "A Theory of Access," *Rural Sociology* 68 (2003): 153–81; Schlager, Edella, and Elinor Ostrom, "Property-Rights Regimes and Natural Resources: A Conceptual Analysis," *Land Economics* 68 (1992): 249–62.

THOMAS SIKOR, JOHANNES STAHL

Proximate Causes.   Proximate causes are the direct or immediate human activities that change land use or land cover. While underlying **driving forces** provide the impulse and context for change, proximate causes relate to the mechanisms of change or the immediate reasons for it. Distinguishing features include closeness (i.e., proximity) to the site of change, and the evident role of human agents in making and implementing decisions to carry out change. These human agents are the channels or media through which the underlying driving forces operate to effect change at particular places and times. They "reflect" the underlying driving forces, according to William Meyer and B. L. Turner II, or give expression to them.

The conventional separation into proximate causes and underlying driving forces implies a hierarchical structuring of the factors determining change in land use and land cover (see figure below).

While this model can be helpful in providing a framework for analysis, it is, like all models, a simplified version of reality. It provides only limited space for factors such as the development of road networks. "Infrastructural development" is shown as a "proximate cause" in the figure, but arguably it has a status and role that are different from those of agricultural expansion or wood extraction. Well-known examples (e.g., from Rondônia in Brazil) have demonstrated an association between the extension of the road network

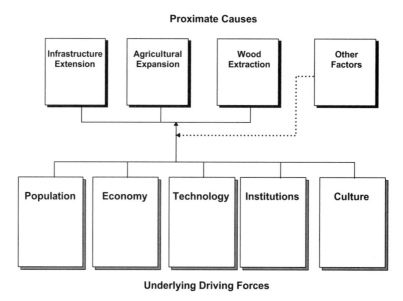

Underlying driving forces and proximate causes.
SOURCE: Simplified after Geist and Lambin (2002)

and forest clearing for pioneer agriculture. In this instance, economic development is manifested in the growth of a road network that permits penetration of the **forest** and hence forest clearance for **agriculture**. In practice, the influences operating on land use and land cover form a complex web: separation into a two-tier structure is somewhat arbitrary.

The framework of underlying driving forces and proximate causes has been used in research on tropical **deforestation** since the early 1990s, but certain parallels can be recognized with "political economy" and behavioral approaches employed, for example, in work on agricultural change in developed countries from the 1970s.

The explicit notion of "proximate causes" is associated most closely with tropical deforestation. Proximate causes in terms of agricultural expansion and wood extraction are frequently cited, but both of these have many variant forms and they are not necessarily mutually exclusive. Down through history, the former has been the main cause of deforestation. Wood extraction on its own would not necessarily lead to the permanent removal of the forest, unless activities such as livestock grazing or **environmental change** such as soil deterioration also occurred.

Agricultural Expansion

Underlying driving forces of population growth and economic development can lead to increasing demand for food and hence expansion of the agricultural area. Usually such expansion is at the expense of the forest. In short, a plausible link can be perceived between population growth and deforestation. The relationship, however, is not deterministic: an alternative response to increasing demand for food could be **agricultural intensification** in existing areas. Also agricultural expansion can take various forms, by no means all of which are closely related to a simple growth in demand for food.

Helmut Geist and Éric Lambin reviewed 152 case studies of deforestation as reported in ninety-five journal papers, and analyzed them in terms of conclusions about causation. Overall, agricultural expansion was found to be by far the most frequent proximate cause of deforestation: they found that it applied in 96 percent of cases. Different authors, using different approaches, have reached similar conclusions about the primacy of agricultural expansion.

Shifting cultivation is a form of agriculture that has often been blamed for, or implicated in, deforestation. The term, however, is variously defined and interpreted, and confusion has arisen in particular between "traditional" shifting cultivation, usually practiced by indigenous peoples as **swidden cultivation** or "swidden fallow" farming, and "colonist" shifting cultivation or **slash-and-burn agriculture**. The former, if practiced in long cycles, does not necessarily result in deforestation, and indeed in some parts of the world people and forests have co-existed in this way for millennia. A mosaic of cleared patches, successional stages, and mature forest can exist in perpetuity. If the cycles shorten, through population growth or some loss of land available to the shifting cultivators, then the mosaic is truncated and full recovery to mature forest cannot be maintained. In other words, traditional shifting cultivation is likely to be a proximate cause of deforestation only if it is de-stabilized through population growth or curtailment of **access**. Geist and

Lambin identify traditional shifting cultivation as a proximate cause in 30 percent of the cases they have reviewed. The **United Nations Food and Agriculture Organization**'s (FAO) Global Forest Resources Assessment 2000 suggests that the contribution of shifting cultivation in terms of area deforested is much less.

Permanent agriculture is a more frequently reported proximate cause of tropical deforestation. Again, however, different forms exist. Divisions are made into large-scale and small-scale farming (or commercial farming and **subsistence agriculture**) and also into pastoral farming and cropping. Clear continental contrasts seem to exist in terms of relative importance of large- and small-scale agriculture: in Africa, "small scale" is far more important than "large scale," while the reverse is true in **Latin America**. There are, however, other possible divisions, including those involving formal land-settlement schemes as opposed to informal settlement. Variations exist both in the underlying drivers and the characteristics of the resulting deforestation.

Forest clearance for **cattle ranching** and soybean production is closely associated with Brazil. Some of the most dramatic images of tropical deforestation relate to these end uses, and as a result it is possible that their relative importance as proximate causes in the tropical zone as a whole has been exaggerated. Geist and Lambin report that only 6 percent of the deforestation cases they examined in Asia and 16 percent in Africa involved cattle ranching, compared with 82 percent in Latin America. This clearly illustrates the essential spatial variability of the causes of deforestation and thus helps to explain why simple, convincing explanations for deforestation have been so elusive. The case of Brazil, however, highlights the role of economic factors as underlying driving forces. The integration of **Amazonia** into the national and international economy has triggered an episode of change in land cover, in the same way that the integration of parts of India into the world economy in the nineteenth century led to the replacement of forest by tea plantations.

Agriculture has also expanded into the forest in some areas as a result of planned land settlement. Examples include the Bolivian lowlands, and areas such as Sumatra and Kalimantan in Indonesia where government-sponsored "transmigration" schemes have settled small farmers in previously unoccupied areas. According to the analysis of Geist and Lambin, this form of **colonization** occurred in around 40 percent of cases, but was much less frequent in Africa than in Asia and Latin America.

If agricultural expansion is the major proximate cause of deforestation, it might be assumed that agricultural retrenchment would also result in change in land use and/or land cover. To some extent this has happened, for example in central Europe and New England. While this change can sometimes be "active" (e.g., involving decisions to establish forest plantations on land released from agriculture), it can also be passive. While a human decision may be taken to discontinue agriculture, no such decision is necessarily involved in relation to forest invasion. The change can therefore be regarded as more "passive," and arguably the proximate causes of it are more complicated.

Wood Extraction

The harvesting of timber from the forest at a rate greater than that of natural increment depletes the forest and is an obvious candidate as proximate cause of deforestation. Logging in general and clearcutting in particular have become symbols, in the eyes of many, of environmental or **land degradation**. Forest exploitation for timber has sometimes been on a "cut and run" basis: in Southeast Asia, for example, a series of countries were affected successively in the second half of the twentieth century, and the result was a reduced forest area and a depleted forest resource. A similar episode occurred in the **United States of America** in the nineteenth century, giving rise to fears of a timber famine. Yet wood extraction need not necessarily result in a change in land cover, and in any case it seems that it rarely acts alone to reduce forest areas.

Wood extraction is reported by Geist and Lambin to have been involved in two-thirds of the deforestation cases they examined, but the frequency ranged from nearly 90 percent in Asia to little over half in Latin America. They subdivide wood extraction into "commercial" and "fuelwood," the former (for trade) being almost twice as important numerically as the latter (for domestic use). Again, the continental pattern shows strong contrasts; this ratio is inverted in the case of Africa.

The underlying driving forces differ between the two segments, with economic development presumably being a major driver of commercial extraction and population growth more directly significant in relation to extraction as fuelwood. It can probably also be concluded that the ecological impact of these two types of extraction varies according to forest type.

Commercial extraction of timber on a large scale requires roads (or railways). The construction of a communications network increases the accessibility of the forest and permits its penetration by small settlers. They may then proceed to clear further areas of forest for agriculture. In other words, wood extraction and agricultural expansion may combine to bring about removal of the forest in a synergistic way. This illustrates the general point that combinations of proximate causes, as well as of underlying driving forces, are usually more significant than single factors. According to Geist and Lambin's analysis, single-factor causation applied in fewer than 10 percent of cases, with the remainder involving two or more causal factors. They use the notion of tandems, such as "infrastructure-agriculture" or "logging-agriculture," to convey the message that combinations are more important than single causes.

Infrastructural Expansion

Serious problems of analytical logic are encountered when the role of infrastructure such as roads is considered. Infrastructural development permits or facilitates agricultural expansion and wood extraction, but it is debatable whether it can be viewed as a "proximate" cause. It can certainly be a permissive and/or intermediate causal factor, but it is questionable whether it can be regarded as the same category as agricultural expansion or wood extraction. Nevertheless, "transport extension" is reported by Geist and Lambin to have applied in nearly two-thirds of deforestation cases overall, and over three-quarters of those in Latin America. They also confirmed that it was very frequently involved in combination with agricultural expansion and wood extraction.

This factor in recent times has been closely associated with road building, but on a longer time scale, the advent of cheap forms of shipping was probably of major significance. Forests have tended to disappear more readily in accessible coastal lowlands than in the interiors of land masses. This pattern may be related to patterns of **land quality** such as soil fertility, but accessibility and ease of export have almost certainly also been significant.

Proximate Causes and Deforestation

Combination and contingency are key themes. The clear conclusion from the work of Geist and Lambin is that multiple causation is far more common than single-factor causation. Less immediately obvious but equally significant is the notion that the role of a process such as agricultural expansion, in terms of deforestation, depends on the setting. This setting includes both locational aspects or **initial conditions**, in relation to transport networks, and biophysical and social characteristics. In short, some forest types are more easily removed than others, and some social climates are more likely to be associated with small-scale agricultural expansion, for example, than others. A further complication is that **feedback** influences can and do link proximate causes and underlying driving forces: neither is fixed or independent.

"Proximate Causes" as an Investigative Instrument

Thus far, the hierarchical framework of underlying driving forces and proximate causes has been used mainly in analyses of tropical deforestation. A critique of that framework can readily reveal limitations of a logical and/or operational nature even in relation to that type of change in land cover. As yet, experience with the explicit application of the framework in other types of change in land use and land cover is limited, and therefore a general evaluation of the approach cannot be reached. It may be, however, that adaptation of this framework, or the development of an alternative one, may be required for the analysis of changes other than tropical deforestation. Modification or adaptation may be needed even if understanding of tropical deforestation is to deepen.

Leaving aside factors such as climatic change, most alterations to land use and land cover are the result of human **decision-making**. It follows that an understanding of them requires an understanding of the context of the decision. Various schematizations of land-use decision environments have been attempted, but most of them have been more useful as pedagogic tools than as research instruments. Most of them involve two or more levels, such as the behavior of the individual decision-maker and the "environment" in which (s)he operates. Attempts have been made to advance understanding of land-use change both at the "behavioral" level and that of prevailing regional political economy. The linking of the two levels is logical and desirable, but it has proved difficult to achieve in practice, and it raises wider philosophical issues of structure and agency. The significance of human agents is sometimes acknowledged in work on tropical deforestation, but their full incorporation into the "underlying proximate" framework has not yet occurred. Even if that "immediate" level were added, and the framework further modified to accommodate intermediate levels, problems would remain. The simplicity of the framework would be compromised and linearity of the underlying cause-effect model sustained.

In recent years, actor-network theory has attracted much attention in the social sciences. It avoids the problem of a hierarchical structuring of causality, and its use in land-use analyses can circumvent intractable problems of whether infrastructure, for example, is a different category of factor from wood extraction. It also provides the means of avoiding issues such as the question of whether there are fundamental differences between active and passive factors. On the other hand, while actor-network theory can help in promoting an understanding of a system, it does not yield results akin to those of the statistical analyses that have flowed from most work on land-use/cover change in recent years. It may therefore be less likely to produce results that can effectively inform policies relating to issues such as tropical deforestation.

There is nothing inherent in models of the "underlying driver/proximate causes" type that logically restricts their application to situations of linearity or of independence of variables. Indeed in relation to the latter, work with them has emphasized tandems or other combinations. Nevertheless, they do not necessarily point to the significance of non-linear relationships and **thresholds** of change, and it remains to be seen whether they can be adapted to cope with reversals of trends such as that associated with the **forest transition**.

The understanding of change in land use and land cover raises epistemological and well as practical issues. The quest for "explanation" that can be expressed in apparently simple (and preferably) numerical form has at best been only partly successful. It has been located within a context that can perhaps be described as traditional and reductionist—within a belief that effects can be explained by causes and that causes can be isolated and that they exist, in chain-like form, at different levels. The alternative, some would claim, is the despair that comes from acknowledging that the complexity of the real world is irreducible and inexplicable. The notion of "proximate causes" clearly lies in the first position rather than in the second: it is a notion that is problematic, for a variety of reasons, but to say that it has limitations is not to deny its helpfulness in contributing to a fuller understanding of change.

See also: **Agent-Based Model; Anthropocene; Biodiversity; Biomass; Degradation Narrative; Desertification; Exotic Species; Extensification; Forest Anomalies; Mediating Factor; Population Dynamics; Pristine Myth; Secondary Vegetation; Turnover; Vulnerability.**

Further Reading

Geist, Helmut J., and Lambin, Éric F., "Proximate Causes and Underlying Driving Forces of Tropical Deforestation," *BioScience* 52 (2002): 143–50; Meyer, William B., and B. L. Turner II, eds., *Changes in Land Use and Land Cover: A Global Perspective*, Cambridge, UK: Cambridge University Press, 1994; Meyer, William B., and B. L. Turner II, "Human Population Growth and Global Land-Use/Cover Change," *Annual Review of Ecology and Systematics* 23 (1992): 39–61.

ALEXANDER S. MATHER

Public Policy.   A course of action plan for that dimension of human activity that is regarded as requiring governmental or societal regulation or intervention. As Thomas Dye once summarized, public policy is focused on "what governments do, why they do it, and what difference it makes." It is concerned with how public issues and problems such as the impacts of land-use/cover change come to be defined and constructed, and how they are placed on the political and policy agenda. According to Stuart Nagel, the study of "the nature, causes, and effects of public policies" requires frameworks, metaphors, and models for thorough policy analysis. The clear distinction between the public and the private began to blur from the late nineteenth century onward. Almost all areas of social life that political economists have regarded as private were penetrated by public policies, among them sectors as diversified as housing, agricultural **colonization**, rural welfare, conservation, and urban planning.

The closeness of public policy to issues of land-use/cover change is a reminder that the use of land is a highly political activity or **arena of land conflict**. By implication, especially **land-use planning** or **land-use policies** in *strictu sensu* indicate a belief that determination of land use should not be left to chance, to individual landowners, or to the market alone. Governments have the duty to address those aspects of society and economy that individuals cannot handle in an appropriate, society-embracing way. Although **private property** is one sacred cow in modern societies, it is the private market that has yielded often undesirable and sometimes disastrous results, among them loss of **biodiversity**, **land degradation**, air and water pollution, and **urban sprawl**, reinforced by free riding behavior of individuals. The seed of modern public policy is that society should aim to improve the techniques of governing a capitalist system by making the environment less vulnerable. An intelligent management is needed for guaranteeing more efficiency in the public sector, like the promotion of intergovernmental cooperation to better manage global **common-pool resources**.

One of the consequences of governments taking responsibility for reconciling public and private interests was the development of bureaucracy as a more rational form of organization. Max Weber underlined that the growth of bureaucracy was due to the process of an encompassing "rationalization" in industrial society and that a bureaucrat who characterized a rational, neutral professional ethics would—different from the politician—best answer the purpose to serve the public interest. During the last decades of the twentieth century, this distinction between the roles and ethics of a politician and a bureaucrat has decreased, as argued by representatives of the "new right" such as Friedrich Hayek and Milton Friedman, who refer to the failures of public policy caused, among other reasons, by the blur of the public and private sector and a disappearance of a clear distinction between the sphere of the market versus the state.

Whereas it has been the state as prime actor designing and implementing public environmental policies, **globalization** resulted in the renaissance of local policies becoming more and more important in a world in which cities and regions competed for investors on the one hand and consumers on the other. The strategic management of places is becoming a major task for local policymakers who need to strongly promote their region or city. Local policymakers know best how to promote their locality in the optimum fashion. Federal policies offer an important and useful framework for land management, but the main impetus for efficient and **sustainable land use** and management is expected to derive from local policies. Due to globalization, the strategies and roles of governments at the regional, state, and federal levels have greatly changed during the last two decades, resulting in a reinvigoration of local policymaking.

During the last years, public policy analysis has increasingly been based on comparative research relying on methods such as statistical analysis of several case studies across countries or a more focused comparison of a specific policy area, for example, environmental or entrepreneurship policy, within a selected number of comparable countries. Policy analysis is challenged by complex interactive processes and mechanisms related to policymaking that are often intertwined and not clearly separated from each other. Public policy analysis has to take into account processes such as elections, or roles such as of politicians, parties, and interest groups who have a major impact on policymaking. The complexity of policy analysis increases if subliminal context variables are taken into consideration, such as business environment, access to information, or freedom of expression, which affect policymaking to a high degree. In this regard, public policy-related **cultural factors** such as public attitudes, values, and beliefs rank high among the many pressures or underlying **driving forces** of land change, both local and regional and acting in various combinations in different geographical locations of the globe. For example, among factors associated with tropical **deforestation** are public unconcern and missing basic values with respect to the environment, in particular dominant frontier mentality, prevailing attitudes of nation-building, modernization and development, and low public morale. Likewise, and in terms of **landholding**, medium- and large-scale landholders' main motivation very often is to achieve the largest economic benefits through their better connection to markets, usage of industrial inputs, and hired labor force. In these groups, land-use **decision-making**, and particularly deforestation, are contingent upon the profit they will obtain from expanding arable land for commercial crops or **pasture** for beef cattle production. They will, therefore, deforest more in cases in which the prices of commodities are higher than their production costs. The latter is likely to happen when they benefit from prices of support, subsidized and cheap credit, preferential export taxes, favorable exchange rates for export commodities, and low price of inputs. Especially the medium- and large-scale enterprises have in diverse contexts benefited from policies that have for different reasons tended to provide them additional institutional rents (i.e., cheap credit, access to roads, etc.). On the other hand, next to same policies have negatively biased small farmers and limited them in expanding their **access** to assets. Nonetheless, an example of how public policy can enhance low deforestation, in a larger region titled the **arc of deforestation**, stems from **Amazonia**. As in most other states of Brazil, government policy and discussions in the state of Acre have often centered around a developmental strategy involving forest products. Acre, however, shows a contrasting reality because the dominance of extractive economies in this area is based primarily on rubber extraction, and rubbertappers resisted cattle ranchers' invasions and lobbied the federal government for the creation of extractive reserves in the late 1980s and early 1990s. It was **non-timber forest products** extraction that has proved to be more environmentally sound, also tending to favor social equity. Nevertheless, pressures on deforestation persist due to the implantation of some areas in pasture.

For land-use **tradeoffs** in developing countries, Steve Vosti and Thomas Reardon have proposed analyses to be organized around a "critical triangle" of public policy goals that include productivity growth, **poverty** alleviation, and sustainable resource use. Some other policy analysts extended this approach to include **human health**, and a collection by David Lee and Christopher Barrett includes a large number of examples of approaches to quantitative policy analysis of tradeoffs among the "critical triangle" for cases of **agricultural intensification**. In terms of the sociopolitical properties of land use at the national-scale level, in particular **China** is in a crucial moment of its history in finding a balance among **food security**, **economic growth**, and ecological **conservation**.

One of the biggest challenges the world has ever faced is climate change. Its potentially destructive potential has forced the world community into a multilateral process of negotiations, to prevent at least the most disastrous consequences and to slow down the speed of climate change. Global warming and policy issues are increasingly intertwined. The controversies among national and local policymakers all over the world circle around new or at least re-designed international trade agreements, questions of national sovereignty versus global governance, and ideological debates about future strategies for the promotion of economic growth. In this context, the role of public policies has increased to enforce control over the interests of the private sector, as well as governments affecting **global environmental change** in a negative way. Only by the common efforts of the world community and a global policy-driven strategy will the challenge of climatic change be deactivated. Environmental groups supported by sympathizers, for example, in foundations and in the federal government, aim to restrict and phase out the use of fossil fuels as sources of energy. Such measures would reduce emissions of **greenhouse gases** into the atmosphere. International climate policy is based on the **Kyoto Protocol**, which calls on industrialized nations to carry out, within one decade, drastic cuts in greenhouse gas emissions that stem mainly from the burning of fossil fuels. In November 2004, the Russian Duma ratified the protocol, so that the accord was finally brought into force. As of November 2004, the agreement had been ratified by 127 countries representing 61 percent of emissions, which is a success based on an enduring and determined climate policy. The United States, although a signatory to the protocol, has neither ratified nor withdrawn from the protocol.

See also: Agenda 21; Carson, Rachel Louise; Community Involvement; Consumption; Decentralization; Economic Restructuring; Integrated Assessment; Investments; Land-Use System; Multiplier Effect; Property; Reforestation; Scenario; Swidden Cultivation; Tragedy of Enclosure; United Nations Convention to Combat Desertification; Vulnerability.

Further Reading

Audretsch, David B., Grimm, Heike, and Charles Wessner, eds. *Local Heroes in the Global Village: Globalization and New Entrepreneurship Policies*, New York: Springer Science & Media Inc., 2005; Dye, Thomas R., *What Governments Do, Why They Do It, What Difference*

It Makes, Tuscaloosa: University of Alabama, 1976; Friedman, Milton, *Capitalism and Freedom*, Chicago: University of Chicago Press, 1969; Hayek, Friedrich A., *The Road to Serfdom*, Chicago: University of Chicago Press, 1994; Lee, David R., and Christopher B. Barrett, eds., *Tradeoffs or Synergies? Agricultural Intensification, Economic Development and the Environment*, Wallingford, UK: CAB International, 2001; Nagel, Stuart S., *The Policy Studies Handbook*, Lexington, MA: Lexington Books, 1975; Parsons, Wayne, *Public Policy: An Introduction to the Theory and Practice of Policy Analysis*, Cheltenham, UK: Edward Elgar, 1995; Platt, Rutherford H., *Land Use and Society: Geography, Law, and Public Policy*, Washington, DC: Island Press, 2004; Rausser, Gordon C., and Rachael E. Goodhue, "Public Policy: Its Many Analytical Dimensions," in *Handbook of Agricultural Economics, Volume 2B: Agriculture and Food Policy* (Handbook in Economics no. 18), eds. Bruce L. Gardner and Gordon C. Rausser, Amsterdam: Elsevier Science, 2002, pp. 2057–102; Repetto, R., and M. Gillis, *Public Policies and the Misuse of Forest Resources*, Cambridge, UK: Cambridge University Press, 1988; Vosti, Steve A., and Thomas Reardon, eds., *Sustainability, Growth, and Poverty Alleviation: A Policy and Agroecological Perspective*, Baltimore, MD: Johns Hopkins University Press, 1997; Weber, Max, *The Protestant Ethic and the Spirit of Capitalism*, New York: Charles Scribner's Sons, 1930.

HEIKE GRIMM

R

Ramsar Convention on Wetlands. An abbreviation of the "Convention on Wetlands of International Importance Especially as Waterfowl Habitat," which is an intergovernmental treaty adopted in February 1971 in the Iranian city of Ramsar. The convention came into force in 1975 and provided a global framework for the **conservation** of **wetlands**. In December 2004, the Ramsar convention was ratified by 143 contracting parties, and 1,399 wetlands have been designated for inclusion in the "Wetlands of International Importance" list, covering about 1.2 million square kilometers. Originally, the convention emphasized the conservation and wise use of wetlands to provide habitat for waterbirds, but all aspects of conservation and sustainable use are now considered with a broadened perspective. However, the designation of "Wetlands of International Importance" is often conflicting and debatable, as preliminary identification of national priorities in each country is required, as well as international agreements in the case of transfrontier wetlands, shared water systems, and species.

See also: **Coastal Zone; GLOBWETLANDS; Millennium Ecosystem Assessment.**

Further Reading

The Ramsar Convention Secretariat [Online, December 2004], The Ramsar Convention on Wetlands Web Site, www.ramsar.org.

LAURENCE HUBERT-MOY

Reforestation. Forest **regrowth** that constitutes an increasing fraction of the earth's cover and performs a crucial role in mitigating the harmful effects of **global environmental change**. There is thus a critical need to better understand the biophysical and human driving forces associated with reforestation. The term **forest transition** has been widely used to describe changes in forest cover with increased economic development and **industrialization**. The environmental Kuznets curve, on which much of this discussion is based, predicts a rapid increase in **deforestation** during the initial period of economic development for a nation, followed by a turnaround period of net reforestation in economically developed nations. There is increasing evidence now that this trend is not confined to developing countries. A range of socio-economic and institutional **driving forces** have been associated with reforestation, including demographic transitions, **agricultural intensification**, **urbanization**, global collapses in certain agricultural market prices (such as of **coffee**), shifts toward non-wood energy sources, restrictive national policies, **community involvement**, and increased cultural awareness. Factors such as **land-use history** and soil fertility significantly impact the dynamics of reforestation, with the rate of regrowth much faster in areas of shifting cultivation compared to intensively farmed areas and in the fertile tropics compared to the temperate zone. It is important to recognize that the observed rates of forest regrowth are often much lower than the rates of deforestation in preceding periods. Reforested areas can take several decades to return to **biodiversity** levels approximating those in the original old-growth forests, and may never reach these levels in areas where there is large-scale planting of non-native species. **Scale** is a crucial factor impacting the perceived trajectory of change, with reforestation frequently reported as occurring in smaller abandoned clearings, or on tree lots planted by local people to alleviate problems caused by increasing scarcity of wood. Nevertheless, where properly managed, reforestation has multiple beneficial impacts on global environmental change, providing crucial environmental or **ecosystem services** such as **carbon sequestration**, **watershed** protection, habitat for endangered species, and support for **forest**-dependent communities, indicating that there is a pressing need to identify and support the factors associated with the increasing expansion of secondary forest cover across the world.

See also: **Conservation; Degradation Narrative; Exotic Species; Fallow; Forestry; Green Wall Project; Hot Spots of Land-Cover Change; Kyoto Protocol; Land Rehabilitation; Land-Use Policies; Mediating Factor; Property; Public Policy; Secondary Vegetation.**

Further Reading

Fairhead, James, and Melissa Leach, "False Forest History, Complicit Social Analysis: Rethinking Some West African Environmental Narratives," *World Development* 23 (1995): 1023–35; Moran, Emilio F., Brondizio, Eduardo S., Tucker, Joanna M., da Silva-Forsberg, Maria Clara, McCracken, Stephen, and Italo Falesi, "Effects of Soil Fertility and Land-Use on Forest Succession in Amazônia," *Forest Ecology and Management* 139 (2000): 93–108; Nagendra, Harini, Southworth, Jane, and Catherine Tucker, "Accessibility as a Determinant of Landscape Transformation in Western Honduras: Linking Pattern and Process," *Landscape Ecology* 18 (2003): 141–58.

HARINI NAGENDRA

Regrowth.   Conversion of farmland on slopes to **forest** or **grassland** for the aim of restoring **vegetation** and restraining ecological deterioration in **China**, for example. After serious flooding disasters happened to the reaches of the Changjiang and Yellow rivers, the government of China began to realize the urgent necessity to protect and improve the **watershed** environment of the nation's two largest rivers. It is not only a large ecological project, but also a socio-economic project implemented all around western China from the year 2000 onward, covering eleven provinces. As a principle, land with a slope between 15 and 25 degrees should be returned to forests and grassland. It is estimated that the total regressed area is 9.5 million hectares, accounting for 24.9 percent of the total cultivated land in western China. In the mountainous and plateau areas of western China, a tremendous amount of forestland and grassland was impoldered in the past to create arable land to support the growing population. However, water and soil on the land slopes are seriously lost when the slopes are cultivated. The main policy of regrowth is the **conversion** of farmland to forestland (grassland), planting trees, and thus sealing mountain passes. Regrowth will play an important role in the control of environmental deterioration and protection of the environment in China. It has been well implemented, but it is still necessary to keep enough cultivated land for **food security** and to avoid negative impacts such as food shortages.

See also: **Conservation; Decentralization; Fallow; Forest Anomalies; Green Wall Project; Himalayas; Land Degradation; Reforestation; Soil Erosion; Tradeoffs.**

QI LU

Remittances Landscape. An emergent landscape that is driven by the investment of remittances. People migrate in order to ensure their own survival and to improve the economic well-being of their families. Once in their new host countries, migrants often begin to send money home to provide for the basic needs of their families or to support community development projects. In the past ten years, these migrant financial flows, known as remittances, have doubled worldwide.

Remittances can be divided into private and collective remittances. Private remittances are funds that are sent directly from an individual migrant to a family member in his or her country of origin. Collective remittances are sent by a group of migrants to their community of origin to support community and development projects or to contribute to disaster-related recovery efforts. Collective remittances make up approximately 1 percent of total remittances.

According to the World Bank, remittances to eighty countries amounted to nearly $60 billion (in U.S. dollars) in 2000, which is 6 billion more than total net foreign aid from countries of the Organization for Economic Cooperation and Development (OECD). The largest flow of remittances goes to **Latin America** and the Caribbean. Migrant remittances to Mexico in 1996 amounted to $4.22 billion, or fourteen times the total sum of net foreign aid received. Other countries where remittances amount to at least three to nine times the amount of net foreign aid include India, Morocco, Sudan, Croatia, the Dominican Republic, Colombia, and Brazil. In many countries, remittances have replaced agricultural exports to a large extent. Income from **agriculture** as a percentage of the gross national product (GNP) dropped from 21 percent in 1980 to 18 percent in 1997, and remittances as a percentage of GNP rose from 3 percent to 15 percent during the same period. In most South Asian countries, and particularly in Bangladesh, Sri Lanka, India, and Pakistan, migration and remittances are becoming so important that they have become a key focus in development planning.

Many studies reveal a growing flow of remittances from urban to rural areas. Urban migrants from rural households apparently prefer to invest in rural lifestyles and preserve or improve the quality of their rural households rather than to invest in urban areas to which they migrate. However, the implications for land use, land cover, and landscapes of this inflow of capital and the medium- to long-term loss of labor are disputed. The prevailing thought is that remittances are used to cover basic needs but are not invested in agriculture. On the other hand, there are numerous studies from Latin

America, West Africa, South Asia, and elsewhere that show that migration and remittances are also used to improve agricultural or **land-use systems**.

The way remittances are invested depends on the particular situation of the family receiving the money. Nonetheless, in general the largest part of remittances (an estimated 80 percent) are used for **consumption** and household needs (food, clothing, health care, education, etc.). Investment in land, livestock, agricultural inputs, and in-house construction and maintenance is also relatively common but secondary to daily needs. Still less is used for business **investments**, savings, and loan repayments. Agricultural investment—for instance the purchase of improved inputs such as fertilizer and the expansion of irrigation—is more likely to occur in situations where political/economic incentives and environmental conditions are favorable, but typically only after consumption demands that immediately improve family welfare are met. The key factor deciding the way remittances are invested is insecurity; therefore the type of insecurity affects the type of investment.

Despite the impressive increase in remittances worldwide, and the substantial amount of research on migration, there is relatively little knowledge of the impact of remittances on landscapes. At present, no commonly used definition of remittances landscapes seems to exist. One possible criterion that can be applied in search of a definition is the following: a certain type of landscape is being transformed into another type mainly due to the investment of remittances.

So far, the transformations of at least two types of landscape due to the investment of remittances have been identified. The first transformation is that of agricultural landscape to a peri-urban landscape: investment of remittances in house construction replacing traditional dwellings by large, brick and cement homes. Many migrants decide to leave their country in the hope of earning enough money in another country to be able to build a house in their community of origin. This can lead to the emergence of peri-urban landscapes of "cultivated real estate," meaning that land is valued as a safe investment, and a place to demonstrate one's success with a large home, but not as an agricultural investment. This type of remittances landscape dominated by large ostentatious houses to demonstrate achievement and to have a home for retirement has also been described as a "landscape of conspicuous retirement." In a few migrant sending areas, this transformation has included the partial abandonment of cultivated land because remittances substitute agricultural income and are sufficiently large to allow households to abandon an **economic livelihood** based on agriculture, for example, in Bolivia, Ecuador, Mexico, Philippines, Thailand, and India. In the African context, even though remittances are also invested in housing, meeting basic needs continues to be the central preoccupation. Nevertheless, some migrants are able to build homes for their retirement, thereby encroaching on farming land around cities, contributing to the conversion of an agricultural to a peri-urban landscape in Nigeria and Morocco, for example. The second transformation is that of **forest** and agricultural landscape to a pastoral landscape, that is, investment of remittances in livestock to transform forested and agricultural land into **pasture**. The conversion of previously cultivated fields into degraded pasture is also included in this category, with examples stemming from Ecuador, Mexico, the Dominican Republic, and Brazil.

Another type of remittances landscape has been identified, namely the transformation of landscapes affected by natural disasters and civil wars to reconstructed landscapes. Especially in Central America but also in some African countries, much of the remittances sent back by migrants have helped to rebuild countries after civil war or natural disasters in Honduras, El Salvador, Nicaragua, Lebanon, and Somalia, for example. However, reconstructed landscapes are peri-urban, urban, agricultural, or pastoral landscapes that are "restored" to the original type of landscape. As the type of landscape does not change due to the investment of remittances, reconstructed landscapes are not treated here as a third type of remittances landscape. Furthermore, remittances that are invested into agricultural improvements in an agricultural landscape do not change the basic type of landscape and are therefore not included in this typology.

A tentative definition of remittances landscape could thus be the emergence of a certain type of landscape that is driven by the investment of remittances. However, how exactly land cover will be affected by the various ways remittances are invested cannot be generally predicted. The investment of remittances, and therefore the transformation of land use, land cover, and landscapes, will largely be influenced by the cultural, socio-economic, political, and environmental context from which migrants leave and to which they remit. If these conditions are positive, then it is more likely that remittances are invested into agricultural production. Thus, a predominantly agricultural landscape is maintained as opposed to an emerging peri-urban remittances landscape where previously cultivated land may be abandoned and/or turned into real estate. The emergence of a pastoral landscape due to the investment of remittances into livestock can potentially lead to large-scale changes in land cover, in particular through **deforestation**.

See also: **Agricultural Intensification; Decision-Making; Land Abandonment; Land Degradation; Land Rehabilitation; Mediating Factor; Multiplier Effect; Population Dynamics; Urban-Rural Fringe.**

Further Reading

Anonymous, "The Longest Journey: A Survey of Migration," *Economist* November 2 (2002): 1–16; Durand, Jorge, and Douglas Massey, "Mexican Migration to the United States: A Critical Review," *Latin American Research Review* 27 (1992): 3–42; International Institute for Environment and Development, Inquiry on Migration and Development [Online, August 2004], The IIED Web Site, www.iied.org; Jokisch, Brad, "Migration and Agricultural Change: The Case of Smallholder Agriculture in Highland Ecuador," *Human Ecology* 30 (2002): 523–50; McDowell, Christopher, and Arjan de Haan, Migration and Sustainable Livelihoods: A Critical Review of the Literature [Online, August 2004], The IDS Working Paper Web Site, www.ids.ac.uk; Nii Addy, David, Wijkström, Boris, and Colleen Thouez, Migrant Remittances: Country of Origin Experiences— Strategies, Policies, Challenges and Concerns [Online, August 2004], The International Migration Policy Programme Web Site, www.livelihoods.org; Sander, Cerstin, and Samuel Maimbo Munzele, Migrant Labor Remittances in Africa: Reducing Obstacles to Developmental Contributions [Online, August 2004], The World Bank Africa Region Working Paper Series Web Site, www.worldbank.org; Tiemoko, Richmond, Migration, Return and Socio-Economic Change in West Africa: The Role of the Family

[Online, August 2004], The Sussex Centre for Migration Working Paper Web Site, www.sussex.ac.uk.

SILVIA HOSTETTLER

Remote Sensing.   The best practical way of spatially explicit **monitoring** of land-use and land-cover changes on a regional to global scale is from space. It is urgently needed to take data from the sequence of individual moderate- and high-resolution sensors and develop long-term climate data records of sufficient quality for the study of global change. The challenge in creating these records is to account for the evolving instrument design and performance since the 1970s, and necessitates dynamic product continuity.

The foremost observational requirement for land-use/cover change research is to extend the thirty-two-year record of high-resolution **Landsat** observations, which is the indispensable foundation for global land-cover inventories. Availability of the global thirty-meter spatial resolution data that have been collected from Landsat-7 increases the potential for detailed studies of landscape change on regional and global **scales**, a task that only a few years ago was not possible for some regions of the world. The long-term record of high-resolution observations will be continued through the Landsat Data Continuity Mission, which is being developed by the U.S. National Aeronautics and Space Administration (NASA) and the **United States Geological Survey (USGS) Program**. It is important that the data quality and availability from this mission be adequate for land-use/cover-change research, and emphasis needs to be given to instrument calibration, data quality assessment, data acquisition strategy, and data policy. It is also critical that high-resolution observations (ca. thirty meters) become part of the operational environmental satellite suite, enabling data continuity without gaps in the data record.

While high-resolution satellite data are used to map and quantify **land-cover change** at the regional scale, moderate resolution data are used to classify and characterize land cover at the global scale. The TERRA and AQUA satellite products, especially from the **moderate resolution imaging spectrora-diometer** (MODIS), extend the twenty-three-year moderate-resolution data record from the **advanced very high resolution radiometer** (AVHRR), providing global land-cover products, including continuous fields of tree cover and burned area. During the last several years there has been an increasing emphasis on quality assessment and **validation** for MODIS satellite products, which sets an important precedence for future missions. The newly developed MODIS-based system is being used to detect changes in land cover at spatial resolutions of 250 meters, allowing the monitoring of potential **hot spots of land-cover change** over the globe. The continuity of these moderate-resolution data will be sustained by the future National Polar Orbiting Environmental Satellite System (NPOESS) mission, which will have on board the "MODIS-like" Visible Infrared Imaging Radiometer System (VIIRS). Quality of VIIRS observations and products must be assured to enable the measurement suite developed for the MODIS time series to be continued.

In addition to the longest continuous series of observations from Landsat and the AVHRR-MODIS-VIIRS line, there are other very useful optical coarse-to-moderate resolution data that are used for monitoring land surface. The International Satellite Cloud Climatology Project (ISCCP) global data set contains data on surface reflectances and temperatures. It has been based on compilation of all available geostationary satellite observations during the past two decades and could be used for analyzing global surface changes on a coarse spatial resolution. Data from the U.S. geostationary satellites (GOES) covering the Western Hemisphere have been extensively used for monitoring **fires** over the Americas. The above data sets are available at NASA data archives and/or from the National Climate Data Center (NCDC). Night-lights data from the Optical Line Scanner (OLS) on-board the DMSP satellite have been effectively used in monitoring tropical fires and **urbanization**. These data have been collected at the National Geophysical Data Center (NGDC). Another source of global land optical observations during the last few years is the Multiangle Imaging Spectro-Radiometer (MISR) on-board TERRA satellite providing angular reflectance measurements, which have been successfully utilized to characterize surface bi-directional properties and derive landscape composition.

During the last few years, NASA has been investing in hyperspatial resolution data (fewer than ten meters) for mapping land-cover change with data from the commercially operated **IKONOS** system. Results have demonstrated the utility of these data for mapping logging and **forest degradation** and validating high-resolution interpretations. While hyperspatial resolution data are becoming available for local uses, there remain significant hurdles to their adoption, including access to data, cost, and knowledge of the use of geospatial information.

In the New Millennium EO-1 mission, NASA has launched technology demonstration projects to explore the utility of high-resolution thermal (ASTER) and hyperspectral (Hyperion) measurements. The latter provides insight into vegetation cover composition and soil type discrimination.

Lidar technology—the active optical observations—is useful for fine-scale mapping of urban areas and provides elevation data that can be applied directly to existing/traditional applications. Lidar observations also provide information on **vegetation** structure that cannot be directly derived from passive optical data. NASA recently launched ICESat with the Geoscience Laser Altimiter System (GLAS) on board. Potential for using GLAS data for land-cover change studies is being investigated.

In addition to optical measurements, microwave observations have been widely utilized for mapping changes at the land surface. Radar measurements are especially useful in the areas of persistent cloudiness. Among the most useful applications with active microwave observations from space borne radars is monitoring flood events. **Biomass** estimation and classification of **forest**-cover types have produced encouraging results with available radar data.

The **NASA Land-Cover/Land-Use Change (LCLUC) program** has a goal to develop the capability to perform repeated global inventories of land use and land cover from space to identify changes occurring at the land surface. Priority for **change detection** under NASA LCLUC has been given to areas of the world that have been undergoing the most rapid changes and where stresses from human activities are likely to increase most rapidly.

Emphasis has been placed on the exploitation of satellite remote sensing data as the best source of information on the spatial distribution of land cover and rates of landscape change on regional, continental, and global scales. Time series of satellite data described above are analyzed in the LCLUC program to provide a consistent record of global land-cover change; characterize the end states of land-cover modification in regions of high population density; and study the impacts of spatial patterns and past history of land-use conversion on carbon and water cycles, and ecosystem changes—for example, changes in biodiversity, interaction of land-cover change with climate, and the sustainability of ecological services.

See also: **AFRICOVER; Airborne Light Detection and Ranging; BIOME 300 Project; Center for International Earth Science Information Network; Continuous Data; Desertification; Discrete Data; Earth Resources Observation Systems (EROS) Data Center; ENVISAT ASAR; European Remote Sensing (ERS-1/-2) Satellites; Forest Anomalies; Global Burnt Area 2000; Global Land Cover Map of the Year 2000; Global Observation of Forest and Land Cover Dynamics; Global Positioning System; Global Terrestrial Observing System; GLOBWETLANDS; Hot Spot Identification; Land Cover Classification System; Land-Use History; Normalized Difference Vegetation Index; Pattern Metrics; Pixel; SaHEL Land Cover; Socializing the Pixel; TERRA-ASTER; Thermal Band Analysis; Tropical Ecosystem Environment Observations by Satellite (TREES) Project.**

Further Reading

Gutman, Garik, Janetos, Anthony C., Justice, Chris O., Moran, Emilio F., Mustard, John F., Rindfuss, Ronald R., Skole, David, Turner, B. L., II, and Mark Cochran, eds., *Land Change Science: Observing, Monitoring and Understanding Trajectories of Change on the Earth's Surface (Remote Sensing and Digital Image Processing Series 6)*, Dordrecht, NL: Kluwer Academic Publishers, 2004; Rogan, J., "Remote Sensing Data and Methods in Land Cover Change Detection: A Review and Implication for Operational Large Area Monitoring," *Progress in Physical Geography* (in press).

GARIK GUTMAN

Resilience. Capacity for renewal and persistence in a dynamic environment facing disruption and change. In 1973, Crawford Holling first defined the concepts of resilience and stability. Stability is the tendency of a system to return to an equilibrium following a disturbance. "Ecological resilience" measures the amount of disturbance that an ecosystem can absorb before it changes into an alternative stability domain. Crawford Holling distinguished it from the "engineering resilience" that is the ability to return to a steady state following a perturbation. The ecological resilience paradigm includes research on alternative states or transitions. "Social resilience" implies that not only the relevant ecosystem, but also the corresponding human and management systems are able to absorb perturbations such that the system as a whole remains able to sustain (on average) a reasonable flow of benefits over time. "Social resilience" is then the human ability to cope with and adapt to social and **environmental change**, mediated through appropriate institutions. It remains unclear, however, whether resilient ecosystems enable resilient communities. Using the structure proposed by the Resilience Alliance, the concept of resilience in a broad sense has three defining characteristics. It is a measure of the amount of change the system can undergo and still retain the same controls on function and structure, the degree to which the system is capable of self-organization, and the community's ability to build and increase its capacity for learning and adaptation.

The resilience concept is interlinked with the concepts of sustainability and **vulnerability**. **Sustainable land use** exists if the use and management of resources do not lead to the loss of future options. The quality in a resource that renders its use sustainable is its resilience. Vulnerability is the negative side of resilience: when a social or ecological system loses resilience, it becomes vulnerable to change that previously could be absorbed. In a resilient system, change has the potential to create opportunity for development, novelty, and innovation. In a vulnerable system, even small changes may be ravaging. **Land degradation**, to put it very simply, is loss of resilience.

The degree to which the social-ecological system can build and increase the capacity for learning, adaptation, and responding in a manner that doesn't constrain future opportunities is a central aspect of resilience. In the face of the inevitability of changes (but with uncertain impacts), the resilience of the system demands evolving and adaptive strategies. Short-term responses to changes in land-based activities, which are identified as coping mechanisms, are one component of this adaptive capacity. The second component is related to cultural and ecological adaptations and represents long-term adaptive strategies. Coping mechanisms often take the form of emergency responses in abnormal seasons or years. Adaptive strategies, on the other hand, are the ways in which individuals, households, and communities change their productive activities and modify local rules and **institutions** to secure livelihoods. Compilation and analysis of case studies that illustrate adaptive management by local or indigenous people (a process of adjusting management actions and directions, as may be appropriate) are two goals of the Resilience Alliance.

See also: **Agrodiversity; Biodiversity; Disequilibrium Dynamics; Economic Growth; Economic Livelihood; Land-Use System; Mediating Factor; People at Risk; Pristine Myth; Threshold; Tradeoffs; Turnover.**

Further Reading

Adger, Neil W., "Social and Ecological Resilience: Are They Related?" *Progress in Human Geography* 24 (2000): 347–64; Berkes, Fikret, and Dyanna Jolly, "Adapting to Climate Change: Social-Ecological Resilience in a Canadian Western Arctic Community," *Conservation Ecology* 5 (2001): The Ecology and Society Web Site, www.ecologyandsociety.org/vol5/iss2/art18/; Folke, Carl, *Social-Ecological Resilience and Behavioural Responses* (Beijer Discussion Paper 155), Stockholm, SW: The Beijer Institute, 2002; Gunderson, Lance H., Holling, Crawford S., Pritchard, Lowell, Jr., and Garry D. Peterson, "Resilience," in *The Encyclopedia of Global Environmental Change, Vol. 2: The Earth System: Biological and Ecological Dimensions of Global Environmental Change*, eds. Harold A. Mooney and Josep G. Canadell, Baffins Lane, Chichester, UK: John Wiley & Sons, 2002, pp. 530–1; Resilience Alliance, Research on Resilience in Social-Ecological Systems: A Basis for Sustainability [Online, January 2005], The Resilience Alliance Web Site, www.resalliance.org.

NATHALIE STEPHENNE

Rice. An ancient grain, with *Oryza sativa* being by far the most important cultivated rice species worldwide. The origins of rice have been much debated, but the precise time and place of its first development remains unknown. Wild species of rice have been found over a broad area extending south and east from India and south **China** across continental Asia and insular Southeast Asia.

Rice terraces carved out of the hillsides hundreds of years ago by Ifuago farmers on the Philippine Islands, photo taken between 1890 and 1923.
PHOTO: Marion Post Wolcott. [Courtesy Library of Congress]

Rice is presently cultivated in nearly all parts of the world where the soil and climate are favorable. It is the most important **food crop** for human beings and the primary staple for more than half of the world's population. Asia accounts for about 90 percent of production and **consumption** of rice. There are numerous ways of using the crop, including the straw.

Rice is the dominant food crop in the humid subtropics (76 percent of the area under food grains), humid tropics (75 percent), and subhumid tropics (51 percent). It is an important crop in the subhumid subtropics (36 percent) and semiarid tropics (19 percent), but is insignificant in the semiarid and cool subtropics.

There are about 120,000 varieties of cultivated rice. The most important cultivar groups are indica and japonica (or sinica) rice. Indica rice accounts for more than 75 percent of global trade and is grown mostly in tropical and subtropical regions. Japonica rice, typically grown in regions with cooler climates, accounts for about 12 percent of global rice trade. Other types of rice are aromatic rice, primarily jasmine from Thailand and basmati from India and Pakistan; and glutinous rice, grown mostly in Southeast Asia and used in desserts and ceremonial dishes.

Cultivated rice is generally considered a semiaquatic annual grass, although in the tropics it can survive as a perennial. Physical requirements such as warm temperatures and available water determine its geographical distribution. Rice is grown as irrigated rice, rainfed lowland and upland rice, and flood-prone rice. Irrigated rice accounts for 54 percent of the harvested rice area and contributes 76 percent of global rice production. It is concentrated mostly in the humid and subhumid subtropics and humid tropics.

World rice production in 2004 exceeded 600 million tons and has almost tripled in the last four decades. The most important rice-producing countries are China and India, with about 51 percent of the total rice production. Other important countries are Indonesia, Bangladesh, Thailand, and Vietnam.

Demands for rice have steadily increased, mainly due to population growth. In the past centuries, increasing demand for rice was met mainly through increase in production area. In the last four decades, rice's cultivated area further increased by about 30 percent to around 150 million hectares, which is more than 10 percent of total global arable land. However, the increase in rice production area has gradually declined during the last twenty-five years largely due to competition with other land-use types and for environmental reasons. In some countries such as China, land use for rice production has even decreased in recent decades. Growing demands for rice are increasingly met by higher productivity.

Global rice yields more than doubled from 1961 to 2004 and have almost reached 4 t ha^{-1}. These yields are mainly a result of the **Green Revolution** and associated progress in breeding for new varieties and improved crop management, including extensive use of synthetic fertilizers and agrochemicals. Rates of yield increases and presently obtained yields largely vary among countries and regions. The highest yields are obtained for irrigated rice, for example, in China, Indonesia, and Vietnam, among others. High cropping intensity and extensive use of agrochemicals affect **human health** and the environment.

An additional matter of concern is the contribution of intensive rice production to global warming through the release of methane (CH_4) into the atmosphere. Methane, one of the principal **greenhouse gases**, accounts for 15–20 percent of current radiative forcing. Its concentration has more than doubled over the past 200 years. Rice paddy soils are one of the major anthropogenic CH_4 sources. It has been estimated that the 60 percent increase in rice production required by the year 2020 to meet the demands of an increased population may increase CH_4 production by up to 50 percent. In contrast, the **Intergovernmental Panel on Climate Change** (IPCC) has recommended immediate reductions of 8 percent in anthropogenic emissions of CH_4 to stabilize atmospheric concentrations at current levels.

However, rice production is not only contributing to but also affected by climate change. Rice is a C3 plant with a pathway of carbon fixation typically more responsive to CO_2 increase than for C4 plants such as **maize**. Experimental and modeling studies suggest increases in productivity of about 20–25 percent if present CO_2 concentration doubles. However, projected increases in temperature will reduce rice yields so that the combined CO_2 and temperature effects will tend to compensate, which, however, will vary depending on the specific growing conditions and crop management.

The population of Asia is growing at 1.8 percent per year, until the middle of the next century, which will increase the demand for rice by about 70 percent over the next thirty years. Accordingly, rice production has to be raised by another 70 percent over this time period to sustain **food security**. Since expansion of rice-growing areas is largely limited, further increases in yields appear to be the only option to meet the growing demand for rice. However, rice yields in the high-yielding irrigated regions tend to approach a ceiling. Also, most rice is grown in small farms with limited **investment** capital to intensify production. The environmental and socio-economic consequences associated with any attempts to further increase the supply of rice are largely unclear.

See also: **Agricultural Intensification; Biotechnology; Cash Crops; Coastal Zone; Domestication; Cumulative Change; Environmental Change; Land Privatization; Land Rights; Malaria; Paleo Perspectives; Population Dynamics; Turnover; Wetlands.**

Further Reading

Kimball, Bruce A., Kobayashi, Kazuhiko, and Marco Bindi, "Responses of Agricultural Crops to Free-Air CO_2 Enrichment," *Advances in Agronomy* 77 (2002): 293–368; Matthews, Robin, and Reiner Wassmann, "Modelling the Impacts of Climate Change and Methane Emission Reductions on Rice Production: A Review," *European Journal of Agronomy* 19 (2003): 573–98; Parry, Martin L., Rosenzweig, Cynthia, Iglesias, Ana, Livermore, Matthew, and Günther Fischer, "Effects of Climate Change on Global Food Production under SRES Emissions and Socio-economic Scenarios," *Global Environmental Change* 14 (2004): 53–67; Ramankutty, Navin, and Jonathan A. Foley, "Estimating Historical Changes in Global Land Cover: Croplands from 1700 to 1992," *Global Biogeochemical Cycles* 13 (1999): 997–1027; Riceweb [Online, February 2005], The Rice Web Site, www.riceweb.org; United Nations Food and Agriculture Organization [Online, February 2005], The FAO Web Site, www.fao.org; Vergara, B. S., and S. K. de Datta, "*Oryza sativa L.*," in *Plant Resources of South-East Asia Handbook 10: Cereals*, eds. Gerard J. H. Grubben and Soetjipto Partohardjono, Leiden, NL: Backhuys Publishers, 1996, pp. 106–15.

FRANK EWERT

River Basin. Part of the land surface of the earth from which rain and snow drain into a river and which eventually flows into a lake or an ocean. It is also called a "drainage basin" or "catchment area." River basins are defined by natural hydrological boundaries that are usually **mountain** ridges between two neighboring basins. Basins can be divided into **watersheds** or areas of land around a smaller river or stream. In a river basin, all the water, including the wastewater produced by human beings, drains to a large river.

River basins absorb and channel all the runoff from snowmelt and rainfall. All the watersheds within a basin constitute a coherent system through the linking of the river system. By this function, a river basin forms a critical link and an integrator among the hydrologic, ecological, and socio-economic components of a river basin. It also supplies the resources we need and regulates the environment we live in if it is managed in a sustainable way.

Within each river basin, water travels over the surface and across **forest**, farm fields, **grassland**, villages, and city streets, or it seeps into the soil and travels as groundwater on its way to the main river channel. Land cover varies in its evaporation, transpiration, and water retention properties, and then has different hydrologic functions. Change in land use and land cover can have important impacts on the hydrology of river basins.

As water moves downstream, it erodes soil on slopes and carries and redeposits gravel, sand, and silt. This function of rivers makes them a major

An elevated view of Potomac River and Tidal Basin in Washington, D.C., ca. 1920–1950.
PHOTO: Theodor Horydczak. [Courtesy Library of Congress]

force of landscape **modification**. The upper reaches of a river basin are dominated by erosional landforms. Land use in this part is usually at risk of water and **soil erosion**. Sedimentary landforms like alluvial plains dominate on the lower part of a river basin, and land use in this part is usually at risk of flooding when the seasonal discharge exceeds the capacity of the river drainage system.

Water also transfers chemicals, excess nutrients, organic matter, and bacteria. The nutrients carried by river are usually the main resources for aquatic (marine or lacustrine) ecosystems. These functions of rivers make river basins into coupling means between hydrologic and biogeochemical cycles and between land and ocean ecosystems.

As gathering grounds, river basins play an important role in the ocean-air-land-water cycle. Perhaps the most important resources that river basins provide for human beings are fresh drinking water. When wisely managed, river basins can also provide **access** to food, hydropower, building materials, medicines, and recreational opportunities. By acting as natural "filters" and "sponges," well-managed basins play a vital role in water purification, water retention, and regulation of flood peaks. In many parts of the world, seasonal flooding remains the key to maintaining fertility for grazing and **agriculture**.

The function of river basins as a provider of resources and regulator of the environment has long been recognized and taken advantage of by human beings. Probably the greatest achievements and foci of river basin management during the past two centuries have been engineering works for drinking water supply, irrigation farming, hydroelectricity, sanitation systems, and flood-reduction purposes. This tightly focused engineering approach to river basin management, however, has put river basins under threat. Construction of dams and dykes, excessive pumping of groundwater aquifers, long-distance transfers of water from one basin to another, and drainage of **wetlands** disrupt natural hydrologic, biogeochemical, and ecological systems, resulting in mounting conflicts among both human communities and sectors, and a degradation of overall human welfare in the medium to long term.

The present situation of environment and development calls for a more holistic and integrated river basin management. A new **perception** of river basins, namely, river basins as not only water drainage systems but also part of the whole land surface of the globe, is needed with the new approach. Thus, the integrated river basin management should coordinate **conservation**; management; and development of water, land, ecosystem, and related resources across both sectors and human communities.

See also: **Coastal Zone; Land-Ocean Interactions in the Coastal Zone (LOICZ) Project; Siltation; Water-Land Interlinkages.**

Further Reading

Calder, Ian R., *The Blue Revolution: Land Use and Integrated Water Resources Management*, London: Earthscan, 1999; Calder, Ian R., "Hydrologic Effects of Land-Use Change," in *Handbook of Hydrology*, ed. David R. Maidment, New York: McGraw-Hill, 1993, pp. 13.1–13.50; Chorley, Richard J., and Barbara A. Kennedy, *Physical Geography: A Systems Approach*, London: Prentice-Hall, 1971; Global Water Partnership, *Integrated Water*

Resources Management (Global Water Partnership Technical Advisory Committee Background Papers no. 4), Stockholm, SW: Global Water Partnership, 2000.

XIUBIN LI

Russia.   The largest country in the world, comprising 13.1 percent of the earth's land surface. Russia has about one-quarter of the territory of 1.7 billion hectares (of which 1.63 billion hectares are covered by soil and vegetation) in **Europe**, and the rest (behind the Ural Mountains) forms Asian Russia. Russian land is a natural resource of global importance both environmentally and economically. Globally, this land contains the largest deposits of natural gas, diamonds, and wood, and it occupies the second place in oil, coal, and other important natural resources. Russia has 23 percent of the world's **forests** and more than half of the world's bogs (**peatlands** cover about one-fifth of the country's area). The largest wild area of the planet comprises more than two-thirds of the country's land: about 26 percent of the last frontier forests are in Russia. Russian land plays an important role in global biogeochemical cycles: about 350 Pg organic carbon (C) (about 17 percent of the global) are accumulated in the top one-meter layer of Russian soils and bogs, and 40 Pg carbon are stored in living **biomass** (or 2.45 kg C m^{-2} for land covered by **vegetation**). On average, net primary production is estimated to be 4.36 Pg C yr^{-1} or 267 g C $m^{-2} yr^{-1}$, with a clear zonal gradient from arctic desert and tundra (120 g C $m^{-2} yr^{-1}$) to temperate forests and steppe (530 g C $m^{-2} yr^{-1}$). During the recent decades, Russian terrestrial ecosystems (mostly forests and **wetlands**) served as a net carbon sink of 400–700 Tg C yr^{-1}.

The major part of the country falls under a temperate to cold climate, and the zone of cold coniferous forests comprises 70 percent of total land. More than 60 percent of Russian territory is covered by permafrost; **mountains**, desert, and semi-desert regions together occupy almost half of the country's land, and arable lands cover only 7.3 percent. Agricultural lands, particularly in densely populated regions, show negative anthropogenic impacts up to a critical state and cannot be restored in a natural way, that is, only with special ameliorative activities. The area of critical ecological situations in Russia exceeds 250 million hectares or 15 percent of the country's territory. The current condition has been formed historically, but mostly it is a heritage of the seventy-year Soviet period, substantially aggravated by insufficient policies during the last fifteen years of the transition to a market economy.

The most dramatic climate change over the globe is expected to occur in the high latitudes of Russia. In spite of probably increasing productivity of terrestrial biota, negative consequences of climate change (in particular, thawing permafrost and accelerating regimes of natural disturbances) are likely to be more tangible.

Current Land Use and Land Cover

The two-dimensional official Russian land-use and land-cover classification includes seven land-use categories (distribution of land by destination and administration). In millions of hectares in 2002, they are the following: **agriculture**, 400.7; settlements, 18.9; industry, transport, etc., 17.2; specially protected areas, 34.2; forest fund, 1,103.1; water fund, 27.8; and unused land reserves, 107.9; each of the above categories includes different land-cover classes.

Agricultural land (i.e., territories, which are systematically used for production of agricultural goods) comprises 220.9 million hectares, including 123.5 million hectares of arable land. Russia has 0.85 hectares of arable land per capita (less than in **Canada** and Australia) and is seventh in the world in the amount of arable land weighted by potential biological production. Due to estimates by the Russian State Committee on Land Policy, the condition of agricultural lands is not satisfactory. The processes of active degradation of soil cover (increasing **soil erosion**, negative transformation of soil water regime, desertification, contamination, paludification, dehumification and decline of humus horizon, secondary carbonization, losses of nutrients, etc.) are typical for land of almost all densely populated regions. The area of the agricultural and arable lands under erosion threat comprises 64 and 71 percent, respectively. More than 50 million hectares of agricultural land (including 35 million hectares of arable land) suffer from water and wind erosion, basically in regions of steppe and semi-desert zones. It leads to annual decreasing of humus content on arable land at 0.7 t ha^{-1}. About 45 percent of arable land has an insufficient amount of humus, phosphorous (23 percent), and potassium (9 percent). High deficit and negative balance of nutrients are typical for agricultural land as a whole [e.g., the balance of NPK in 1997 comprised -97 kilograms per hectare $(+21.5 - 118.5)$]. **Desertification** is observed in twenty-eight administrative regions out of a total of eighty-two. The first desert in Europe has arisen and is dramatically increasing in Kalmikia and along the lower reaches of the Volga River. Paludification affects 30–50 percent of agricultural lands in the humid zone of European Russia. Areas of salty soil are increasing in many regions, in particular on irrigated land. Many regions suffer from contamination of soil by heavy metals, sulfur, radionuclides, and other industrial impacts. For instance, about 40,000 different accidents with spills of oil took place in western Siberia during only three years (1995–1997). Average production of agriculture is low and measured by 2000 (t ha^{-1}): grain crops, 1.5; sugarbeets, 18.5; sunflower, 0.83; potatoes, 9.7; vegetables, 14.9; etc. Agricultural production decreased by 10–30 percent during the last 10–15 years. The current **land reform** (started in 1991), which aims at the dismantling of the Soviet system of state properties, still did not present substantial results. About 265,000 agricultural farms on 14.4 million hectares (or 0.8 percent of land area of the country) accounted for 2.2 percent of the total agricultural production in 2002. In contrast, subsidiary plots (about 20 million plots on 6.2 million hectares, with an average size of 0.38 hectares) generate 60 percent of all agricultural production, namely 92 percent of potatoes, 77 percent of vegetables, and 59 percent of meat. In 1999, the productivity of each hectare of cultivated land on private plots was 23.5 times that of agricultural enterprises and thirty times that of farms.

Forestland including closed forests and temporarily unforested areas, clearcuts, and burnt areas makes up 51.6 percent of the country's land.

According to the State Forest Account in 2003, closed forests cover 776.1 million hectares or 45.4 percent of the total land area. Close to one-fifth of the forests are classified as protected forests, 7.6 percent are destined mostly for ecological services with prohibited or restricted industrial logging, and 69.4 percent are destined for timber extraction. Due to large areas of protected and low-productive forests, only 45 percent of Russian forests are industrial logging zones. The absolute majority (about 95 percent) of Russian forests are in the **boreal zone**, especially coniferous species (70 percent) dominated by larch, pine, spruce, fir, and stone pine (Russian cedar). Soft-leaved deciduous forests cover 17 percent, mostly secondary birch and aspen. Relatively small areas (2.5 percent) are covered by valuable hardwood deciduous trees. The rest (about 10 percent) mostly presents dwarf pine and other shrubs, in regions where tall trees are not able to grow due to severe climatic conditions. The growing stock of Russian forests (i.e., the total amount of stem wood of all living trees) was 88.3 billion cubic meters in 2003.

Large areas of treeless bogs (154.2 million hectares) are mostly found in peatlands and on peat soils. Surface water (71.7 million hectares) includes 120,000 rivers with a total length of 2.3 million kilometers, and about 2 million freshwater and salty lakes. Special protected areas, corresponding to categories I–II of the World Conservation Union (IUCN) and totaling 34 million hectares, include ninety-five state natural reserves; thirty-five national and natural parks; eleven federal preserves; and other territories that have special environmental, scientific, cultural, esthetic, and sanitation functions. Roads cover 7.9 million hectares, built-up areas and land under construction is 5.5 million hectares, and 1.1 million hectares fall under land that has been destroyed by industrial activity.

Legislatively, about 800 million hectares are indicated as resident regions for indigenous people of the north. Of this area, northern deer **pastures** cover about 300 million hectares. Previous and current policies led to the development of degraded and contaminated large territories there. In all northern pastures, 65 percent are degraded to medium to strong degrees. The amount of fodder from lichen pastures decreased during the last fifty years more than twofold. With more than 30 million hectares, the European north is the most heavily polluted area. Increasing heat contamination on permafrost, and processes of thermocarst and solifluction are widely distributed. Altogether, more than 3 million hectares in central European Russia and in the central Ural Mountains are contaminated by radionuclides due to the Chernobyl nuclear accident in 1986.

Land-Use History

During the last 2,000 years, major features and current conditions of land cover in Russia were formed by gradually increasing and regionally specific anthropogenic impacts resulting in decreasing forest cover. In central and southern regions of European Russia, a special system of agricultural land use (under which forest land was cleared and used for crops for three to ten years, following a recovery of the land for the following forty to eighty years during which **secondary vegetation** encroached upon these areas) led to increasing fragmentation of landscapes and eventually formed large open territories. In the north, human-induced burning of vast areas, aiming at development and improvement of northern deer pastures, also caused a substantial decrease of forest cover. This practice resulted in the 150–200 kilometer southward shift of the treeline and likely formed the current forest tundra zone. By 1700, forest covered about 50 percent of European Russia. Land cover of Asian Russia was regulated by rather stable regimes of natural disturbances, and forest likely covered 65–70 percent of the land there. The documented history of land-use/cover changes in European Russia started in the 1700s and was put on a solid legislative basis after the *Manifesto on General Land-surveying* (1765). During some 200 years (1696–1888), forested areas decreased significantly for all major parts of European Russia. Agricultural areas increased by 127 million hectares (from 7 percent to 27.3 percent). During the first half of the nineteenth century, 150,000–200,000 hectares of forests were cleared annually; during 1860–1890, this rose to about 900,000 hectares per year, while only fewer than 30,000 hectares were planted annually. From 1700 to 1914 (the year of minimal forest cover in European Russia during the last 300 years), the forest cover decreased from 52.7 percent to 35.2 percent. About 67 million hectares of closed forests were converted to other land-use categories during 1700–1910. The reform of 1861, which ended serfdom in Russia, had a pronounced effect on land use. Peasants received 40 million hectares of lands. Arable land increased in 1860–1887 from 88.8 million to 117.2 *dessiatinas* (=2.7 acres), and the average yield increased from twenty-nine *pods* (=16.38 kilograms) in 1861–1870 to thirty-nine *pods* in 1890–1891. **Colonization** of western Siberia and the Far East, which started in the seventeenth century, substantially intensified after the construction of the Trans-Siberian Railroad (1980s) and in particular after Piotr Stolypin's reform (1906–1917). The latter aimed at developing a wide layer of independent agricultural producers, and substantially triggered the migration of population to new lands further east. In 1917, Russia produced 20 percent of the world's grain (the Soviet Union produced only 12 percent in the 1990s). The October Revolution (1917) outlawed **private property**, which then became state property. The history of **agriculture** in the Soviet Union was heterogeneous—from "military communism" (1918–1921) through an era of market-oriented "New Economic Policy" (1922–1927) to collectivization (1927–1937; 99.4 percent of all cultivated land was in *kolkhozes* by 1937). In the following period (up to the 1990s), the Soviet Union generated a rather mechanized collective agricultural production; but simultaneously it was a period of discrimination against the rural population, and humiliation and annihilation of the peasants. In spite of many attempts to introduce sustainable agriculture (e.g., Stalin's so-called plan of transformation of nature, 1948–1953, and a large-scale developing of virgin lands in the 1960s), the Soviet period did not solve the issue of national **food security** and left agricultural lands in an unsatisfactory condition.

Reliable and detailed information on the dynamics of land-use/cover change in Russia is available for the period since 1961, when the first forest inventory was done for the entire country. In spite of the wide occurrence of natural and human-induced disturbances and about 17 billion cubic meters of wood that was harvested in Russia from 1950 to 2000, the area of closed forests increased to 70.5 million hectares (10 percent to the area of 1961) during 1961–2003, basically due to the high natural regeneration capacity of Russian boreal forests, suppression of wild fire (the area of burnt forests decreased from 57.6 million to about 30 million hectares), and **reforestation** activities (in 2003, 18.6 million hectares of planted forests were accounted for). During 1960–1990, annual timber harvest increased from 350 to 450 million cubic meters. The intrinsic features of the Soviet forest industry were the use of unsustainable logging technologies and machinery, and substantial losses of harvested wood. The harvest was provided in the most productive, valuable species and accessible forests. It resulted in the substantial decrease of forest quality. During the last four decades, the area of secondary birch and aspen forests increased by 13 percent, and the area of mature coniferous forests decreased by about one-third.

The severe economic and social decline in territories of the former Soviet Union has dramatically impacted land-use/cover dynamics during the last fifteen years. In agriculture, it caused increasing **land abandonment**, that is, encroachment of trees, shrubs, and **grasslands** (more than 35 million hectares from 1988 to 2003, or an estimated 25 percent of all land under cultivation). In many regions, abandoned land is covered by weeds and becomes a source of crop diseases and pests. In the summer of 1999, thirty Russian regions suffered from locust invasions for the first time during the last fifty years. As a new tendency, the human formation of "narcotic landscapes" is observed in some regions of Asian Russia on degraded and abandoned agricultural lands. Application of mineral fertilizers decreased from eighty-eight kilograms per hectare in 1990 to fifteen kilograms per hectare in 1999 (about thirty-seven times less than in the Netherlands), and organic fertilizers from 3.5 t to 0.9 t ha^{-1}. Any systems measures destined toward improvement of the quality of agricultural lands have not been implemented during the last fifteen years. **Land quality** has been decreasing: for example, during 1990 to 1996, the area of acid soils increased by 14 percent, and the area of soil with low humus content by 5 percent. The forest sector is on a substantial decline. During the last decade, the annual amount of harvested wood covered 120–150 million cubic meters, while the annual allowable cut, that is, the many-year norm of sustainable harvest, is estimated to be 500 million cubic meters. However, harvest mostly decreased in remote regions, and harvested areas are concentrated in regions with developed infrastructure. The amount of harvested wood increased in many dense populated areas. High prices for fuel have caused increased tree harvesting in forest deficit regions. According to estimates of non-governmental organizations, illegal logging is 20–30 percent of the officially reported numbers. The last decade was the warmest in the Russian territories for all the documented history of climate records in Russia. It caused an unprecedented increase in natural disturbances: satellites estimated the average annual area of vegetation **fires** in Russia to be about 10 million hectares in 1997–2003, with half to two-thirds of the fires having occurred on forestland. Outbreaks of dangerous insects were reported on about 10 million hectares in 2001–2002, and a substantial part of these outbreaks happened to be in the northern part of the country, where outbreaks have never been observed in the past.

Climate change might dramatically impact the condition of ecosystems and land cover throughout Russia. Over the twenty-first century thus far, the predicted increase of mean annual temperature varies from +4 to +9°C for different models, scenarios, and regions of Russia. The relative increase of precipitation will likely be smaller than that of warming. It will support desiccation of landscapes in many vast regions of Asian Russia, and will likely provoke a dramatic acceleration of natural disturbances (fire and insects' outbreaks). Melting of permafrost will cause the physical destruction of northern landscapes due to thermokarst and solifluction, thus threatening built-up structures. A substantial increase of anthropogenic pressure on fragile ecosystems at high latitudes can be expected due to dilative processes of exploration and extraction of natural resources. Taken together, all these influences might negatively impact the hydrological regime of vast territories, leading to steppization of landscapes and "green desertification" (i.e., the long-term replacement of forests by grasslands and wetlands). In order to provide the conditions for a transition to **sustainable land use**, Russia needs to develop and implement an anticipatory strategy of adaptation of lands to, and mitigation of the negative consequences of, global change.

See also: **Carbon Cycle; Conservation; Extensification; Hot Spots of Land-Cover Change; Industrial Revolution; Land Rehabilitation; Mineral Extraction; Nutrient Cycle; Salinization; Taiga; Transhumance.**

Further Reading

Kraninka, Olga N., Sun, Guoqing, Shugart, Herman H., Kharuk, Vyacheslav, Kasischke, Eric, Bergen, Kathleen M., Masek, Jeffrey G., Cohen, Warren B., and Doug R. Oetter, "Northern Eurasia: Remote Sensing of Boreal Forest in Selected Regions," in *Land Change Science: Observing, Monitoring and Understanding Trajectories of Change on the Earth's Surface* (Remote Sensing and Digital Image Processing Series 6), eds. Gutman, Garik, Janetos, Anthony C., Justice, Chris O., Moran, Emilio F., Mustard, John F., Rindfuss, Ronald R., Skole, David, Turner, Billie L. II, and Mark A. Cochrane, Dordrecht, NL, Boston, London: Kluwer Academic Publishers, 2004, pp. 123–38; Milanova, Elena V., and Maria Sennikova, Land Use and Land Cover Changes in Russia [Online, March 2005], The Encyclopedia of Life Support Systems Web Site, www.eolss.net; Shvidenko, Anatoly, and Sten Nilsson, "Synthesis of the Impact of Russian Forests on the Global Carbon Budget for 1961–1998," *Tellus* 55B (2003): 391–415.

ANATOLY SHVIDENKO

S

Sahel Land Cover. A characterization of land-use and land-cover trends for ecological regions in the Sahel of West Africa, carried out by the **Earth Resources Observation Systems (EROS) Data Center** of the **United States Geological Survey (USGS) program**, working in close collaboration with the AGRHYMET Regional Center in Niamey, Niger. SaHEL land cover is about mapping and quantifying changes in natural resources through land use and **land-cover changes** in nine countries: Burkina Faso, Cape Verde, Chad, the Gambia, Guinea-Bissau, Mali, Mauritania, Niger, and Senegal. The objective is to characterize land-change trends using historical and current satellite imagery and supporting ground information from three points in time (1965, using declassified CORONA photography; 1985; and 2000 using **Landsat** Thematic Mapper data), and to better understand the socioeconomic and biophysical **driving forces** of regional land-use and land-cover change. For this reason, the USGS-AGRHYMET partnership includes the Institut du Sahel in Bamako, Mali, to include issues related to socioeconomic trends and policy research as they relate to understanding the driving factors of **environmental change**.

See also: Aerial Photography; BIOME 300 Project; History Database of the Global Environment; Land-Use History; Proximate Causes; Remote Sensing; Sudan-Sahel.

Further Reading

United States Geological Survey, Science for a Changing World [Online, March 2005], The USGS Web Site, www.usgs.gov.

THOMAS R. LOVELAND

Sahelian Land Use (SALU) Model. A dynamic simulation model designed to represent regional-scale land-use change processes in the African Sahel region over a time scale of around thirty years backward and forward. Identified from an in-depth literature review of local case studies from the **Sudan-Sahel**, the dominant **driving forces** and processes of land-use change in the region are represented by a combination of simple equilibrium equations and knowledge rules. The model assumes that three land uses are generating the basic resources for the population: natural **vegetation** areas from which fuelwood is extracted, cropland that produces food for subsistence and market needs, and pastoral land that feeds livestock.

In this model, the **exogenous factors** or variables (rural and urban human population, livestock population, rainfall, and cereals imports) drive yearly changes in land-use allocation. For any given year, the land-use demand is calculated under the assumption of equilibrium between the production and

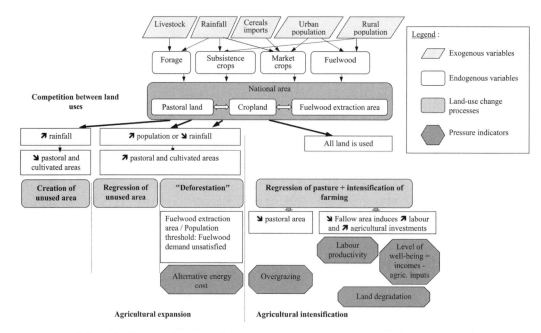

Structure of the Sahelian Land Use Model.

consumption of resources. Thus, the supply of food and energy resources derived from the areas allocated to the different land uses must satisfy the demand for these resources by the human and animal populations, given the exploitation technologies used at a given time. The model simulates two processes of land-use change: agricultural expansion at the most extensive technological level, followed by **agricultural intensification**. Driven by changes in human and animal populations, agricultural expansion leads to **deforestation** or to a regression of pastoral land. Once the expansion of cropland and pastoral land has occupied all unused land, and when fuelwood extraction areas have reached their minimal area, additional demand for **food crops** results mainly in agricultural intensification. This mostly takes place as a shortening of the **fallow** cycle, compensated by the use of labor and agricultural inputs to maintain soil fertility. The model also generates several endogenous pressure indicators, which are symptoms of land-use change and can be used to identify sustainability **thresholds** and dynamic **feedbacks**.

See also: Cash Crops; Endogenous Factor; Extensification; Modeling; Pasture; Subsistence Agriculture.

Further Reading

Stephenne, Nathalie, and Éric F. Lambin, "A Dynamic Simulation Model of Land-Use Changes in Sudano-Sahelian Countries of Africa (SALU)," *Agriculture, Ecosystems & Environment* 85 (2001): 145–61.

NATHALIE STEPHENNE

Salinization. The accumulation of water-soluble salts such as sodium, calcium, magnesium, potassium, sulfate, and chloride within the soil profile.

Salinization occurs naturally in areas characterized by a net upward flux of soil water, that is, mostly in semiarid and arid regions. Land-use/cover change can enhance or cause salinization of soils. However, the major cause of human-induced salinization is improper management of irrigated soils, sometimes called secondary salinization or irrigation salinization. This is considered a major contributor to **soil degradation** in some parts of the world. A related problem is alkalization.

Being a product of chemical weathering, salts are abundant in soils, surface runoff, and ground water. Under humid climate conditions, salts are washed out of soils during the initial phase of soil formation and carried off by streams to the oceans. Therefore, saline soils providing ecological niches are found in humid areas only under very specific conditions, that is, close to the oceans, in areas with saline marine sediments or where saline groundwater reaches the surface. Under semiarid to arid conditions, saline soils occur naturally in low-lying areas and depressions receiving additional water and solutes from the catchment area. As the water evaporates, the solutes are left behind and accumulate. Higher grounds are usually not affected, because salts are either not mobilized or drained. The rise in salt concentrations in the soil decreases the osmotic potential of the soil and diminishes the plants' ability to take up water. This results in drought damages to plants long before salt concentrations reach a toxic level.

Salinization is an immanent climate-induced hazard for irrigation farming in semiarid to arid environments, irrespective of the political or economic system. Its major cause is improper design and management of irrigation schemes (irrigation salinization). This includes the excessive application of water due to inefficient water distribution systems that eventually cause the groundwater table to rise within the reach of capillary rise, the use of brackish or saline surface or groundwater, and the neglect of proper drainage systems. Another cause of salinization is the removal of **vegetation** in drylands, again causing a rise of groundwater and enhancing evaporation and accumulation of salts in the upper soil horizons.

Although flushing of soils is a potential method of regenerating saline soils, it is considered economically unfeasible and ecologically doubtful. Therefore, salinization has to be considered an irreversible process of soil degradation. Drawing together different estimates from the 1990s, the Global Environment Outlook (GEO 3 report) states that over 30 percent of all irrigated land (2.55×10^6 km^2) is affected by, and about 10 percent severely degraded from, salinization, leading to the abandonment of nearly 4 percent of the total irrigated area annually. As the total irrigated area continues to rise, this implicates a strong pressure on other land resources.

See also: **Agricultural Intensification; Coastal Zone; Cotton; Desertification; Environmental Change; Land Rehabilitation; Watershed; Water-Land Interlinkages; Wetlands.**

Further Reading

Umali, Dina L., *Irrigation Induced Salinity: A Growing Problem for Development and the Environment* (World Bank Technical Paper 215), Washington, DC: The World Bank, 1993; United Nations Environment Programme [Online, March 2005), http://www.unep.org/geo/geo3/], *Global Environment Outlook 3: Past, Present and Future Perspectives,*

London: Earthscan, 2002; United States Salinity Laboratory [Online, March 2005], The USSL Web Site, www.ussl.ars.usda.gov.

JUSSI BAADE

Sauer, Carl Ortwin (1889–1975).   American geographer and professor at the University of California–Berkeley (where he built a distinguished graduate school) who is widely credited for reinvigorating cultural and historical geographies and unifying physical and human geographies in his time. Seeing geography as inseparable from human history inasmuch as the earth is profoundly affected by humanity, he creatively worked the concepts of anthropology, archeology, and sociology into geography through studies of the human occupation of a specific region over intervals of historic time. Enduring influence of his cultural landscape studies, as manifest in such notions as "cultural agency," "cultural hearth," and "historically accumulative effects" remains a countercurrent to the prevalent view of seeing "clocklike" processes of land-use/cover change amenable to computer **modeling**.

See also: **Domestication.**

Further Reading

University of Colorado–Boulder, Department of Geography, Geographers in the Web [Online, December 2004], The Carl Ortwin Sauer Web Site, www.colorado.edu/geography/giw/sauer-co/sauer-co.html.

KE CHEN

Savannization.   The process whereby a particular land cover is converted to savanna, either naturally or as a result of human activity. The word savanna is used to describe a landscape with a continuous herbaceous layer, composed of grasses and forbs, with scattered woody vegetation that includes trees and shrubs. However, the exact composition of the herbaceous layer and the degree of woody vegetation cover varies across savannas. Although both natural and anthropogenic savannas exist today, the origin of specific savannas is difficult to pinpoint in many cases. The process of savannization has occurred throughout the world, converting forested areas or areas along a forest-savanna into savanna. This process has been highly debated in many areas where savannas may be of natural origin but are sustained by human activities, including **deforestation** and regular burning of the landscape. Savannization also refers to a purposive management strategy in arid and semiarid areas.

Ecosystems across the world have exhibited change in **vegetation** composition, configuration, and health due to multiple forcing factors, many of which are human in origin, whether directly (e.g., deforestation, **agricultural extensification**) or indirectly (e.g., via increased competition for water, impacts on local and regional climate). Savannas occupy 20 percent of the earth's surface and show marked change in vegetation composition, function, and **biodiversity**. Savannas are typically defined by two components: a continuous grassy or herbaceous layer and a discontinuous woody component. Many of the differences in traditional and newer models of tree/grass ratios in savannas have been researched of late, which is important, as tree density is often used as a surrogate for determining whether an area is **grassland**, savanna, or woodland. The transition of a given area into one of these land-cover types has been hypothesized to be potentially controlled by physical factors (e.g., temperature, precipitation, soil texture), biologically meaningful factors (such as plant-available moisture and available nutrients), and disturbance factors (fire, flood, trampling, and herbivory/grazing). These ecosystems and their ecological fluctuation or stability are considered important because of their high levels of biodiversity (faunal as well as floral) and socioeconomic benefits. Globally, savannas and grasslands contribute approximately one-third of net primary productivity.

Savannas typically have a climate consisting of alternating wet and dry seasons. The prominent dry season results in dormant grasses that become susceptible to **fire** in contrast to a wet season during the year. These areas include tropical Africa, Asia, Australia, and both North and **South America**.

Savannization has been offered in some cases as a tool for combating **desertification**; in this case savannization occurs by turning deserts into savanna by dryland-adapted afforestation of rangelands. This process helps to control erosion by promoting greater ground cover and changing water pathways, whether the water source is irrigation or rainfall, and has been successfully implemented in Israel. Soil **conservation** and concomitant water protection result in sustainability of greater vegetative land cover that is also of higher forage quality. Grazing, then, is more productive and allows for fire suppression through controlled grazing. It is hypothesized that impacting the water budget through changing water storage and overall evapotranspiration promote a positive **feedback** between available moisture and subsequent microclimate conditions, though these relationships have not been exhaustively modeled to date. Thus, savannization has two differing connotations: as a management strategy for fighting desertification (converting desert to savanna) and as a seminatural process involving either woody encroachment into grasslands or woody mortality in woodlands.

Desavannization, the process of converting savanna into another ecosystem type, is also occurring around the world. Both North American and African savannas face impending woody encroachment, whereby the woody species density eventually exceeds the level associated with savannas and moves into that of woodland. Desavannization may also occur when woody species are systematically eliminated, leaving primarily grasses and forbs, turning the savanna into a grassland.

Most savannas operate in a climatic context (precipitation and temperature) whereby very small changes can trigger wholesale **conversion** into or out of woodlands and grasslands. These shifts, documented in both North America and Africa, may also occur in response to disturbance regimes. The two primary disturbance regimes affecting savannization and desavannization are fire and grazing. Fire and/or grazing convert woodlands to savanna, and can also convert savannas to grasslands if frequent or severe enough. Fire suppression and grazing limitations have the opposite effect, allowing grasslands to segue into savanna and savanna to convert to woodland. **Land-use policies** concerning fencelines and grazing management or fire suppression therefore have immediate, direct, and substantial impacts on savannas and their rates of areal growth or decline.

While fire and grazing have negative and positive short-term effects on aboveground productivity, composition, and diversity, their independent and interactive effects on woody plants are fundamental to production and maintenance of savannas and grasslands. However, the effects of season of disturbance have rarely been examined and as such are especially inviting for analysis via an intra-annually rich satellite time series. Similar to fire, grazing effects on herbaceous plant species composition can be dependent on season. However, depending on grazing intensity, this effect can be less than that caused by fire, selectivity of grazers, or climate.

A third disturbance regime, flooding, has also been found to impact savannas in Africa and South America. For example, in the Okavango Delta of Botswana, floodwaters from the Angolan highlands each year spread through the low-energy river system into nearby grasslands, savannas, and woodlands.

As flooding extent increases over several years, savanna can be lost to grasslands or gained in woodlands due to woody species mortality. When flooding extent is lower than normal for several consecutive years, savannization occurs in the grassland areas while desavannization occurs in the more distal woodlands.

Studies of seasonal effects of disturbance, particularly fire, are fraught with difficulties. The causes of post-disturbance response of a plant community may be due to direct (interruption or impact on phenological phase) or indirect (**modification** of conditions in post-disturbance environment) mechanisms and are difficult to identify and therefore model. A coarse prediction of post-disturbance compositional outcomes may be achieved by allocating plant species to guilds, also known as functional groups. However, this method may be flawed due to both the variation within guild and the variation of species phenology year to year and place to place. Furthermore, any direct or indirect effects of disturbance may be obscured by both the variation in response time of some components of the vegetation and the possible overriding effect of climate on this ecosystem. This suggests that any examination of plant community responses to ecological disturbance must have high replication across time and space to extract local geographic and short- and long-term effects, again suggesting a gap that an earth **remote sensing** approach could potentially help fill.

The vast majority of the literature on savannization and desavannization tends to treat those processes as either wholesale sweeping change or species substitution. Little of the work considers the spatial nature of the conversion into or out of savanna ecosystems. That is, the spatial process by which savannas morph has not been sufficiently studied. For example, few studies of savannization employ **pattern metrics** to better describe the configuration of savanna landscape elements. When savannization occurs, does it move along a frontier, advancing into its new territory at the shared ecotone or edge? Or does savannization occur more through percolation and subsequent patchy **colonization** of woody species, fragmenting the former ecosystem until it reaches a new background or dominant land-cover type of savanna? The answers lie in comparing the way savanna species (particularly woody species) propagate into new areas, and the global inventory of savanna woody species is vast.

There is some controversy as to whether defining savannas (as separate from woodlands or grasslands) can be performed solely on the basis of woody species density (whether stem count or canopy closure). Furthermore, disentangling anthropogenic effects from natural effects has proven difficult. For example, lightning-induced fires may be natural in origin but start fires that burn at higher temperatures with greater fuel loads because of human fire suppression. However, recent work comparing savannas from Australia, Africa, and North America suggests that more than many other ecosystem types, savannas may in fact be incredibly similar across widely varying conditions. The implication for global savanna studies is that the processes at work on differing savanna species around the world may in fact function very similarly, making research in savannization applicable to savanna landscapes worldwide. As the threat of global warming and population-driven fire frequency increase, even portions of **Amazonia** have been identified as likely targets for natural savannization in the next century, adding to the increase in the world's savanna ecosystems.

See also: **Degradation Narrative.**

Further Reading

Archer, Steve, Boutton, Thomas W., and Kathy A. Hibbard, "Trees in Grasslands: Biogeochemical Consequences of Woody Plant Expansion," in *Global Biogeochemical Cycles in the Climate System,* eds. Ernst-Detlef Schulze, Martin Heimann, Sandy Harrison, Elisabeth Holland, Jonathan Lloyd, Ian C. Prentice, and David Schimel, San Diego, CA: Elsevier Academic Press, 2001; Belsky, A. Joy, "Tree/Grass Ratios in East African Savannas: A Comparison of Existing Models," *Journal of Biogeography* 17 (1990): 483–9; Bourlière, François, ed., *Ecosystems of the World 13: Tropical Savannas,* New York: Elsevier, 1983; Scholes, Robert J., and Steve R. Archer, "Tree-Grass Interactions in Savannas," *Annual Review of Ecology and Systematics* 28 (1997): 517–44.

KELLEY A. CREWS-MEYER, AMY L. NORMAN

Scale. Scale is the spatial, temporal, quantitative, or analytic dimension used to measure and study objects and processes. Two important components of scale are extent and resolution. Extent is the magnitude of a dimension used in measuring, whereas resolution is the precision used in this measurement. For example, if a map consists of square **pixels**, then the amount of land portrayed by the collection of pixels determines the extent, and the size of the individual pixels determines the map's resolution. In this case, the length of the side of the pixel is the measurement of the map's resolution, which is also known as the grain.

Categorical scale is the level of detail of the categories in an analysis. For example, in classification of land cover, a hierarchy may be defined in which many detailed categories can be aggregated into fewer coarser categories, which can be aggregated further into even fewer, more general categories.

The word scale frequently causes confusion because it has many meanings. The most common source of confusion derives from the fact that some scientists use simple descriptors such as "small" and "large," while they simultaneously fail to specify which version of the definition of scale they are using. If a scientist uses the word scale to mean "extent," then "small scale" means a small study area. If a scientist uses scale to mean the ratio between the map and the ground (e.g., 1/1,000), then "small scale" means a large study area for a given size of paper map. If a scientist uses the word scale to mean resolution or categorical detail, then the size of the scale is independent of extent. There would be less confusion if scientists used the adjective "fine" to describe the small grain of highly detailed resolutions and "coarse" to refer to large grains of less-detailed resolutions. The adjectives "small" and "large" should be reserved to describe extent.

The operational scale is the scale at which a certain process (e.g., a land use decision) manifests on the landscape. Land use and cover changes are the results of many interacting processes. Each of these processes operates over a range of scales in space and time. Land-use decisions are made at different levels of social organization (e.g., the individual or community level). Level refers to level of organization in a hierarchically organized system and is characterized by its rank ordering in the hierarchical system. Other examples of levels include the organism, ecosystem, landscape, social system, nation, and globe. If the operational scale does not correspond with the scale of the data, then the scientist might not be able to determine the correct processes. **Land-use systems** rarely produce a single scale that can be regarded as correct or optimal for measurement and prediction, because landscapes usually demonstrate many interacting processes. An optimal scale of analysis for a specific data set might exist where predictability is highest; however, that scale will not necessarily be the same for other analyses. Therefore, it is desirable to determine the scale of the analysis from a careful examination of the phenomenon, rather than using *a priori* scales of observation.

The interaction of processes at various scales and the mismatch between operational and data scales are common. These can cause results of empirical analysis of land-use change to be sensitive to the selection of both extent and resolution. In cases where an investigation attempts to use data that are readily available, the extent and resolution of the analysis are sometimes determined by factors that are independent of the phenomenon. For example, if the phenomenon of interest is **deforestation**, then the available data are frequently in the form of satellite images, which are available in pre-determined extents and resolutions that are functions of the path and sophistication of the satellite used to create the image. If the scientist adopts the satellite image as the study area, then both the extent and the resolution are dictated by the format of the available data, not the phenomenon. Scientists are developing tools to measure the degree to which changes in scale influence the analysis of the phenomenon. This challenge is closely related to geography's famous and unsolved "Modifiable Areal Unit Problem," which states that statistical results are sensitive to changes in the areal unit. Numerous examples in land-use change research have indicated that the analysis of the relation between land-use change and its **driving forces** should not be restricted to a single scale. Multiscale and multilevel techniques are better able to unravel the complex interactions and **feedbacks** over scales that determine the functioning of the land-use system. Scale remains one of the major challenges of geographic analysis, and is consistently ranked as a top research priority by the University Consortium on Geographic Information Science.

See also: Auto-Correlation; Environmental Change; Integrated Assessment; Land Cover Classification System; Land-Use History; Mediating Factor; Millennium Ecosystem Assessment; Pattern Metrics; Pixelizing the Social; Remote Sensing; Socializing the Pixel; Validation; Vulnerability.

Further Reading

Gibson, Clarke C., Ostrom, Elinor, and Toy-Kyeong Anh, "The Concept of Scale and the Human Dimensions of Global Change: A Survey," *Ecological Economics* 32 (2000): 217–39; Openshaw, Stan, *The Modifiable Areal Unit Problem*, Norwich, UK: GeoBooks, 1984; Pontius, Robert G., Jr., "Statistical Methods to Partition Effects of Quantity and Location During Comparison of Categorical Maps at Multiple Resolutions," *Photogrammetric Engineering & Remote Sensing* 68 (2002): 1041–9; Verburg, Peter H., and You Qi Chen, "Multi-Scale Characterization of Land-Use Patterns in China," *Ecosystems* 3 (2000): 369–85; Verburg, Peter H., de Groot, Wouter T., and A. (Tom) Veldkamp, "Methodology for Multi-Scale Land-Use Change Modelling: Concepts and Challenges," in *Global Environmental Change and Land Use*, eds. A. J. (Han) Dolman, A. (Jan) Verhagen, and C. A. (Ilse) Rovers, Dordrecht, NL, Boston, London: Kluwer Academic Publishers, 2003, pp. 17–51.

ROBERT GILMORE PONTIUS JR., PETER H. VERBURG

Scenario. A hypothetical, possible future pathway that describes dynamic processes and represents sequences of events over a period of time, including causally related states, **driving forces**, events, consequences, and actions. The word scenario comes from the dramatic arts. In the theater, it is an outline of the plot; for a movie, a scenario sets forth details relevant to the plot. Its formal intellectual roots trace back to the Manhattan Project six decades ago when world-renowned nuclear physicists explored and evaluated the possibility that the energy buildup from a full-scale explosion of the hydrogen bomb might ignite a devastating deuterium reaction in the skies and the oceans. In the 1950s, Herman Kahn and Anthony Weiner at the Rand Corporation used the concept of scenarios in a series of strategic studies for military planning purposes. In the corporate world, scenarios were refined at Royal Dutch/Shell by Pierre Wack in the 1970s and 1980s, and Shell became a leader of the scenario approach to business planning. The word scenario came to the attention of the general public in 1972 with the publishing of *The Limits to Growth* by Dennis Meadows and colleagues. Today, scenario development is used in a variety of different contexts ranging from political decision-making, to business planning, to local community management, and to global environmental understanding.

Garry Peterson and colleagues provide a good first point of entry to define a scenario. They propose to position scenarios along two axes: the degree of uncertainty and the degree to which a system can be controlled. As depicted in the figure, scenarios are useful in systems where uncertainty is high and controllability low. Most definitions of scenarios agree with this assumption. However, the highest level of agreement is on what a scenario is not.

A scenario is not a prediction, understood as the best possible estimate of future developments; neither is a scenario a forecast, the best estimate from a particular method or model. Last, a scenario is also not a projection, as it depends on assumptions about drivers and leads to "what if" statements.

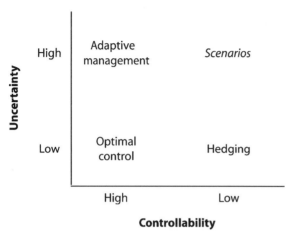

Degree of controllability and uncertainty of the system being studied. Adapted from Petersen et al. (2003).

A first definition of scenarios states that they are hypothetical sequences of events constructed for the purpose of focusing attention on causal processes and decision-points. Alternatively, scenarios can be defined as plausible, challenging, and relevant stories about how the future might unfold that can be told in both words and numbers, as done by the Millennium Ecosystem Assessment.

A large diversity of types of scenarios has been developed, and various attempts to classify them have been undertaken. For example, Philip Van Notten and colleagues propose a classification based on degree of complexity, exploration versus decision support, and degree of formalization. Other useful classification criteria include forecasting versus backcasting, or whether participatory methods are employed. Within the framework of land-use/cover-related research, it is particularly the division between qualitative and quantitative scenarios that is of interest.

Qualitative scenarios are usually in the form of narrative storylines and involve stakeholders during the development. Stakeholders can be involved in two fundamentally different manners: in-depth interviews and group sessions, and workshops. The advantages of interviews are larger sample size, larger reproducibility and replicability, and thus arguably a higher "scientific" value, although the latter can be disputed. Nevertheless, there are important advantages of a workshop setting that outweigh the possible drawbacks. A group setting enables stakeholders to create a common vision; it can be a stimulus for processes of social interaction and learning; it is the ideal setting for development of complex, multidisciplinary scenarios; and it might enable viewpoints that might not have been discovered in individual interviews. Given that participatory processes are at the heart of the development of qualitative scenarios, an important goal is usually the process itself. Stimulating discussion among stakeholders and initiating long-term participation in the **decision-making** process are often as important as the actual results of the exercise. The final output is usually in the form of a storyline, although similar methods are increasingly being employed for qualitative model building, of which the output can be semi-quantitative, for example.

Quantitative scenarios are almost always created to serve as inputs into **modeling**. They therefore rarely serve as products of their own. Of particular interest are area-based scenarios that are linked to spatially explicit models. Key questions are less related to how the future might unfold and more to where these changes will take place. Typical elements of area-based scenarios are the location of national parks, migration, or the construction of new roads. The focus of the model and therefore of the scenarios is on the distribution of land use. The availability of data can potentially limit the scope and richness of quantitative scenarios.

A particularly pressing issue is establishing the link between qualitative outputs from employing participatory methods and quantitative (i.e., data-demanding) models. Many of the recent (global) scenario efforts—such as those of the **Intergovernmental Panel on Climate Change**, the **United Nations Environment Programme**, and the **Millennium Ecosystem Assessment**—developed narrative storylines that were subsequently quantified and linked to models. However, in most documented cases, the link between the stories and models remains rather rudimentary, and both products are developed in parallel but separate exercises. Among the novel methods that have been employed is the use of **agent-based models** that can be directly parameterized by stakeholders.

Another type of scenario that has increasingly been developed in recent years is the multiscale scenario. VISIONS and MedAction are good examples of how multiscale scenarios can be developed using stakeholders and how decision support systems (DSS) can be used to quantify them. The Millennium Ecosystem Assessment has advocated a multiscale approach as well in its conceptual framework that serves as the guidelines for a large number (more than thirty) of sub-global assessments. The ultimate goal is to link a multitude of local studies through a meso-level assessment, which includes scenarios, to four global storylines.

Historically, land-use (change) scenarios are mostly developed to serve as inputs for land-use change models. Therefore, most land-use scenarios are quantitative rather than qualitative, although recently participatory methods are increasingly being employed to develop qualitative scenarios. As most currently available scenarios are linked to a model, they logically follow the framework of that model. The architecture of many land-use models consists of two parts: a large-scale, non-spatial part where total quantities of each land-use type are projected for a number of scenarios; and a small-scale spatially explicit part where those quantities are subsequently allocated. Land-use scenarios therefore often have a place-based component.

Projections of large-scale developments are usually related to processes that are slower and act over large areas like climate change, macro-economic developments, or institutional factors like the expansion of the European Union. Depending on the spatial extent of the study area, these non-spatial aspects can be related to quantitative storylines. In most land-use models, these large-scale developments are a direct input into the allocation part of the model.

The future changes of all dynamic drivers of land-use change that are spatially explicit are part of any land-use scenario. Typical examples are the construction of a new road—linked to accessibility, the institutionalization of a national park, or migration patterns—linked to population density. Often, these place-based assumed changes will heavily influence resulting land-use patterns.

Increasingly, scenarios are being recognized as one of the best tools in participatory approaches, like stakeholder workshops and focus groups.

Discussion on possible futures is an excellent way to avoid conflict, increase awareness, and start a process of mutual learning. It is therefore to be expected that the role of qualitative scenario development will increase in the coming decade. At the same time, DSSs have rapidly become easier to use, also by laypeople. Top-down, bottom-up iterative procedures linking locally developed narratives and DSSs that can be discussed during series of stakeholder workshops could be a great step forward.

See also: Change Detection; Integrated Model to Assess the Global Environment; Land-Use Planning; Participatory Geographical Information System; Socializing the Pixel.

Further Reading

Alcamo, Joseph, Kok, Kasper, Busch, Gerald, Priess, Jörg, Eickhout, Bas, Rounsevell, Mark, Rothman, Dale, and Maik Heistermann, "Searching for the Future of Land: Scenarios from the Local to Global Scale," in *Land Use and Land Cover Change: Land Processes, Global Impacts*, eds. Éric F. Lambin and Helmut Geist, Berlin: Springer, 2006; Kahn, Herman, and Anthony Weiner, *The Year 2000: A Framework for Speculation on the Next Thirty-Three Years*, New York: McMillan, 1967; Meadows, Donella H., Meadows, Dennis L., Randers, Jørgen, and William W. Behrens III, *The Limits to Growth: A Report for the Club of Rome's Project on the Predicament of Mankind*, New York: Universe Books, 1972; Peterson, Garry D., Cumming, Graeme S., and Stephen R. Carpenter, "Scenario Planning: A Tool for Conservation in an Uncertain World," *Conservation Biology* 17 (2003): 358–66; Rotmans, Jan, Van Asselt, Marjolein B. A., Anastasi, Chris, Greeuw, Sandra C. H., Mellors, Joanne, Peters, Simone, Rothman, Dale S., and Nicole Rijkens-Klomp, "Visions for a Sustainable Europe," *Futures* 32 (2000): 809–31; Wack, Pierre, "Scenarios: Uncharted Waters Ahead," *Harvard Business Review* 63 (1985): 72–89; Xiang, Wei-Ning, and Keith C. Clarke, "The Use of Scenarios in Land-Use Planning," *Environment and Planning B: Planning and Design* 30 (2003): 885–909.

KASPER KOK

Secondary Vegetation.   A transition state in which vegetation is in the process of recovering to a more mature state of primary vegetation following natural or anthropogenic disturbances. Much of the world's **vegetation** has been altered by a variety of anthropogenic forces. Natural events such as forest **fire**, landslides, and flooding also cause significant **modification** to vegetated landscapes, after which vegetation begins to recover. Following these types of disturbances, vegetative cover goes through a process of recovery to a more mature condition.

Secondary vegetation differs from primary vegetation in terms of species composition and vegetative structure. An early state of succession usually consists of hardy species that survive well under harsh conditions. These pioneer species can gradually change the ecological conditions that make the environment suitable to other species unable to initially colonize disturbed areas. For example, following forest clear-cutting, the arrival of new pioneer species can provide enough shade to allow the growth of species adapted to low sunlight conditions.

Some literature refers to secondary vegetation as a condition that specifically follows human disturbance while other literature includes both human and natural disturbances as precursors to secondary vegetation. The more inclusive definition has been used here. Specific examples of secondary vegetation include forest **regrowth** following agricultural or **land abandonment**, catastrophic fire events, or clearcutting. In addition, secondary vegetation is a key component of the cycle of **swidden cultivation**, in which secondary growth is a phase in a cycle of agricultural production. Thus, the term secondary vegetation applies even though the system will never actually reach a mature vegetated state.

Landscapes modified by human actions such as timber harvesting, pasturing, and **slash-and-burn agriculture** can revert to a more natural vegetated state after abandonment, but this process takes many years. The intensity and duration of disturbances affect the rate of vegetation recovery. An important part of the growth of secondary vegetation is the presence of a seed bank from which new vegetation growth can develop. Areas that have been under a disturbance regime such as cattle **pasture** for many years may have lost the seed bank necessary for the regrowth of tree species and thus take many decades to return to what can be considered a mature vegetated state.

It should be noted that a primary vegetation state is not a static endpoint target. Even in mature vegetated areas, species composition can change within local areas due to natural events such as tree falls. A large contiguous area of **forest** can be considered in a mature state, but within the larger forest there are local areas with different species compositions and ecological conditions. Thus, the **scale** and context of a vegetated area are important in considering whether the term secondary vegetation is applicable. Forests subject to small-scale disturbances such as natural tree falls or low-intensity selective cutting are generally not considered secondary vegetation.

See also: Amazonia; Biomass; Boreal Zone; Driving Forces; Ecological Colonization; Fallow; Land-Use History; Pristine Myth; Proximate Causes; Reforestation.

Further Reading

Chokkalingam, Unna, and Wil De Jong, "Secondary Forest: A Working Definition and Typology," *International Forestry Review* 3 (2001): 19–26.

THOMAS P. EVANS

Siltation.   The accumulation of fine primary or flocculated particles (clay, silt, sand, and organic matter) in ditches, rivers, lakes, and other water bodies. Siltation is a natural subprocess of sediment transport occurring when flow velocity of water carrying suspended sediment is reduced below a particle or floc size depending on **threshold**. Alluvial plains, **wetlands**, and many landforms in the **coastal zone** result from siltation. Land-cover and land-use changes and alternations of watercourses (e.g., building of locks and dams) have a strong impact on rates of instream and downstream sediment transport and siltation.

A major concern is enhanced siltation of waterways, estuaries, and reservoirs due to **deforestation** and **soil erosion** in the catchment areas of water bodies. This threatens aquatic habitats; reduces water depth and flow capacity in waterways, eventually hindering navigation and increasing the flooding hazard; and diminishes storage capacity of flood control, fresh water, and hydropower reservoirs. The World Commission on Dams estimates that

Aswan High Dam in southern Egypt, March 2001 (before completion of the dam in 1970, 110 million tons of silt was deposited by the annual flood of the Nile, which is now trapped behind the dam).

PHOTO: Space Shuttle. CREDITS: Image Science and Analysis Laboratory, NASA-Johnson Space Center, 18 Mar. 2005 (Earth from Space—Image Information). http://earth.jsc.nasa.gov/sseop.efs

0.5–1 percent of the world's freshwater storage capacity in small and large reservoirs is lost to siltation annually. Land **conservation** measures reducing runoff, soil erosion, and sediment yield to watercourses help to reduce siltation of water bodies in the long term. Short-term measures to mitigate siltation include dredging and flushing of waterways and reservoirs.

See also: **Land Degradation; River Basin; Water-Land Interlinkages; Watershed.**

Further Reading
International Sedimentation Initiative [Online, March 2005], The ISI Web Site, portal.unesco.org; World Commission on Dams, Dams and Development: A New Framework for Decision-Makers [Online, March 2005], The WCD Web Site, www.dams.org/report.

JUSSI BAADE

Slash-and-Burn Agriculture.   A colloquial as well as a pejorative term often referring to the method of clearing and preparing land that is common among swidden cultivators. Slash-and-burn can also refer to the practice of using **fire** simply to clear **forests** for permanent cultivation, **pasture**, or further development. This practice was employed on a massive scale during the European settlement of the frontiers of the eastern **United States of America**.

Some scientists argue that sustainable **swidden cultivation** is being replaced by unbalanced, unsustainable slash-and-burn agriculture because the population of swidden cultivators has expanded so rapidly that their traditional

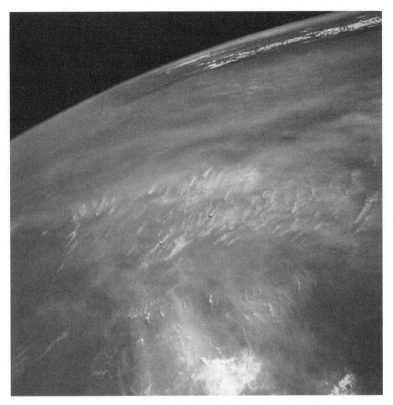

Major Smoke palls due to fires for land clearing by slash-and-burn practices in the humid tropical forest in the state of Rondônia, western Brazil, in September 1984.

PHOTO: Space Shuttle. CREDITS: Image Science and Analysis Laboratory, NASA-Johnson Space Center, 18 Mar. 2005 (Earth from Space—Image Information). http://earth.jsc.nasa.gov/sseop.efs

practices can no longer support their numbers. Under these circumstances, traditional cultivators reduce the length of **fallow**, leading to unsustainable slash-and-burn agriculture.

Norman Myers tries to distinguish between what he calls "shifting" and "shifted" cultivators. The first are traditional peoples who practice **sustainable land use** with a sufficiently long fallow period for soil fertility to regenerate. The second, the shifted cultivators, are small-scale farmers who find themselves squeezed out of traditional farmlands and feel they have no alternative but to pick up machete and matchbox and head for the forests as the last unoccupied lands available. Norman Myers estimate these peoples, who might be called slash-and-burn agriculturalists, as numbering 200–600 million.

Christine Padoch and colleagues point out that the problem with this distinction is that identifying particular forms of swidden as ideal (swidden cultivators) and others as problematic (shifted cultivators or slash-and-burn agriculturalists) is that many site- and situation-specific variations from those considered to be ideal may often be misinterpreted as shortcomings in technical expertise or agricultural performance.

See also: **Agricultural Frontier; Agricultural Intensification; Alternatives to Slash-and-Burn Programme; Colonization; Degradation Narrative; Extensification; Land-Use History; Secondary Vegetation; Tragedy of Enclosure.**

Further Reading

Fox, Jefferson, "How Blaming 'Slash and Burn' Farmers Is Deforesting Mainland Southeast Asia," *Asia Pacific Issues* 47 (2000): 1–9; Myers, Norman, "Tropical Forests: The Main Deforestation Fronts," *Environmental Conservation* 20 (1993): 9–16; Padoch, Christine, Harwell, Emily, and Adi Susanto, "Swidden, Sawah, and In-Between: Agricultural Transformation in Borneo," *Human Ecology* 26 (1998): 3–20.

JEFFERSON FOX

Socializing the Pixel.   To "socialize the pixel" and to "pixelize the social" are to discern information embedded within spatial data on land-use/cover change that is directly relevant to the core themes of the land-change science agenda, by making **remote sensing** (RS) and **geographical information systems** (GIS) relevant to the social, political, and economic issues driving land-use/cover change. While pixel refers to a unit of observation associated with satellite data on land cover, the term is used more broadly here to encompass all spatial analysis on land-use/cover change, even if the data are from other sources. Remote sensing, both data and image processing, and analysis through GIS have played crucial roles in the development of land-use/cover change models in both data creation and analysis.

During the course of the past ten years, significant progress has been made in acquiring spatial land-cover data sets from remotely sensed data, conceptualizing the basic geographic and environmental processes that are associated with land-use change. Numerous spatially explicit and heterogeneous land-use change models now exist, spurred by the vast amount of spatial land-use/cover data that are now available. The research agenda of the **Land-Use/Cover Change (LUCC) project** (and its associated groups) has demanded an interdisciplinary approach to **modeling**, with major contributions made by landscape ecologists, geographers, anthropologists, political scientists, economists, and demographers.

While the goal of both "socializing the pixel" and "**pixelizing the social**" is to better understand the spatial pattern of land-use/cover change, there are broad generalizations of the differences between the two, as envisioned here. To "socialize the pixel" is to develop land-change models that move from the RS imagery beyond its use in the natural sciences and toward the concerns of the social sciences per se. These models usually have as the unit of observation the satellite data **pixel** and include explanatory variables that can be "seen" from the remotely sensed data and calculated using GIS, such as distance measures, other spatial biophysical variables (e.g., soil, slope, elevation); and occasionally socio-economic "drivers," such as population or other sociodemographic characteristics, usually measured at some aggregate level, such as village or state.

To "pixelize the social" is to apply social science theories to spatial data in order to develop an underlying structural model that seeks to explain the human behavior that generates these patterns of land-use/cover change.

These models usually take as the unit of observation the land under the control of an individual decision-maker and directly link the decision-maker to specific pixels. Of course, as research has progressed, and as more "socialization" has been done in the "socialize the pixel" realm and more "pixelization" has been done in the "pixelize the social" area, the boundaries between the two, rightly so, have become blurred.

Landscape ecologists were early developers of spatially explicit models of land-use change used to predict changes in spatial patterns of the landscape. The early models were simple grid-based models that merely calculated the percent change of each land-cover type during a time period and predicted future changes by assuming that these proportionate changes remained constant over time. More sophisticated models were then developed that estimated these changes as a function of other explanatory variables and not just simply a function of previous land-use changes. While many of these models had sophisticated treatment of ecological relationships that affected or were a result of land-use/cover change, they were very simple with respect to human behavior.

Subsequent models developed by geographers and other social scientists included some variables that captured some aspects of human behavior. For example, distance to urban centers, and variables that reflect the biophysical heterogeneity of the landscape have been commonly included for economic reasons. Distance to urban centers matters because the accessibility to markets affects transportation costs of getting agricultural goods to markets; whereas biophysical features of the landscape, for example certain soils being preferred for agricultural use, will affect the choices of land managers. Typically, these models estimate the effects of such explanatory variables as slope, soil, elevation, aspect, location, and population measures on different types of land-use/cover change. The results of these estimated models have often been intuitive. For example, in models that focus on **deforestation**, the effects of the explanatory variables are often as expected: the higher the elevation, the smaller the probability of deforestation; the further a pixel is from the road, the less likelihood of deforestation; the closer a pixel is to a market or a village, the greater the probability of deforestation.

However, there are issues of aggregation and **scale** with linking these diverse data sets. For example, in a single model, the size of the unit of observation, the satellite pixel, will often be of a different scale than the data for the explanatory variables. Slope and elevation data are often derived from a digital elevation model that can be at a different scale, and soils can come from maps at an entirely different scale, while any socio-economic census data will be at some other governmental entity scale, such as state or town, the size of which is often determined by population levels. Therefore, decisions must be made in how to best link these data for modeling purposes. For example, a model that includes socio-economic census data must somehow allocate those values of those variables to the individual pixels, such as assigning all pixels in a particular village the same value for population density, even though, in reality, not all pixels are likely impacted equally by the individuals living in the village. As a result, the estimated effect of such variables will only be the average effect on land-use change for the scale at which the variable is measured.

These estimated models are often subsequently used to simulate the effect of different **scenarios** on land-use change, such as to simulate the effect on future land-use change of a moratorium on logging or road building or the effects of **urbanization**, abolition of national parks, extension of national parks, **soil erosion**, or crop disease at certain elevations. However, for simulation or predictive purposes, all models are limited by the variables that are included in them and how closely these variables are linked to individual **decision-making** on land-use change. For example, in some models, the only simulations that can be performed are changes that are imposed on the explicit features of the landscape, such as building a new road or prohibiting certain land uses, as there are no explicit variables directly linked to human behavior. As a result, the impact of an agricultural policy change, such as a subsidy change on a farmer's decision to farm his land, cannot be predicted. The only way to simulate such a policy with these models would be to assume the land-use decision of the farmer in response to the policy.

Another consideration of the "socializing the pixel" models is that the unit of analysis is either an individual pixel or some aggregation of landscape units, rather than the individual decision-maker. For this reason, having information on the boundaries of individually owned land parcels, rather than just the boundaries between two dissimilar land-use pixels, is greatly preferred. For example, individuals owning large land parcels may react differently to a policy than those with small land parcels. Distinguishing the effects of a policy change among large and small landowners is important and only possible if ownership boundaries are known.

See also: Cadastre; Driving Forces; Forest Anomalies; Landholding; Land-Use System; Pattern to Process; Property.

Further Reading

Geoghegan, Jacqueline, Pritchard, Lowell, Jr., Ogneva-Himmelberger, Yelena, Chowdhury, Rinku Roy, Sanderson, Steven, and B. L. Turner II, "'Socializing the Pixel' and 'Pixelizing the Social' in Land-Use and Land-Cover Change," in *People and Pixels: Linking Remote Sensing and Social Science*, eds. Diana Liverman, Emilio Moran, Ronald Rindfuss, and Paul Stern, Washington, DC: National Academy of Science Press, 1998, pp. 51–69; Irwin, Elena G., and Jacqueline Geoghegan, "Theory, Data, Methods: Developing Spatially-Explicit Economic Models of Land Use Change," *Agriculture, Ecosystems and Environment* 84 (2001): 7–24; Verburg, Peter H., Kok, Kasper, Pontius, Robert Gilmore, Jr., and A. (Tom) Veldkamp, "Modelling Land Use and Land Cover Change," in *Land Use and Land Cover Change: Local Processes, Global Impacts*, eds. Éric Lambin and Helmut Geist, Berlin: Springer, 2006.

JACQUELINE M. GEOGHEGAN

Soil Degradation. The loss or deterioration of soil properties and soil ecological and socio-economic functions due to direct or indirect human interventions (sometimes more narrowly defined as the reduction of soil productive capacity due to human action). Natural changes of soil properties and soil functions are called soil formation. Soil degradation is often used as an indicator for **land degradation**, and sometimes soil degradation and land degradation are used synonymously. Because of the close relationship between soils and land cover, it is a major concern under changing land use and land cover.

Situated at the interface of the lithosphere, atmosphere, biosphere, hydrosphere, and anthroposphere, soils play a vital role in supporting life and human well-being. Soils filter and store water, plants root in soils providing nutrients and water, soils sequester carbon, and most human activities take place on soils. Under natural conditions, soils are in a dynamic equilibrium with climate and **vegetation**. In most cases, human utilization of soils includes the removal of the natural vegetation, resulting in a deterioration of the dynamic equilibrium and ecological soil functions while fulfilling socioeconomic functions. This seems to be an unavoidable conflict of interests.

Processes contributing to soil degradation include a wide range of often-inadvertent events ranging from local contamination from leakages to the worldwide erosion of soils accompanying some agricultural management practices. Frequently, processes contributing to soil degradation are related to each other in positive **feedback** mechanisms. **Soil erosion** and soil organic matter depletion are a good example.

The Global Assessment of Soil Degradation (GLASOD), up to date the only consistent global evaluation of human-induced soil degradation, distinguishes four main types of soil degradation: water erosion, wind erosion, chemical deterioration (including loss of nutrients and organic matter, **salinization**, acidification, and pollution), and physical deterioration (including compaction, sealing and crusting, waterlogging, and subsidence of organic soils). Taking into account the degree and the relative extent of degradation, severity of soil degradation was assessed on a national level by expert judgment.

On the global **scale**, 15 percent of all soils accounted for $(130 \times 10^6$ km$^2)$ were found to be degraded. Strongly and extremely degraded soils considered irreclaimable cover an area of 3×10^6 km^2, mostly in Africa and Asia. The most important processes are soil erosion by water and by wind, causing degradation on 56 percent and 28 percent of the degraded soils, respectively. Chemical deterioration affects 12 percent and physical deterioration 4 percent of the degraded area. The three most important causative factors are overgrazing, **deforestation**, and agricultural mismanagement, which are held responsible for soil degradation on 92 percent of the affected area. Soil degradation of the remaining area is due to overexploitation of the vegetation cover for domestic use and bio-industrial activities.

See also: **Anthropocene; Carbon Sequestration; Cotton; Degradation Narrative; Desertification; Modification; Monoculture; Pasture.**

Further Reading

EEA (European Environment Agency), Environment in the European Union at the Turn of the Century (Chapter 9), Copenhagen: EEA, 1999 [Online, March 2005], http://reports.eea.eu.int/92-9157-202-0/en; Natural Resources Conservation Service, World Soil Resources [Online, March 2005], The NRSC Soils Web Site, soils.usda.gov/use/worldsoils; Oldeman, L. R., Hakkeling, R. T. A., and W. G. Sombroek, *World Map of the Status of Human-Induced Soil Degradation: Global Assessment of Soil Degradation (GLASOD)*, 2nd ed., Wageningen, NL: International Soil Reference and Information Center, Nairobi, Kenya: United Nations Environment Programme, 1991.

JUSSI BAADE

Soil Erosion. The accelerated displacement of soil or parent material by natural drivers, that is, water, wind, and gravitational processes, following

Soil erosion in croplands and remaining chimney of abandoned farm house in Greene County, Georgia, April 1939.
PHOTO: Marion Post Wolcott. [Courtesy Library of Congress]

Soil erosion gullies on farm land in Caswell County, North Carolina, October 1940.
PHOTO: Marion Post Wolcott. [Courtesy Library of Congress]

human interference with the landscape. The natural displacement of soil or parent material lowering hills and mountains and filling up basins since millions of years ago is referred to as (geological) erosion or denudation. Soil erosion is a major driver of **soil degradation**, contributes to the siltation and pollution of water bodies, and might impair air quality. Climate, soil properties, and topography determine the type of, and **vulnerability** to, soil erosion. However, land use, land cover, and land management are the prime factors controlling the extent and severity of soil erosion.

Soil erosion is a complex phenomenon resulting from human interference with landscape properties that control the evolution of landforms. Interacting in a complex manner, climate, soil properties, and topography set the scene for soil erosion to determine the (prevailing) type of soil erosion, that is, soil erosion by water or by wind and the vulnerability of a landscape to soil erosion. For example, sloping topography is a prerequisite for soil erosion by water, but not for soil erosion by wind; the steeper a slope, the higher the risk of severe soil erosion by water. Sandy soils are very susceptible to soil erosion by wind, but less so to soil erosion by water. Humid climate conditions hinder soil erosion by wind, while semiarid areas are subject to soil erosion by both wind and water. The third type, soil erosion by gravitational processes, is often associated with soil erosion by water, like shallow land sliding at the edge of gullies formed by water.

Climate, soil properties, and topography set the scene, but human land use and land management cause soil erosion and control its extent and severity. Vegetation cover and soil management are crucial. Any disturbance of the vegetation cover protecting the soil against the shear stress exerted by wind and flowing water or the impact of raindrops detaching soil particles can trigger soil erosion. Deterioration of soil properties (which usually accompanies the disturbance of the vegetation cover), like compaction or loosening of the soil surface, reduction of soil organic matter content, and soil infiltration and water storage capacity, increases the susceptibility to soil erosion.

This implies that a wide range of direct and indirect human activities can cause soil erosion. The most common and fateful ones are **deforestation**, overgrazing, and tillage-based **agriculture**. However, in fragile environments even the trampling of foot paths can result in soil erosion. Therefore, soil erosion has to be considered an inevitable consequence of human utilization of the earth's surface, at least to some extent.

Soil erosion results in onsite and offsite damages. Onsite damages include the loss of minerogenic material, soil organic matter, and nutrients, contributing to an often-gradual degradation of soil depth and soil fertility. Ultimately, this degradation might reach a point where the land-use practice causing the soil erosion is brought to an end. Strongly eroded and abandoned field sites presently either under **forest** cover, as in some sites in the northern loess zone of **Europe**, or already bare of soil as in some places in the Mediterranean, highlight the ultimate onsite consequences of soil erosion. Offsite damages result from the deposition of material mobilized by soil erosion. The covering of adjacent fields or roads with blown or washed-out sediment, the **siltation** of waterways and reservoirs, and the input of nutrients and pesticides into aquatic habitats are issues of concern.

The fact that **vegetation** cover and soil management are crucial for soil erosion implies the opportunity to control soil erosion. Basically, activities to mitigate soil erosion are characterized by either reducing disturbance of plant cover and soil properties or even improving plant cover, both in time and space (e.g., **fallow** plant cover, fertilization of grassed areas); and strengthening soil structural stability and soil hydraulic conductivity by reducing or even giving up tillage operations. More technical and more costly solutions include terracing of sloping fields, implementing subsurface drainage, and treating fields with soil stabilizing chemicals. The basic principles have been well known for decades. However, the utilization of soil-conserving land-management measures often meets resistance due to traditional, practical, and economic reservations. However, discussions of **carbon sequestration** in soils might create a new momentum for the application of soil **conservation** measures, which not only increase soil organic matter content, but also help to mitigate the soil erosion problem.

Soil erosion is a problem controlled by several factors acting at the field scale and might therefore vary from field to field. This makes it rather difficult to assess soil erosion on smaller scales, that is, regions, nations, continents, or the whole world. Often, models are used to estimate the soil erosion risk based on climate, soil properties, and topography. However, the disadvantage is that land use and land management, that is, the factors determining the actual severity of soil erosion, are not taken into account. The Global Assessment of Soil Degradation (GLASOD) estimated the global dimension of soil erosion based on expert judgments from all major countries. According to GLASOD, soil erosion by water (including mass movements) and by wind (including overblowing as an offsite effect) are the most important processes causing soil degradation: 56 percent of areas with degraded soils are deteriorated from soil erosion by water and another 28 percent from soil erosion by wind. The results have been criticized for several reasons; and taking into account the considerable changes in land use over the past decade, the data are certainly in danger of becoming obsolete. However, the GLASOD assessment remains to date the only consistent global evaluation of the actual status of soil erosion.

See also: **Agricultural Revolution; Cash Crops; Cassava; Degradation Narrative; Desertification; Environmental Change; Hot Spots of Land-Cover Change; Land Degradation; Land-Use System; Monoculture; Nutrient Cycle; Water-Land Interlinkages; Watershed.**

Further Reading

Favis-Mortlock, Dave, Soil Erosion Site [Online, March 2005], The Soil Erosion Web Site, www.soilerosion.net; Morgan, Roy, *Soil Erosion and Conservation*, 3rd ed., Oxford, UK: Blackwell, 2005; Oldeman, L. R., Hakkeling, R. T. A., and W. G. Sombroek, *World Map of the Status of Human-Induced Soil Degradation*, *Global Assessment of Soil Degradation* (GLASOD), 2nd ed., Wageningen: International Soil Reference and Information Center (ISRIC), and Nairobi, KE: United Nations Environment Programme (UNEP), 1991 [Online, March 2005], www.isric.nl/Docs/Glasod.zip.

JUSSI BAADE

South Africa.   Lying at the southern tip of Africa, the present-day landscape of South Africa reflects the country's colonial history and particularly the racial policies pursued during the twentieth century. Like the rest of **southern Africa**, hunter-gatherer hominids have been present in the area for over a million years, and pastoralism and cultivation have been important land uses for about 2,000 years.

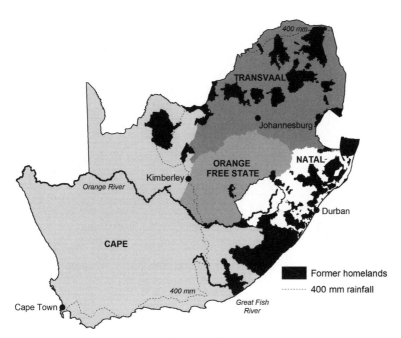

The pre-1944 political units of South Africa.

Rainfall in South Africa decreases from east to west, forming a dry western half (<400 mm per year), and a wetter eastern half and southern coastal plain (400–1500 mm p.a.) (see figure titled "The pre-1944 political units of South Africa"). Most of the region receives rainfall in summer, except for the extreme southwest, which receives winter rainfall. The dry western region is covered by shrublands (the succulent and the Nama Karoo) and the fine-leafed acacia savannas of the Kalahari region. The wetter eastern half consists of **grasslands** and either broad-leafed savanna on sandy infertile soils, or acacia savannas on more fertile clay soils. The southwestern tip was originally covered by sclerophyll thicket, known as *fynbos*; both the eastern grasslands and the *fynbos* are now substantially transformed by **agriculture**. Small, isolated patches of **forest** occur along the **mountain** escarpment in the southern and eastern parts of the country.

By 1900, approximately 3.5 percent of the total land area of South Africa was cultivated; 96.2 percent was covered by natural **vegetation**, and 0.3 percent consisted of urban settlements (see figure titled "Changes in land use"). At the end of the twentieth century, 85 percent of South Africa was still covered by natural vegetation, while 14 percent had been converted to planted crops and 1.3 percent to urban land uses. Approximately 1.3 million hectares (8 percent of the cultivated area) was under irrigation. The untransformed area in 2000 consisted of 5 percent formally protected land, while 75 percent was used for activities such as livestock ranching and, increasingly, nature-based tourism ventures. Approximately 5 percent of South Africa was covered by natural vegetation but in a degraded state. About 68 percent of the country consisted of freehold land, 14 percent was communal tenure, and 18 percent was owned by the state. South Africa is currently a net food

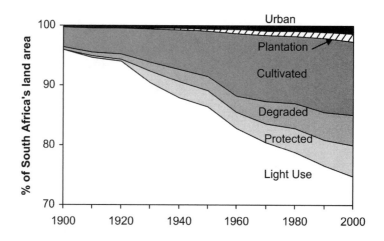

	Light use	Protected	Degraded	Cultivated	Plantation	Urban
1900	95.94	0	0.41	3.37	0.02	0.26
1910	94.65	0.32	0.62	3.97	0.1	0.34
1920	94.02	0.32	0.85	4.29	0.16	0.37
1930	90.51	1.88	1.42	5.52	0.25	0.42
1940	87.87	2.7	2.06	6.56	0.31	0.5
1950	86.39	2.72	2.41	7.52	0.37	0.6
1960	82.8	2.79	2.68	10.37	0.69	0.67
1970	80.52	3.02	3.81	11	0.81	0.83
1980	78.76	4.03	4.26	11.14	0.94	0.87
1990	76.55	4.39	4.64	12.33	1.01	1.08
2000	74.76	5.3	4.95	12.23	1.47	1.29

Changes in land use over the twentieth century

exporter in most years, and produces a wide range of food, timber and fiber, and horticultural, cosmetic, and medicinal crops.

European Settlement (1652–1910)

During this period, land-cover impacts in South Africa closely tracked the growth and spread of the human population. The Dutch East India Company set up a trading post in 1652 in present-day Cape Town to provide fresh food for the company's ships that rounded the Cape on their way to East Africa and Asia. The settlement expanded over the next 150 years, and by the early 1800s, approximately 26,000 European farmers had settled the wetter region along the southern coast between Cape Town and the Great Fish River (see first figure). These immigrants engaged primarily in **subsistence agriculture** and produced little for export. During this time there were also dispersed pockets of subsistence cultivation in the interior east of the 400-millimeter rainfall line, which had existed for up to 1,500 years, in the areas settled by various Bantu-speaking African tribes. The largest settlements were those of the Zulu and Xhosa people along the east coast, and the Sotho, Ndebele, and Venda people in the northeastern regions of present-day South Africa. To the west of the 400-millimeter rainfall line (the limit of cultivation) were the sparsely distributed Khoikhoi (also called the San), who were hunter-gatherers and pastoralists.

Upheavals in the political, economic, and social structures of the Bantu-speaking communities in the first half of the nineteenth century led to widespread migrations and dramatically altered ownership and land-use patterns. The period is known as the *mfecane* or *difaqane* and was brought about by a combination of factors including population growth, depletion of natural resources, and devastating drought and famine. The slow southward migration of the Bantu people in Africa over the preceding 2,500 years had reached its southern limit, and pressure for land increased. The introduction of **maize** (*Zea mays*) from the Americas in the eighteenth century via the Portuguese in Mozambique had enabled a growth in the population, as maize produced more food per unit land or labor than the main indigenous staples sorghum and millet. However, maize requires more water than local grains, and the onset of a severe ten-year drought starting in about 1800 had devastating impacts as people fought for supplies of grain and cattle. Large numbers of people were displaced, and changes in military organization took place. Warfare continued over the following decades and led to the establishment of a number of Bantu states with large armies, the most successful and aggressive being that of the Zulu kingdom. To secure foodstuffs and protect themselves from the Zulu marauders, several defensive states formed in the 1820s and 1830s, most notably the Swazi and Sotho kingdoms, which form the core of the modern-day countries of Swaziland and Lesotho.

Between 1795 and 1806, the British captured the Cape Colony from the Dutch. The British attempted to alleviate the land problems of the expanding European population in the southwestern Cape by sending imperial armies to fight against the Xhosa in the region of the Great Fish River, expelling most of the African population in that area by about 1820. In order to stabilize the frontier, about 5,000 British immigrants assisted by the British government were settled on smallholder farms in the region at this time. British rule and the increasing number of British settlers, however, engendered growing hostility from the Boer community (farmers primarily of Dutch, German, and French descent). Beginning in 1836, about 12,000 people, Boer families together with their servants, left the eastern Cape frontier and traveled by ox-wagon into the high-lying interior. This movement, known as the "Great Trek," led to the eventual establishment of two republics (the Orange Free State in the 1840s and the South African Republic, later known as the Transvaal, in the 1850s) in the northeast of present-day South Africa (see first figure). The economic mainstay of the Boer republics during the first thirty years of their existence was wildlife, and especially a trade in hides and ivory. The abundant herds of large ungulates were eliminated by 1870. By this time approximately 50,000 migrants of European descent had settled in the two republics, practically all living in rural areas and, like the African people among whom they settled, making a living from small-scale subsistence agriculture, extensive pastoralism, and hunting. During the period from 1800 to 1865, the total European population increased twelvefold from about 20,000 to nearly 250,000 (180,000 in the Cape Colony, 18,000 in Natal, and 50,000 in the Boer republics).

The diffusion of British and Boer farmers into the interior had two major impacts. First, the Europeans had firearms and shot large numbers of indigenous mammals, so that the animals' populations were substantially reduced.

Herbivores, and particularly the megaherbivores, play an important role in ecosystem processes. Their removal, together with altered **fire** regimes, has been blamed for the "bush encroachment" problem that has rendered an estimated 2.5 percent of the savanna region of South Africa uneconomical for livestock ranching. Second, modern technologies, particularly the establishment of boreholes in the late nineteenth century, enabled farmers to introduce large numbers of livestock into previously inhospitable parts of South Africa, such as the arid karoo area. The marked decline in livestock numbers in many of these areas in the early twentieth century, following an initial peak, has been taken as evidence of **land degradation** in these environments due to early overstocking.

The discovery of diamonds in 1867 and large deposits of gold in 1886 initiated the shift to an economy based on manufacturing and service industries. Mineral discoveries attracted large numbers of European immigrants and African migrants in the late nineteenth century. The diamond mines of Kimberly brought about a doubling of the European population in the Cape Colony between 1865 and 1900, to reach a total of 400,000. Similarly, in fewer than fifteen years, the gold mines in Johannesburg attracted a population of 75,000 Europeans and 100,000 African laborers, mostly from the Portuguese-ruled areas of Mozambique. The growth of the urban population necessitated the establishment of surplus-producing commercial agriculture and the first large-scale commercial timber plantations to provide support timber for the mines. The discovery of gold in the South African republic was the major factor leading to the South African War (1899–1902), in which the British captured the Boer republics, and eventually led to the formation of the Union of South Africa in 1910, establishing the country's present-day boundaries. The **urbanization** of thousands of Africans led to the establishment of large multiracial communities in the new industrial cities, competition between the African and European populations for jobs, and increasing bargaining strength on the part of the Africans that was seen as a threat to the supply of cheap labor for the mines. These threats prompted the introduction of the first pass law in the Cape Colony in 1872, limiting the mobility of and economic opportunities for Africans. This formed the foundation for subsequent discriminatory laws that had far-reaching effects on the South African society, economy, and environment throughout the twentieth century.

The Period of White Dominance (1910–1994)

Following the unification of South Africa in 1910, several pieces of racially discriminatory legislation were introduced, the most important of which was the Natives Land Act of 1913. This act restricted land ownership for Africans (4 million people; two-thirds of South Africa's population) to 7.5 percent of the country's land area, and only on a customary rather than a freehold basis. Although this was increased to 13 percent of South Africa in 1936, the legislation had significant impacts, and the effects on land-cover are still clearly visible today (see image). The population densities in these "homeland" areas were five to ten times greater than in the rural areas in the rest of South Africa. At least 1.5 million people were forcibly removed from white-owned areas and resettled in the homelands over the course of the twentieth century. Most of the population were peasant farmers, but the small **landholdings**; insecure tenure; and the lack of capital and **access** to technology, advice, and markets made it very difficult to make a living from farming. When apartheid ended in 1994, the average African farm was 5.2 hectares in size, while the average white-owned farm was 1,300 hectares. The result was that most of the economically active sector of the population worked as poorly paid migrant laborers in the cities of "white" South Africa, leaving the elderly and children to manage the land. The consequent shortage of able-bodied labor in the homelands, together with the lack of capital for agricultural inputs and conservation, led to deteriorating land productivity in many of these areas. Gathering wood, dung, and crop residues to provide fuel for cooking and heating further reduced soil fertility. Large numbers of livestock, kept primarily as a form of capital investment rather than for meat, were a significant factor in the denudation of vegetation in these areas. The so-called "betterment" schemes, implemented from the 1940s to the 1970s, aimed to address land degradation in these areas, but largely failed as they did not address the underlying socio-political problems.

Land degradation has also been an issue of significant concern in white-owned commercial farming areas since the late nineteenth century. Efforts

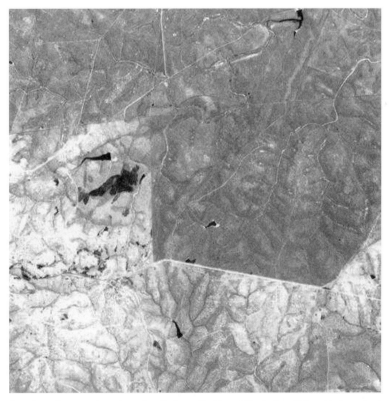

Land sat image showing the Krüger National Park (north) and denuded former homeland area (south).

CREDITS: Tobias Landmann, Council for Scientific and Industrial Research, South Africa

to combat land degradation took the form of government subsidies for conservation works, legislation, agricultural extension, and farmer study groups, and were generally successful at improving the **conservation** status of these areas. The effects of drought and **soil erosion** received particular emphasis. The practice of corralling domestic livestock at night was identified as a major factor leading to soil and vegetation degradation, and consequently a program of exterminating jackals and other predators was initiated in the early twentieth century. Cheap fencing material was provided to create paddocks for rotational grazing management, and water points were provided so that livestock did not have to walk long distances. Particularly influential was the work of John Acocks, who believed that the margins of the Karoo shrubland had been expanding into the grasslands since colonial times as a result of overgrazing by commercial small-stock farmers. The expanding Karoo hypothesis, subsequently found to be unsupported by evidence, heavily influenced policy development until the 1980s, specifically interventions such as the Stock Reduction Scheme (1969–1978) and the National Grazing Strategy (1985). Soil conservation and veld management acts were introduced in the mid-1900s and required landowners to, among other things, obtain permission for the cultivation of virgin soil and steep slopes, to prevent waterlogging and **salinization**, to restore eroded land, to control the number and type of livestock kept on the land, and to control wild fires and invader plants.

The government's racial policies led to growing international isolation from the 1950s until the end of apartheid in 1994. Consequently, agricultural policies increasingly emphasized food self-sufficiency. White farmers received privileges such as loans from the government's Land Bank, labor law protection, and crop subsidies. Marketing boards were established to stabilize the production of many crops, and various drought subsidies and rebates existed to keep white farms economically viable. These policies were partly responsible for the 4-million-hectare (3 percent of South Africa's land area) increase in cultivated area that occurred between 1955 and 1965. Large increases in crop yields of white-owned commercial farms occurred in the 1960s, resulting from the introduction of higher-yield cultivars, pesticides, fertilizers, and irrigation. In contrast, the productivity in the homeland areas throughout the twentieth century stayed at the same level it was in the early 1900s.

The New Democracy (1994–Present)

The end of apartheid brought the need to reverse the discriminatory laws around **land rights** and **land tenure**. Two **land reform** programs have been instituted: a legally driven restitution process to return land that was confiscated in terms of the 1913 and 1936 Land Acts; and a market-based redistribution process to significantly increase the share of South Africa under freehold tenure by Africans. The marketing boards have been phased out, and agricultural produce is now freely traded. Currency fluctuations are a significant factor impacting the economic viability of especially export agriculture. The total area under cultivation is unlikely to expand much due to the climatic limitations of the country. On the other hand, nature-based tourism is growing annually at approximately five times the rate of the agricultural sector, resulting in a significant shift out of pastoral land uses and into tourism ventures, changing the policy emphasis in the country. According to the United Nations medium-variant projections, the total population of South Africa is expected to stay at approximately the current size over the next half-century, but become increasingly urbanized. This is likely to bring an increasing demand for protein as people grow wealthier, which is likely to be met by increasing use of feedlots in livestock production.

See also: **Cattle Ranching; Colonization; Degradation Narrative; Green Revolution; Mineral Extraction; Spatial Diffusion.**

Further Reading
Acocks, John P. H., "Veld Types of South Africa," *Memoirs of the Botanical Survey of South Africa* 8 (1953): 1–128; Biggs, Reinette, and Robert J. Scholes, "Historical Changes in Natural Capital in South Africa: An Approach Using Changes in Biodiversity," in *Restoring Natural Capital in the 21st Century: Emerging Countries of the South*, eds. Sue J. Milton and James Aranson, Washington, DC: Island Press, 2005; Biggs, Reinette, and Robert J. Scholes, "Land-Cover Changes in South Africa 1911–1993," *South African Journal of Science* 98 (2002): 420–4; Byrnes, Rita M., ed., *South Africa: A Country Study* [Online, September 2004], Federal Research Division, United States Library of Congress, Country Study Web Site, http://countrystudies.us/south-africa; Dean, W. Richard J., and Ian A. W. MacDonald, "Historical Changes in Stocking Rates of Domestic Livestock as a Measure of Semi-Arid and Arid Rangeland Degradation in the Cape Province, South Africa," *Journal of Arid Environments* 26 (1994): 281–98; Fairbanks, Dean H. K., Thompson, Mark W., Vink, Dineke E., Newby, Terence S., van den Berg, Hendrik M., and David A. Everard, "The South African Land-Cover Characteristics Database: A Synopsis of the Landscape," *South African Journal of Science* 96 (2000): 69–82; Grossman, David, and Mark V. Gandar, "Land Transformation in South African Savanna Regions," *South African Geographical Journal* 71 (1989): 38–45; Hoffman, M. Timm, and Ally Ashwell, *Nature Divided: Land Degradation in South Africa*, Cape Town, South Africa: University of Cape Town Press, 2001.

REINETTE (OONSIE) BIGGS, ROBERT J. SCHOLES

South America.   This continent covers an area of approximately 17.8 million km^2; the west of South America is dominated by the **Andes** Mountains, which form the continental divide between the Atlantic and Pacific oceans, and a coastal belt extending from Tierra del Fuego to Panama; while the east is dominated by the Amazon (5.87 million km^2), Orinoco (966,800 km^2), and Paraguay-Paraná (3.1 million km^2) **river basins**, which drain 55 percent of the continent, and the Guiana and Brazilian highlands, and smaller **watersheds** in the north and in southern Argentina, which flow into the Caribbean and South Atlantic, respectively.

Extending from approximately 11° north to 56° south, South America's climate, soils, and **vegetation** are diverse. Tropical humid, dryland, midlatitude, polar, and **mountain** climate zones are represented on the continent. The main climatic influences are the Intertropical Convergence Zone, the Southern Hemisphere air masses, and the orographic effect of the Andes. Natural vegetation and non-irrigated agricultural land uses are in dynamic equilibrium with climate. Short-term perturbations, particularly the El-Niño Southern Oscillation (ENSO), affect normal climate patterns.

The **Global Land Cover Map of the Year 2000** map for South America indicates that a little over 75 percent of South America was covered by natural vegetation with varying levels of disturbance. Four types of **forests** cover 46.4 percent of the land area (humid tropical, 35.5 percent; dry tropical, 8.3 percent; flooded tropical, 1.5 percent; and temperate, 1.1 percent), **grasslands** and shrublands cover 18.1 percent, and steppe vegetation covers 6.9 percent. South America is globally important in terms of **biodiversity**. Five of the world's twenty-five biodiversity hotspots—the Atlantic forest, **Brazilian Cerrado**, central Chile, Choco-Darien-western Ecuador, and tropical Andes—cover 945,130 km^2 (5.3 percent) of the continent, of which 24.5 percent is protected. The expansion of protected areas for **conservation** has been an important driving force in land-use change during the 1980s and 1990s.

Migration and urban population growth have been important land-use change drivers in the region since the eighteenth century. Population reconstructions for the **BIOME 300 project** show low densities (<5 persons/km^2) during the eighteenth century with exception areas around Rio de Janeiro, Santiago de Chile, and Lima. During the nineteenth century, these cities provided foci for population growth. The high population densities around Lima pushed northward along the Pacific coast and southward into the Andes. By 1900, high population densities existed from Lima to northern Venezuela along the coast and in the Andes (particularly major cities such as Bogotá) and southward to Cochabamba. High population densities existed in the northeast Brazil drought triangle (Alagoas, Paraíba, and Pernambuco states) by 1850, and by 1900 high population densities extended from Maranhão (northeast Brazil) along the Atlantic coast to Buenos Aires province (Argentina). The early spatial distributions of population are reflected in the contemporary distribution of people. Low densities (<5 persons/km^2) in much of **Amazonia** are surrounded by an arc of high population density that extends from eastern Venezuela through Colombia, Ecuador, Peru, Bolivia, northern Argentina to Buenos Aires, and then northward along the Atlantic coast to northeast Brazil. A second belt of relatively dense population extends from Buenos Aires through Córdoba and Mendoza provinces to central Chile.

Approximately 266 million South Americans people live in urban areas, and urban populations are particularly high in Argentina, Chile, Uruguay, and Venezuela. **Urbanization** is an important land-use change driver, even though urban areas account for only about 0.1 percent (11,400 km^2) of the land area. Large, rapidly growing urban areas (e.g., Buenos Aires, Lima, Rio de Janeiro, and São Paulo) continue to be locally significant drivers through **urban sprawl**, by the demands exerted on rural areas in terms of goods and services, and by attracting rural out-migrants.

Migration to South America, particularly from **Europe** and **Japan**, from the mid-nineteenth to early twentieth centuries, underpinned much of the continent's population growth. Many migrants settled in the **Southern Cone region**, and the area under cropland expanded considerably. The subtropical grasslands (pampas) of Argentina and Uruguay were converted to large farms from the mid-nineteenth century to rear cattle (for meat and hides) and cereals for export. The Argentine pampas now mainly comprise highly mechanized commercial farms, while in Uruguay the pastoral economy is stagnating. Patagonia saw development of large sheep estancias at the same time at the expense of steppe vegetation. The grasslands and shrublands of interior and northern Argentina have been converted more recently for sugarcane, citrus, and soybean cultivation. Large estancias producing beef, **maize**, sugarcane, and **cotton** characterized southern and eastern Paraguay from the mid-nineteenth century, but recently soybean cultivation has expanded and the dry thorn shrublands and woodlands of northern Paraguay have become areas of **colonization**. According to the **United Nations Environment Programme**, most of the continent's drylands were undergoing **desertification** at the end of the twentieth century.

Southern Brazil also witnessed much immigration in the mid-nineteenth and early twentieth centuries, and the region's sub-tropical and temperate forests (the Atlantic forest) were converted for **cattle ranching**, **coffee**, and soybeans. The loss of the Atlantic forest extended southward into Paraguay, Uruguay, and Argentina, and northward along the Brazilian coast. The staggering decline in **tropical humid forests** (16 million hectares between 1990 and 1997, a rate of 0.38 percent per annum) is generally a more recent phenomenon. The majority of these forests in South America are found in the Amazon basin, which covers the Brazilian Legal Amazon and parts of Bolivia, Peru, Ecuador, Colombia, and Venezuela. Significant land-use conversions to these sparsely populated forests began about 150 years ago in **Amazonia** with a succession of boom-and-bust cycles based on **non-timber forest products** (most notably rubber), and demands for land from mining, timber extraction, and livestock. However, most land-use conversion has taken place in the last thirty to fifty years, and its underlying driving forces have been the development of regional economic markets (e.g., MERCOSUR), integration into global economic markets, human **population dynamics** (especially the need to resettle the rural and urban poor), and infrastructure development as part of national development planning. Much of this has taken place in an **arc of deforestation** in eastern and southern Brazil; in western Amazonia in Bolivia, Peru, Ecuador, and Colombia; and in southern Venezuela. Major drivers have been legal and illegal timber extraction, **mineral extraction**, hydroelectric power generation, the formation of cattle ranches, and the expansion of croplands. The conversion of forests to **pastures** has been a major land-use change in Brazil since 1960s, and continues to be so even though government subsidies have been withdrawn and conversion has proven unsustainable in the medium-term. Many pastures are being converted to soybean cultivation. In western Amazonia, there have been efforts to settle people at the western margins of the Amazonian lowlands, along the Carretera Marginal de la Selva. A number of deforestation hot spots occur in this area, some of which are coca production zones. Western Amazonia and foothills of the Andes from Lake Maracaibo to northern Argentina are major hydrocarbon production zones. Oil and gas exploitation have led to localized **deforestation** and provided **access** to settlers. Much of the montane tropical forest of the Andes is protected for nature conservation, but mid-altitude forests in Colombia and Peru have been extensively cleared for coffee. The temperate forests and alpine ecosystems of the southern Andes have been conserved mainly for watershed protection and/or tourism or converted during the twentieth century to commercial forests with introduced species such as Monterrey pine (*Pinus radiata*). In central Argentina and Chile, significant amounts of land have been converted to viticulture.

See also: **Agricultural Frontier; Agricultural Intensification; Economic Growth; Exotic Species; Hot Spots of Land-Cover Change; Land Degradation; Malaria; Poverty; Tropical Dry Forest.**

Further Reading

Denevan, William, *Cultivated Landscapes of Native Amazonia and the Andes*, Oxford, UK: Oxford University Press, 2001; Eva, Hugh D., Belward, Alan S., De Miranda, Evaristo E., Di Bella, Carlos M., Gond, Valéry, Huber, Otto, Jones, Simon, Sgenzaroli, Matteo, and Fritz Steffen, "A Land Cover Map of South America," *Global Change Biology* 10 (2004): 731–44.

ANDREW MILLINGTON

Southeast Asia, Mainland. Mainland Southeast Asia includes the nations of Cambodia, Laos, Myanmar, Thailand, and Vietnam. The region harbors an immense wealth of natural resources, including globally important stocks of **forests** and biological diversity as well as many major river systems. It is also home to a rich heritage of indigenous cultures. Little is known about land-use and **land-cover change** in the region before the early 1800s when the British began to map the boundaries between their newly acquired territories in Myanmar and Siam. John Richards and Elizabeth Flint documented land-use and land-cover change in mainland Southeast Asia between 1880 and 1980. They concluded that agricultural expansion and, to a lesser extent commercial logging, caused substantial land-use change in the region during this century. The greatest single demand for land **conversion** came from pioneer peasant farmers clearing land for wet rice paddies, as the sparsely inhabited river deltas of the region became **rice** granaries. During this period, temporary crops such as **maize** and **cassava** increased by 330 percent in area to become 17 percent of the total land areas. Permanent crops such as rubber increased by a factor of 6.5 during the century but still represented only 1.5 percent of the total land area in 1980.

Over the last five decades, countries that make up the region have been under vastly different economic and political regimes; and these differences influence land use and land cover in the region today. Thailand has had an open market and democratic government. Cambodia, Laos, and Vietnam are socialist states that have differed in the timing and ways in which they opened their economies to world markets and provided private usufruct rights to natural resources. Myanmar remains a closed economic and political system. Cultural differences in the ways in which different ethnic groups use lands, their customary trading practices, and their relationships with other groups affect how they use land and their responsiveness to different government polices and market pressures.

Forest Cover

Forestry statistics in this region are notoriously unreliable and differ by country, time collected, and collection methods. In order to ensure that the data are as comparable as possible, this discussion is limited to statistics of the **United Nations Food and Agriculture Organization** (FAO) for 1990 and 2000. These data (see table titled "Forest cover") suggest that Cambodia,

Forest cover in 1990 and 2000 (total and as percent of national land area)

	1990	2000				
	Forest Area (ha × 1,000 and %)	Natural Forest (ha × 1,000)	Annual Change (ha × 1,000)	Total Forest (ha × 1,000 and %)	Annual Change (ha × 1,000)	Annual Change (ha × 1,000)
Cambodia	9,896 / 56	9,245	90	9,335 / 53	−56	−0.6
Laos	13,088 / 57	12,507	54	12,561 / 54	−53	−0.4
Myanmar	39,588 / 60	33,598	821	34,419 / 52	−517	−1.4
Thailand	15,886 / 31	9,842	4,920	14,762 / 29	−112	−0.7
Vietnam	9,303 / 29	8,108	1,711	9,819 / 30	52	0.5
Total	87,761 / 44	73,300	7,596	80,896 / 40	−686	−0.8

Source: FAO 2001.

Laos, and Myanmar maintain approximately 50 percent forest cover. Annual rates of forest loss in these countries during this period ranged from 0.4 percent in Laos to 1.4 percent in Myanmar, with the greatest annual reduction of forest cover occurring in Myanmar. In 2000, Thailand and Vietnam maintained approximately 30 percent forest cover, with Thailand losing approximately 0.7 percent annually and Vietnam gaining approximately 0.5 percent annually. In recent years, Thailand and Vietnam have undertaken strong efforts to reverse the rapid rates of **deforestation** of previous years. This has occurred through both natural regeneration of previously logged areas as well as the establishment of forest plantations (FAO statistics include rubber as forest plantations).

Croplands

Extensification of cultivated areas continued in various parts of the region throughout the last five decades although most alluvial lowland areas had already long been occupied. In 1961, rice accounted for nearly two-thirds of all arable land in the region (29.3 million hectares). By 2002, this had increased to three-quarters of all arable land (36.5 million hectares), ranging from 131 percent in Vietnam (because of double-cropping) to 53 percent in Cambodia (see table titled "Changes in arable and rice land").

Northeast and eastern Thailand were **hot spots of land-cover change** over the last half-century. Harald Uhlig reported a 47 percent decrease in forest cover in this area and even 75 percent loss in the provinces of Chon Buri and Rayong as spontaneous settlers cleared roughly four to five million hectares of previously little-used but not necessary forested uplands. During this period the proportion of cultivated land under rice in the northeast fell from 96 to 76 percent; the total number of farming holdings increased by over 36 percent; and the total cultivated area increased by 53 percent as maize, kenaf (*Hibiscus sabdariffa*), and cassava were planted. This expansion was driven by export prices as well as numerous loopholes in legal codes and the political difficulty of enforcing their provisions; and inadequacies in survey and land registration permitted continued expansion.

Government-sponsored land settlement schemes were not as large as those found in Indonesia and the Philippines but did occur. In Thailand some

Changes in arable and rice land between 1962 and 2002

	Arable Land 1961 (ha × 1,000)	Arable Land 2002 (ha × 1,000)	Rice Land 1962 (ha × 1,000)	Rice Land 2002 (ha × 1,000)	Rice as Percent of Arable Land 1962	Rice as Percent of Arable Land 2002
Cambodia	2,840	3,700	2,286	1,966	0.80	0.53
Laos	640	800	590	690	0.92	0.86
Myanmar	9,900	9,550	4,654	6,200	0.47	0.65
Thailand	10,400	16,800	6,540	9,990	0.63	0.59
Vietnam	5,550	5,700	4,889	7,485	0.88	1.31
Total	29,330	36,550	18,959	26,331	0.65	0.72

Source: IRRI 2004.

241,000 settler families were placed on about 700,000 hectares under various programs. In Vietnam, state-sponsored resettlement of lowland Kinh settlers in the central highlands caused extensive alienation of lands that minority peoples formally occupied. After 1975, large tracts of forestland were incorporated into state farms and state enterprises on which the newly arrived Kinh were settled. More recently, waves of spontaneous migrants, often themselves ethnic minorities (Tay, Nung) from land-scarce provinces in the northern **mountains**, have displaced local minorities from better lands located near roads. Intensification has affected different agricultural or **land-use systems** in different ways and to different degrees. Wet-rice systems, whether irrigated or rainfed, have undergone the highest degree of **agricultural intensification**. Irrigation is the major method of intensifying production and has generally encouraged demand for labor and capital outputs. Irrigated rice accounted for about 26 and 32 percent of the land area in Myanmar, Thailand, and Vietnam in 1976 and 1996, respectively. In 1996 irrigated rice represented 20 percent of all rice land in Thailand and 51 percent of Vietnam (see table titled "Irrigated rice land").

The greatest expansion in rice production has been in Vietnam, especially in the southern delta region following the abandonment of land collectivization in 1989 and the introduction of free markets. Vietnam now exports 1–1.5 million tons of rice per year and ranks at the world's third-largest exporter. Average yields in most southern provinces have reached six tons per hectare per season, against a level of 1.5 tons per hectares before decollectivization. On the Plain of Reeds, previously a region of low-yielding floating rice production, intensification has seen rice replaced by maize grown for sale as animal feed, as livestock raising has expanded markedly.

In Thailand, supplementary wet season irrigation and drainage have been implemented on a large scale since 1960. The expansion of irrigation farming, and in the Chao Phraya Delta improved drainage, has typically been accompanied by the use of double cropping, high-yielding varieties, mechanization of tillage, the use of fertilizers, insecticides, and other pesticides. In the Chiang Mai valley in northern Thailand, triple cropping began in the late 1960s. By 1980, two-fifths of the land was triple cropped in glutinous rice (main season), rice or vegetables, and non-glutinous rice rotation.

This was possible due to reliable water supplies, partly from tube wells; and the use of high-yielding varieties for the second and third crops, together with heavy fertilizer use, raising main-crop yields to 5.5 tons per hectare.

Rainfed cropping systems have become more intensive by increasing the number of crops obtained from the same fields and the yields of single-crop systems. In northeast Thailand, for example, where rice grows from August to December, it has been mixed with crops such as sesame, peanuts, cassava, **tobacco**, and vegetables. In Vietnam, rice-rice rotations give lower net returns than rice-vegetable (eggplant, cowpea, squash) rotations because of high material and labor inputs used for rice, and so cowpeas or peanuts have been added to rice-rice rainfed rotations.

In upland areas of mainland Southeast Asia, **swidden cultivation** has been the dominant land-use practice for at least a millennium. Jefferson Fox and John Vogler summarize eight sites where they documented changes in land cover in upland areas across the region over the last fifty years. They found that at the level of secondary forest **regrowth** after swidden cultivation, land cover remained stable across the region throughout this fifty-year period. This level of abstraction, however, hid major changes that occurred within the sub-categories of **secondary vegetation** as well as in the size and number of land-cover fragments. These studies suggested that the land-use change that occurred was due to intensification of swidden cultivation and the introduction of **cash crops**.

Such intensification is marked by reduced **fallow** to the extent that in some areas cultivation is now semi-permanent. In the Mae Chaem and Mae Sarieng **watersheds** of northern Thailand, the fallow period of the former long fallow swidden cultivation systems of the Karen and Lua people has now been reduced to five years or less. However, because without external inputs the productivity of upland rice falls drastically, many farmers have switched to non-rice crops such as soybeans, maize, cabbage, and other cash crops, accompanied by heavy use of chemical fertilizers, herbicides, and pesticides. In northern Vietnam, the amount of land devoted to swidden cultivation decreased in the 1990s as farmers devoted more labor and capital to permanent agricultural fields.

Cash cropping has been promoted throughout the region, but particularly in Thailand and Vietnam. Cash crops include temperate vegetables such as tomatoes, carrots, green beans, potatoes, and cabbage, as well as cut flowers, and subtropical fruits such as persimmon, apricot, plum, lychee, and longan. These crops exploit the uplands' biophysical advantage—its temperate climate—for off-season and high-value commercial crops for lowland markets and export.

Irrigated rice land between 1980 and 1996 (total and as percent of all rice land)		
	Irrigated Rice 1980 (ha × 1,000) and (%) Rice Land	Irrigated Rice 1996 (ha × 1,000) and (%) Rice Land
Cambodia	** / **	** / **
Laos	** / **	** / **
Myanmar	873 / 18%	1,535 / 27%
Thailand	2,006 / 22%	1,894 / 20%
Vietnam	2,250 / 40%	3,572 / 51%
Total	5,129 / 26%	7,001 / 32%

Source: IRRI 2004.

Agribusiness companies are having increased influence on farming systems in many parts of the region. These companies are interested in purchasing, and sometimes processing, agricultural products for both domestic and export markets. In recent years there has been enormous expansion of various forms of **contract farming**, offering farmers alternative ways to earn income. For example, in Chiang Mai province, there are many canning and food processing plants that subcontract farmers to deliver products that meet particular grade specifications. Inputs are prescribed and grid-pricing systems of payments used for crops such as soybeans, green beans, and bamboo shoots. Farmers in some areas of Thailand and Vietnam have begun to allocate a substantial part of their land to plantation crops such as sugarcane. Others have allocated land to **coffee** on contract to township sugar mills and coffee producers.

Although intensification has been a dominant process in recent decades and although the agricultural workforce continues to grow, areas of **disintensification** as well as areas of production stability can be found. Disintensification does not imply active withdrawal of labor from agriculture so much as the use of off-farm, non-agricultural work to supplement subsistence. This has involved some swidden cultivators but many more lowlanders. This process has been particularly strong in Thailand although it has occurred elsewhere as well. In Thailand, few farmers rely solely on **agriculture** to meet their needs. Rural households are dividing their time between farm and non-farm activities, constructing livelihoods that are increasingly hybrid, both spatially and sectorally. Whether these changes mean that rural people will eventually become permanent city dwellers or not is not yet clear. They may choose to live in rural areas and to split their time between farm and non-farm activities.

See also: **Agro-Industry; Biodiversity; Colonization; Conservation; Forestry; Hot Spot Identification; Land Reform; Land Rights; Malaria; River Basins; Subsistence Agriculture.**

Further Reading

Fox, Jefferson, and John Vogler, "Land-Use and Land-Cover Change in Montane Mainland Southeast Asia," *Environmental Management* 36 (2005): 394–403; Hill, Ronald D., "Stasis and Change in Forty Years of Southeast Asian Agriculture," *Singapore Journal of Tropical Geography* 19 (1998): 1–25; International Rice Research Institute, World Rice Statistics [Online, December 2004], The IRRI Web Site, www.irri.org; Rerkasem, Kanok, "Uplands Land Use," in *Social Challenges for the Mekong Region*, eds. Mingsarn Kaosa-ard and John Dore, Chiang Mai, Thailand: Social Research Institute, Chiang Mai University, 2003, pp. 323–46; Richards, John F., and Elizabeth P. Flint, "A Century of Land-Use Change in South and Southeast Asia," in *Effects of Land Use Change on Atmospheric CO_2 Concentrations: South and Southeast Asia as a Case Study*, ed. Virginia H. Dale, New York: Springer Verlag, 1993, pp. 15–68; Rigg, Jonathan, and Sakunee Nattapoolwat, "Embracing the Global in Thailand: Activism and Pragmatism in an Era of Deagrarianization," *World Development* 29 (2001): 945–60; Uhlig, Harald, "Spontaneous and Planned Settlement in Southeast Asia," in *Agricultural Expansion and Pioneer Settlements in the Humid Tropics*, eds. Walther Manshard and William B. Morgan, Tokyo: United Nations University Press, 1988, pp. 7–43; United Nations Food and Agriculture Organization, Global Forest Resources Assessment 2000, Rome: FAO, 2001.

JEFFERSON FOX

Southeast Europe. Encompassing the whole Balkan Peninsula excluding Greece, post-communist Southeast Europe (SEE) is the least-developed region in **Europe**. This was caused by the heritage of a century-long and successive occupation by the Ottoman Empire (fourteenth to nineteenth century); and then by its inconvenient geo-economic position; and, after World War II, by the impact of a rigidly planned communist economy based on the exhaustion of all natural and human resources. Since the disintegration of Yugoslavia in the early 1990s, SEE has comprised seven countries, with a total area of 613,000 km^2, that is, larger than France and slightly smaller than Texas (about 6 percent of Europe). Four countries are post-Yugoslavian republics (Slovenia is considered part of **east-central Europe**). Among them, the most populous country is Serbia and Monte Negro (10 million inhabitants), and the least populous is Macedonia (2 million). Bosnia and Herzegovina, and Croatia each have about 4.5 million inhabitants. Within all SEE, the most populous country is Romania (22 million), with Bulgaria and Albania holding populations of 8 and 3 million inhabitants, respectively. In 2002, SEE had about 55 million inhabitants, which is 8.5 percent of the European population (with a population density of about eighty-eight persons square kilometers, i.e., far less than east-central Europe).

All countries are maritime except Macedonia. Romania and Bulgaria lie at the coast of the Black Sea, separated by the Danube River. The Danube enters the Black Sea through an ecologically valuable marshy delta in the boundary territory between Romania and Ukraine.

Geologically, SEE belongs to the Balkan region/peninsula. The Sava River and the lower Danube create its northern boundary; the western border is created by the indented island-rich seacoast of the Adriatic Sea, and the shoreline of the Black Sea forms the eastern border. SEE is geologically one of the youngest regions in Europe, as a part of the tertiary Alpine-Himalayan folder zone (with earthquakes being typical there). The surface is predominately mountainous. Plains and lowlands comprise only about 20 percent of the total area. Large lowlands are situated in eastern Croatia, in central-northern Serbia (surrounding Beograd and Vojvodina) and in Romania and Bulgaria along the Danube River. The Dinaric Alps stretch along the Adriatic coastline, and therefore their western slopes are very sheer. Only in Albania and at part of the Croatian seashore can one find coastal lowlands. Large karst areas are typical. **Mountains** of the Alpine-Carpathian system and the Transylvanian Plateau form the natural backbones of Romania and Bulgaria. The Danubian lowland fills the north of Bulgaria; and the south holds the highest ranges of the Balkans (Rodopi, Rila, and Pirin at 2,900 meters). In the Dinaridy region, stony rendzins, litosols, and so-called *terra rossa* prevail, with kambisols in lower parts.

In terms of climate, SEE lies at the dividing line between the temperate and Mediterranean subtropical climates. The majority of the SEE region belongs to the temperate climate zone. The inland area belongs to a zone of warm fluvial continental climate (Cfb type) with high temperature amplitudes throughout the year. Its opening to the north causes an inflow of cooler air from the Panonian basin in east-central Europe. A Mediterranean etesian climate with dry summers is typical for the Adriatic coast. It enables intensive recreational uses and large areas of wine growing.

In inland parts of the SEE, **forests** cover about one-third of the territory, and forest types vary with altitude from oak to beech with mixtures of maple and hornbeam (up to 1,800 meters in the south). This mountainous landscape is typical for its so-called medieval Walachian **colonization**—with dispersed farms at mountainous **pastures** (cattle, mutton), which survived until recent times. At the highest level, forests are made up of conifers, pine (mainly in Bosnia and Herzegovina), and spruce (in Serbia). The treeline ranges from 1,800 to 2,100 meters.

Original oak forests in the Mediterranean region were almost entirely extinguished by logging, especially during the rule of the medieval Republic of Venice, then by grazing (goats, sheep), and by **fire**. Oak forests were replaced by a macchia type of **secondary vegetation** (laurel, myrtle, frygana, and others), followed by evergreen oak forests in a few spots. **Deforestation** led to significant **soil erosion** and denudation, especially on steep slopes of the Dalmatian coast and its islands, where rocky ground of gray limestone emerged even (e.g., "polje," with lakes in southern Serbia). There, **reforestation** is impossible. The Balkan landscapes as well as that of the Apennine peninsula represent examples of the longest human impact on landscape configuration in Europe.

At the dawn of the **Industrial Revolution**, traditional agriculture, cattle grazing, and handicrafts dominated in the central and southern regions, whereas factories emerged in the north, and early capitalistic relationships developed already in the eighteenth century. In 1804, today's Croatia, as well as the former Hungarian Transylvania (from 1920 onward, part of Romania), were under the rule of the Habsburg Monarchy. Especially in the case of Croatia, this "historical start" over the rest of SEE has an imprint until nowadays. **Agriculture** and industry were most developed in Croatia and Serbia at the end of the nineteenth century (including Slovenia). Between the 1870s and 1912, almost all countries that were under Ottoman rule gained partial and consequently full independence, with Romania and Bulgaria (due to Russian pressure upon the Ottoman Empire) exerting control over the Bosporus and Dardanelle straits. After World War I, the kingdom of Serbs, Croatians, and Slovenes was established; and, after World War II, changed into a federal state called Yugoslavia, which gradually collapsed after 1990.

A remarkable agrarian character remained in the whole of SEE. In the Serbian lowland region of Banat, amelioration of land transformed marshlands into arable lands in the eighteenth century, triggering an intensive cultivation of cereal and **cash crops**. At the end of the nineteenth century, Croatia and Serbia were the most economically advanced regions. However, still after World War I, Yugoslavia remained an undeveloped agrarian country with subtle industry and less-developed agriculture—in comparison with Czechia, Hungary, and Slovenia in east-central Europe, for example. The same applies to Romania. However, due to extensive cereal cultivation in the fertile Walachian lowland, Romania became the fifth-largest producer of **maize** and the ninth-largest producer of **wheat** in the world before World War II.

After World War I, land ownership in the entire SEE was atomized into a big number of small farms and small number of great estates. The collectivization of agriculture took place from late 1940s until the late 1950s. A lot of cooperatives and state farms were established. The organizational structure

East Central and Southeast Europe: Land Use 1961–1990–2002 in Percent

Land category	1961		1990		2002	
	ECE	SEE	ECE	SEE	ECE	SEE
Arable land	49.2	35.0	45.3	33.0	43.2	31.4
Permanent cultures	1.8	2.6	1.8	2.7	1.6	2.3
Permanent grassland	14.0	19.9	12.9	21.3	13.1	19.7
Agricultural land	65.0	57.5	60.0	57.0	56.8	53.4
Forests-woodland	25.2	31.6	28.0	31.9	29.6	31.7
Other areas	9.8	10.9	12.0	11.1	12.5	14.9
TOTAL	100.0	100.0	100.0	100.0	100.0	100.0

Source: Statistical database of the UN Food and Agricultural Organization (http://www.fao.org).

of agriculture from the communist period partially survived after 1990. In the late 1990s, almost no new and capitalist-style corporate farms existed. Individual small farms continued cultivation on 50–60 percent of the agricultural land, in Albania even on 80 percent. Socialistic **industrialization** deepened the regional differences between urbanized regions and countryside. It also reinforced the deep differences among former Yugoslavian states, having endangered the integrity of Yugoslavia from 1980 onward. An agricultural crisis set in, especially in poor mountainous regions, with small farms not able to survive in many cases, so that depopulation of these regions followed. In the fertile Walachian lowland of Romania, the relatively important mining of oil and natural gas had a bad environmental impact.

In 2003, the average gross domestic product (GDP) (purchasing power parity per capita) was about the half of that of east-central Europe (i.e., about $6,390). These differences are reflected also in the average share of agriculture in total GDP, which is about four times higher than that of east-central Europe: 12 percent. Together with the low technological and organizational level of agriculture, these data indicate a prevailing rural character. The share of agricultural land in the total territory in 2002 was highest in Romania (62 percent) and Croatia (56 percent), and lowest in Albania (40 percent). Similar trends apply to arable land. Croatia and Macedonia (28 percent and 25 percent, respectively) hold the highest share of permanent **grassland**. The landscapes of SEE are largely wooded, with huge regional differences in the type of forests, though. Bosnia and Herzegovina and Macedonia hold the largest forest cover (44 and 35 percent, respectively), with Albania at 34 percent, Bulgaria at 33 percent, Romania with 27 percent, and Serbia and Montenegro having 28 percent.

The above table presents overall data of different developments in land use in SEE for the last thirty years (including the influence of transformation through capitalism). It is remarkable that east-central Europe has a much higher share of arable land than SEE. On the other hand, SEE shows a higher proportion of permanent **grassland**, is much more forested, and shows also a higher share of other areas (e.g., lakes and denudated rocky areas). Trends of land-use changes after 1989 in SEE differ a little from the development in east-central Europe. The area of arable land decreased significantly because mechanisms of **land rent** started to work again, and as a consequence, less-fertile soils have been abandoned and mostly transformed to permanent grasslands. The share of permanent grassland increased during 1961–1991, but decreased in 1991–2002. No changes occurred in forests and woodlands. In contrast, "other areas" increased considerably after 1990, probably as a consequence of **soil erosion** and **land abandonment** due to a large decrease of arable land.

See also: **Agricultural Revolution; Economic Restructuring; Land Reform; Mineral Extraction.**

Further Reading

Jelavich, Barbara, *History of the Balkans*, Cambridge, UK, New York: Cambridge University Press, 1983; Kann, Robert A., *A History of the Habsburg Empire, 1526–1918*, Berkeley, CA: University of California Press, 1974; Magocsi, Paul R., *Historical Atlas of East-Central Europe*, Seattle, WA, London: University of Washington Press, 1993; Sugar, Peter F., *Southeastern Europe under the Ottoman Rule, 1354–1818*, Seattle, WA: University of Washington Press, 1977; United Nations Food and Agricultural Organization, Statistical Database [Online, December 2004], The FAO Web Site, www.fao.org; U.S. Central Intelligence Agency, The World Factbook [Online, December 2004], The CIA Publication Web Site, www.cia.gov/cia/publications; Zagoroff, Slavcho, and Stanley P. King, eds., *The Agrarian Economy of the Danubian Countries, 1933–1945*, Stanford, CA: Stanford University Press, 1955.

LEOŠ JELEČEK

Southern Africa.   The region south of approximately 15° S, covering the modern countries of Namibia, Botswana, Zimbabwe, southern Mozambique, **South Africa**, Lesotho, and Swaziland. Southern Africa has been influenced by anatomically modern humans for about 200,000 years, and their hominid ancestors for over a million years. For most of the period, human impacts on the environment were limited to the effects of hunting, gathering, and the use of fire. About 2,000 years ago, Bantu agropastoralists arrived in southern Africa, bringing domesticated livestock, crops, and metal-working technology, which substantially increased human pressures on the landscape. The arrival of the Europeans about 400 years ago and their settlement of the region's interior over the past 150 years coincided with the industrial era and formed the basis for extensive transformation of the region during the twentieth century.

Almost 90 percent of southern Africa is covered by natural **vegetation**, although only about 8 percent of the region lies in formally protected areas. Most of the non-protected area is used for livestock grazing, although a growing fraction is being converted to nature-based tourism activities. Cultivation presently accounts for approximately 9 percent of the land area, and urban land uses for less than 2 percent. About half the region is covered by savanna; the remainder consists primarily of arid shrublands in the southwest and **grassland** in the southeast (see figure titled "The natural vegetation of southern Africa"). Most of the region has summer rainfall with a single peak between October and April; the exception is the extreme southwest, which experiences winter rainfall. Namibia, Botswana, and the western interior of

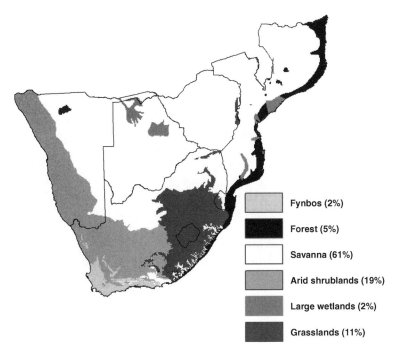

Fynbos (2%)

Forest (5%)

Savanna (61%)

Arid shrublands (19%)

Large wetlands (2%)

Grasslands (11%)

The natural vegetation of southern Africa, still covering almost 90 percent of the area, and a primary determinant of land use in the region.

South Africa are hot and arid; Zimbabwe, southern Mozambique, and the eastern parts of South Africa mostly have a semiarid or dry subhumid climate.

Early Human Influence

The San, also known as the Khoikhoi, bushmen, and a variety of other names, have inhabited southern Africa for at least 30,000 years. They were hunter-gatherers whose impact on the land cover was qualitatively little different than that of ancestral humans. **Fire** is thought to have been used by hominids for more than 1.5 million years for cooking and warmth. The San also used fire for hunting: large mammals were attracted to purposely burning specific pieces of land at certain times. The San harvested plants such as the marula (*Sclerocarya birrea*), !nara melon (*Acanthosicyos horridus*), rooibos (*Aspalathus linearis*), and buchu (*Agathosma betulina*), which continue to be highly valued today and in some cases are now grown commercially for regional and international markets. The pre-historic population density of the San is unknown, but probably low. The total hunter-gatherer population at the time of European arrival in the mid-seventeenth century has been estimated at about 50,000 people, translating to a density of about five people per 100 square kilometers.

The agropastoral Bantu-speaking people arrived from the north over a period of hundreds of years, starting about 2,500 years ago and reaching the southern Cape in present-day South Africa perhaps 1,000 years ago. They brought with them sheep and cattle, which had been domesticated in Asia, and millet, domesticated in North and West Africa. They also knew how to work metal, particularly iron, and make ceramic pots. As a result, their capacity and need to hunt large mammals (including carnivores) and to cut trees for fuel, building timber, and cultivated lands was much greater than that of the San, whom they progressively displaced into the more marginal arid southwestern parts of the region. With the arrival of the Bantu, a portion of the wild herbivore population was replaced by domestic livestock, and small areas of natural vegetation were cleared for cultivation. Bantu population densities in the pre-colonial period are estimated at 0.5–5 people per square kilometer, that is, at least an order of magnitude greater than those of the San.

Africa seems to be unique in that the growing human population did not lead to a rapid demise of large wildlife populations, but co-existed with large numbers and a wide diversity of other mammals. In pre-colonial times, people may have had a controlling influence on the populations of slow-breeding megaherbivores such as elephants, which in turn can have significant local impacts on vegetation. People certainly had the capacity to hunt elephants successfully, and a trade in ivory has existed for at least a millennium on the east coast of Africa.

Hunting, pastoralism, cultivation, and other human activities in the pre-colonial era had consequences for land cover, certainly locally, and probably also regionally. For example, the Tswana people on the edges of the Kalahari Desert lived in settlements of several tens of thousands of people each, covering a hundred hectares or more of stone walling. The oral history, records of early European explorers, and archaeological evidence agree that the settlements would each last a few decades before moving to another location. The reason for moving may have included depletion of grazing and wood in the vicinity of the town, depletion of the water supply, and/or a buildup of pathogens.

European Settlement and the Current Era

The first permanent European settlement in southern Africa was established in 1652 in the area of present-day Cape Town, but significant settlement of the interior took place only after 1835. The technology and political trade links that the Europeans brought and the start of the industrial era provided the basis for the largest-scale transformation of the region's land cover to date, taking place over the course of the twentieth century. A measure of the **land-cover change** that has occurred is that the regional population at the time the European settlers arrived in the mid-seventeenth century was probably no more than a few million people, with a similar number of livestock, and cultivating an area of under a million hectares. In 2000, the population was approximately 80 million, with 75 million livestock (cattle, sheep, and goats) and 35 million hectares of cultivated land. The following figure gives an estimate of the location and extent of areas presently under different forms of land use.

Pastoralism

Today, livestock production is the land use that occupies the largest fraction of southern Africa, and domestic livestock (especially cattle) are the

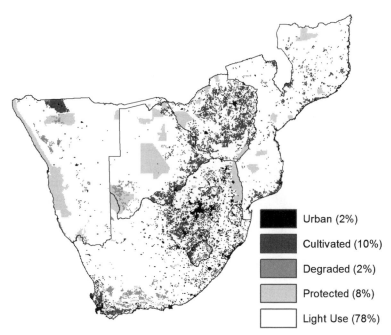

Urban (2%)

Cultivated (10%)

Degraded (2%)

Protected (8%)

Light Use (78%)

The major land uses in southern Africa. Most of the untransformed "light use" areas are used for livestock grazing and, under certain circumstances, may become degraded.

largest consumers of vegetation in the region. While livestock have been present throughout southern Africa for approximately 2,000 years, the numbers and breeds changed significantly during the twentieth century. Unlike in many parts of the world, livestock production in southern Africa seldom involves a deliberate transformation of the land cover. Livestock are almost exclusively raised on natural vegetation, taking the place of large indigenous herbivores. Pastoralism in southern Africa therefore generally has a fairly low impact on the indigenous flora and fauna. Only if livestock numbers are very high relative to capacity of the natural vegetation to support them does a change in land cover take place. In these cases, the natural vegetation persists, but in a degraded form—typically, vegetation production (i.e., biomass and cover) is reduced to about half of the non-degraded state, often accompanied by significant changes in the dominant species of plants, and sometimes by a structural change to greater or less dominance by woody plants, and accelerated loss of soil.

The major increase in livestock numbers in southern Africa followed **industrialization** toward the end of the nineteenth century, aided by the drilling of boreholes to provide surface water, immunization against disease, and feed supplementation. The first significant changes in livestock numbers took place in South Africa with a "boom" in merino sheep, introduced in the Karoo and eastern Cape areas in about 1870. Their numbers grew to a peak and then declined by almost half in the 1920s, possibly because the vegetation cover was permanently altered from a grassy to a shrubbier, less palatable form. A European fashion demand for feathers and the **domestication** of ostriches led to an ostrich population boom in the southern Cape during the early twentieth century. This boom was revived in the late twentieth century, but this time driven by a demand for ostrich meat. In the drier parts of the region (receiving below 450 millimeters of mean annual rainfall), sheep and goats are the dominant livestock. Donkeys are important draft animals in poorer communities throughout the region, but make up a small proportion of the domestic livestock biomass.

Cattle are mainly kept in the wetter (greater than 450 millimeter annual rainfall) savanna and grassland areas of the region. In southern Africa, cattle are not kept only for meat: especially in traditional societies living on communal lands, cattle are an important way of accumulating assets and demonstrating wealth and status. Cattle numbers in southern Africa increased more gradually than those of sheep and goats, reaching approximately their current levels in the 1960s. In Botswana, the Kalahari Desert region, which had no permanent water, was opened to **cattle ranching** by the drilling of deep boreholes during 1970–2000. The low-lying parts of Zimbabwe, Mozambique, and northeastern South Africa were unfavorable for cattle due to the presence of the tsetse fly, which carries the *Trypanosoma* protozoans that cause *nagana* in cattle and sleeping sickness in humans. The tsetse fly was eradicated from most of this area after the invention of DDT, from about 1940 onward, so that cattle ranching is now an important land use in these areas.

Deforestation and Land Degradation

Deforestation in its narrow sense has never been an extensive process in southern Africa, because less than 1 percent of the land surface supported tall, closed forest, even in pre-colonial times. The settler demand for large-dimension timber for ships, buildings, wagons, and later mine and railway infrastructure quickly depleted the local sources (mainly along the southern and eastern coast of South Africa), which had halved in extent by 1850.

Deforestation in the broader sense, that is, reduction in woody cover of the landscape (most of southern Africa is partly wooded), had been going on at a gradual rate for millennia. For example, it is calculated that one *kraal* (a pen for corralling cattle, built by every household and necessary to protect the livestock from predators) consumes 150 tons of wood. During the twentieth century, two divergent land-cover patterns emerged with respect to woody cover in areas with a savanna climate (a hot, wet summer and a warm, dry winter; about 60 percent of southern Africa). Those lands occupied by European cattle ranchers showed a marked increase in woody plant biomass, often to the point of becoming uneconomical for ranching use. This phenomenon, known as bush encroachment, has also been observed following European settlement in Australia, Texas, and Argentina, and has been attributed partly to changes in the prevailing fire regimes. Both the intensity and frequency of fires are believed to have been reduced, resulting in a rapid increase in shrub and tree cover, which further suppressed fires.

On the African communal lands, on the other hand, there was a marked loss in tree cover over the twentieth century. By the 1950s, land ownership by Africans in South Africa, Zimbabwe, Namibia, and Mozambique was restricted by law to a minor part of the land area (13 percent in South Africa, the so-called homelands; about 45 percent of Zimbabwe, the "Tribal Trust

Lands"; and 27 percent of Namibia). **Poverty**, high densities of people and livestock, and a breakdown in traditional **institutions** of land management led to widespread clearing, overgrazing, and **soil erosion**. The consequences on land cover were dramatic and will remain visible for decades after the repeal of the land-ownership laws in the 1980s and 1990s. Large areas of the "homelands" are today effectively peri-urban settlements or **remittances landscapes**, economically sustained by social and wage remittances from urban areas.

Fires collectively burn an area of about 3.7 million km^2 per year in the region of Africa south of the equator (a third of the land area, mostly in savannas and grasslands). At the present time, more than three-quarters of fires in southern Africa are ignited by people, mostly in order to manage ecosystems for grazing, honey, and agriculture. While fires may lead to substantial changes in land cover for a period of days to weeks, and in the very long term (centuries) have shaped entire landscapes, in the annual to decadal time frame they are typically "included disturbances" that do not change the land cover. Fires are a natural and required feature of southern Africa's ecology. Most savannas and grasslands become unusable for grazing if fire is excluded. Unless the total amount of **biomass** burned changes over time, there is no net emission of carbon dioxide—the carbon dioxide emitted by wildfires is reabsorbed within weeks to years by the regrowth of the vegetation. Humans have probably been the principal source of ignition of wildfires in Africa throughout the past million years, and there is little evidence that the overall extent of fires in southern Africa has changed substantially over this period.

Cultivation

Taken over the full extent of southern Africa, crop cultivation is the land use that has had the biggest impact on the terrestrial biota of the region. Most of the cultivated land is in South Africa (36 percent of total cultivated area in the region) and Zimbabwe (29 percent) in large commercial farms (although in Zimbabwe these are currently being reclaimed by the government as part of its land redistribution policy).

Large-scale cultivation in southern Africa started in the early twentieth century. In South Africa, the area under cultivation grew steadily from the end of the South African (Anglo-Boer) War in 1902. A major phase of expansion occurred in the 1960s that was related to the introduction of fertilizers, new cultivars (particularly of **maize**), mechanization, and irrigation. The area under cultivation reached a plateau in the 1970s, when most of the economically arable land was under cultivation, and has declined slightly since agricultural subsidies were reduced. Cultivation in Zimbabwe began to expand from the subsistence level in the 1920s. The major expansion in the 1960s was driven by an influx of white settlers and new agricultural technologies: **tobacco**, hybrid maize, and agricultural lime in particular. Redistribution of this land to smaller-scale African farmers with few technical and financial resources, starting in the 1990s, will have as-yet-unknown land-cover consequences.

The cultivated area in the other countries in southern Africa is currently small in comparison and is largely limited by climatic and soil constraints.

Subsistence agriculture, while not occupying extensive areas, plays an extremely important role in the livelihoods of impoverished rural communities. At the start of the twenty-first century, most of the economically arable land in southern Africa was under cultivation, and a range of **scenarios** shows little further transformation. The major remaining potential for agricultural transformation lies in Angola, Zambia, and northern Mozambique, north of the area covered in this article.

Beginning in the early twentieth century, and accelerating after 1950, substantial areas (currently 1.5 million hectares in South Africa and 0.15 million hectares each in Swaziland and Zimbabwe) of moist highland grassland were converted to exotic timber plantations (eucalyptus, pine, and Australian *acacia* species) for the provision of mine-support timber, lumber, poles, tannin, and paper pulp. Further expansion was slowed in the 1990s by recognition of the deleterious consequences of plantations on **biodiversity** and streamflow in planted **watersheds**.

See also: **African Trypanosomiasis; Colonization; Conservation; Green Revolution; Hot Spots of Land-Cover Change; Land Rights; People at Risk.**

Further Reading

Biggs, Reinette, and Robert J. Scholes, "Land-Cover Changes in South Africa 1911–1993," *South African Journal of Science* 98 (2002): 420–4; Dean, W. Richard J., and Ian A. W. MacDonald, "Historical Changes in Stocking Rates of Domestic Livestock as a Measure of Semi-Arid and Arid Rangeland Degradation in the Cape Province, South Africa," *Journal of Arid Environments* 26 (1994): 281–98; Geist, Helmut J., "Causes and Pathways of Land Change in Southern Africa during the Past 300 Years," *Erdkunde* 56 (2002): 144–56; Hoffman, M. Timm, "Human Impacts on Vegetation," in *Vegetation of Southern Africa*, eds. Richard M. Cowling, David M. Richardson, and Shirley M. Pierce, Cambridge, UK: Cambridge University Press, 1997, pp. 507–34; Scholes, Robert J., and Reinette Biggs, eds., *Ecosystem Services in Southern Africa: A Regional Assessment*, Pretoria, South Africa: Council for Scientific and Industrial Research, 2004, available online at www.maweb.org/en/subglobal.safma.aspx.

REINETTE (OONSIE) BIGGS, ROBERT J. SCHOLES

Southern Cone Region. Southern part of **South America**, which includes Argentina, Chile, and Uruguay, and consists of subtropical and temperate ecosystems (see figure). The southern half of this region, from approximately 40° to 55° S in Argentina, is known as Patagonia. While geographically isolated, the principal land uses in the Southern Cone are linked directly to global markets in agricultural and forest products. Regional land changes affect the viability of local production systems, such as sheep ranching, and a range of environmental services, including **biodiversity** and the **carbon cycle**. Despite these connections to regional and earth systems, the Southern Cone has received relatively little attention from global change and sustainability science. Lacking are integrated studies of the relationships between identified land changes and climatic, local biophysical, political-economic, social, and cultural conditions.

Significant European **colonization** of southern South America did not occur until the nineteenth century, late in global terms. Today it remains a

relatively sparsely settled region of the world; however, its temperate **forests** and **grasslands**, the two land covers emphasized here, have been dramatically altered by human action. Throughout the Southern Cone there has been extensive forest clearing and thinning, and the introduction of non-native flora and fauna (e.g., via forest plantations). Intensive sheep and cattle ranching have led to **soil erosion**, changes in species composition, and falling productivity in the region's native grasslands. The range of biophysical, socioeconomic, and political factors driving these kinds of land changes in other parts of the world are represented in the Southern Cone as well. Also represented are societal attempts to slow human impact, most notably through conservation reserves.

Forest Change

Grass prairies dominate the central and southern parts of Uruguay, and the northern part of the country is covered by subtropical forest and shrub formations, although only 5.72 percent of Uruguay's land area is forested. These forests are being actively managed, however, and there was a 4 percent loss of total forest cover between 1990 and 2000. Almost 90 percent of Uruguay is devoted to **agriculture** or **cattle ranching**.

As of 2000, Argentina had 34,648,000 hectares of forest (12.7 percent of the country's land area), with 926,000 hectares in plantations. According to data of the **United Nations Food and Agriculture Organization** (FAO), there was a net loss of 285,000 hectares of forest at an annual rate of change of −0.8 percent from 1990–2000. Forest change in the country is not a new problem. In the late nineteenth century, the government sold large blocks of land in the country's northern Chaco region to timber companies and cattle barons. Subsequently, most of the forests there were clearcut and now remain as desert scrub.

In 2000, Chile had 15,536,000 hectares of forest (20.7 percent of the country's land area), with 2,017,000 hectares in plantations. Forest-cover change shows a loss of 20,000 hectares from 1990–2000, at an annual rate of −0.1 percent, according to FAO data. Analyzing Chilean forest change in more depth gives a sense of the kinds of forest change dynamics witnessed in the larger region.

Human impact on Chilean forests is most severe in the coastal and lowland zones of the south-central part of the country (35.6°–42.3° S). Endangering the forests is a mix of **proximate causes**, including uncontrolled logging, plantation forestry (usually using introduced species of pine and eucalyptus), farming, the rearing of livestock, invasive species, **fire**, and agro-industrial pollution. There is continuing road building, such as the controversial Southern Coastal Highway between Valdivia and Puerto Montt, which has been re-routed to help protect coastal rainforest. Plantation forestry was initiated in 1940 and has been subsidized by the government since 1974 (Forest Law 701), which has led to some 2 million hectares of *Pinus radiata* being planted between 36° and 39° S latitude. Now less than 45 percent of forest cover in Chile is mature native forest.

Among the underlying **driving forces** of forest change in Chile are policies enacted under the dictatorship of Augusto Pinochet (1973–1990). As part of a suite of neoliberal policies, Pinochet targeted native forests and the intensive exploitation of natural resources nationwide. This focus led to incentives for establishing forest plantations as well as the implementation of a policy in 1985 to sell native forestland, much of which was quite cheap compared to industrial forests elsewhere in the world, and that attracted the attention of multinational corporations. At the same time, the Chilean government's oversight of the logging industry has been weak, with lack of enforcement, **monitoring**, and appropriate fines. Government policy that facilitated logging combined with market forces in 1987 when there was a boom internationally in demand for Chilean wood chips and pulp. The result was the large-scale conversion of native forests to plantations of **exotic species**, principally by multinational corporations, which led to what has been described as the "tragedy of the Chilean forest."

The prominent degradation of forests in south-central Chile hides some more positive trends, however. First, many industries in the country are becoming certified by ISO 14,001 (requiring an environmental management system) and the Forest Stewardship Council. Second, the Ley Marco Environmental Framework Law of Chile, which was created in 1994 and put into effect in 1997, mandates a system of environmental impact evaluation (SEIA). Akin to the environmental impact assessment component of the U.S. National Environmental Policy Act (NEPA), SEIA requires comprehensive assessment of the environmental impacts of development projects. Third, **forestry** shows signs of movement toward more sustainable methods. For example, government regulations in both Chile (since 1974) and Argentina (more recently) require a shelterwood system of forest management (regionally called a *corte de protección* or protection cut). Southern beech (*Nothofagus* spp.), which dominates the forests of Patagonia, has patchy cycles of regeneration due to shade intolerance and the natural mortality of trees; a shelterwood system protects the overstory and mirrors natural disturbance and regeneration processes.

Finally, as is the case for all of **Latin America**, there is a growing embrace of nature **conservation** in the Southern Cone. Many of the national parks in Argentina and Chile were established because of the desire to establish a presence along the long-disputed international border between the two countries rather than for nature conservation. But in Chile, decree-law no. 18,362 created the national public system of protected areas (SNASPE) in 1984, which seeks to protect ecosystem and evolutionary processes as primary goals. One half of the area of Chile's southernmost Region XII is under protection, and 40 percent of the forests in the Argentinean **Andes** of Patagonia are under public protection as well. These protected areas do not include private reserves, such as Estancia Yendegaia in Tierra del Fuego. These figures are somewhat misleading, however, given that most of the protected areas in Patagonia are remote, have low species richness and endemism, and often lie above 600 meters in altitude, consisting of ice and unvegetated terrain.

There is also controversy about some of the nature conservation areas. For example, American Douglas Tompkins has purchased extensive landholdings in Chile, including the 320,000-hectare Pumalin Park in the country's Palena province, which he established as a large private nature reserve. Many Chileans resent that the reserve, owned by a foreign national and extending from the Argentinean border to Chile's Pacific coast, cuts the country in half. In another example of controversy, on the site of a failed logging project by a U.S.-based corporation, the investment company Goldman Sachs announced in September 2004 the establishment of a 280,000-hectare nature reserve in Tierra del Fuego, Chile. Environmentalists are celebrating the protection of the area's forest; however, the decade-long battle over what was, by most accounts, one of the most innovative sustainable logging initiatives to date highlights the difficulty in balancing the environment and development in the region.

Grassland Change

For the most part, too far south for intensive agricultural production, Patagonia's native grasslands have been the site of sheep ranching since the late nineteenth century. Sheep numbers peaked in the mid-twentieth century, after which a combination of market shifts and degraded rangelands resulted in a steady decline. Shrub encroachment and **soil degradation** are widespread phenomena in arid and semiarid rangelands around the world. **Land degradation** affects most of Patagonia and is linked primarily to over-stocking of the rangelands. The landscape has been altered so dramatically that it is difficult to determine pre-European composition and characteristics of the vegetation. In the Patagonian steppe, shrubs provide relatively poor forage for livestock compared with the region's perennial grasses; shrub steppe is found to have more severe erosion and higher water loss than in grass dominated areas; and the poor soil characteristics of the degraded shrub steppe may inhibit the reestablishment of perennial grasses, making attempts to reverse degradation processes difficult.

Despite Patagonia being the most threatened landscape in Argentina and Chile, there are no reserves in either country to protect grasslands. David Aagesen identifies four key human driving forces in the degradation of Pata-gonian rangelands: ranchers' lack of concern about soil conditions, skewed **land tenure** that results in small landholding size (which affects economies of scale and rangeland management strategies) and public lands where there tends to be more severe change, the lack of a diversified regional economy, and top-down and largely ineffective government policy. Human impact on grasslands has been exacerbated by both the strong winds that dominate the region and also by dry periods in the twentieth century. Fire history has also shaped current forest and shrubland patterns across Patagonia.

Intensive Agriculture and Ranching

The humid subtropical zone of eastern Argentina and Uruguay has moder-ate temperatures and precipitation throughout the year. Land covers range from prairie in the pampas of Argentina to a mixed forest-grassland cover in Uruguay. The pampas, along with the western Chaco region of Argentina and the Central Valley of Chile, are the most prominent agricultural zone in the Southern Cone. Often described as having unsurpassed agricultural potential, the pampas of Argentina are an extensive fertile plain that extends in an arc of hundreds of kilometers around Buenos Aires. Given its rich and deep topsoil, and a geographic location that historically has been more accessible to international markets than the rest of the Southern Cone, start-ing in the last quarter of the nineteenth century the region was initially devoted almost entirely to **pasture**, grain cereals, and **maize**. After a bust in the livestock industry in the 1910s, other agricultural uses, such as **cotton** and fruits, were added.

Once the site of over 300 grasses and other forage plants, development of the area involved widespread replacement of these native species with introduced pasture grasses, alfalfa, and feed grain. By the 1950s, overstocking of the rangelands and other poor land management practices, wind and water erosion, and the region's irregular climate (drought affects more pro-foundly the introduced grasses and crops than it does native varieties) had resulted in severe land degradation, such as soil erosion and weed expansion. The unequal distribution of land tenure meant that smallholders were pushed to more marginal lands in the western pampas. More recently, con-tinuous cultivation and double cropping have also led to soil degradation. Despite these changes, however, most of the pampas in Argentina are not severely degraded. But after forests were cleared from the western edge of the Chaco, an agricultural boom occurred in the 1960s that focused initially on grains, beans, and oil seeds. Subsequently, drought, wind erosion, and poor land management practices resulted in severe land degradation.

The Central Valley of Chile (approximately 32°–38° S) has a Mediterranean climate and some of the richest soils in South America. The region, much of it irrigated, is the country's agricultural heartland; there is widespread mining activity, in particular, exploiting the area's extensive copper reserves; and the zone is highly urbanized. Twentieth-century land change in the area is profound.

See also: **Agricultural Intensification; Desertification; Forest Degradation; Mineral Extraction; Urbanization.**

Further Reading

Aagesen, David, "Crisis and Conservation at the End of the World: Sheep Ranching in Argentine Patagonia," *Environmental Conservation* 27 (2000): 208–15; Armesto, Juan J., Tozzi, Ricardo, Smith-Ramirez, C., and Mary T. K. Arroyo, "Conservation Targets in South American Temperate Forests," *Science* 282 (1998): 1271–2; Food and Agriculture Organization of the United Nations, Global Forest Resources Assessment 2000, FAO Forestry Paper no. 140, Rome [Online, April 2004], The FAO Web Site, www.fao.org; Gustavo Zarrilli, Adrián, "Capitalism, Ecology, and Agrarian Expansion in the Pampean Region, 1890–1950," *Environmental History* 6 (2001): 561–83; Neira, Eduardo, Verscheure, Hernán, and Carmen Revenga, *Chile's Frontier Forests: Conserving a Global Treasure*, Washington, DC: World Resources Institute: Comité Nacional Pro Defensa de la Fauna y Flora, and the Universidad Austral de Chile, 2002; Parizek, Bernardo, Rostagno, Cesar M., and Roberto Sottini, "Soil Erosion as Affected by Shrub Encroachment in Northeastern Patagonia," *Journal of Range Management* 55 (2002): 43–8; Pauchard, Anibal, and Pablo Villarroel, "Protected Areas in Chile: History, Current Status, and Challenges," *Natural Areas Journal* 22 (2002): 318–30; Sawers, Larry, "Income Distribution and Environmental Degradation in the Argentine Interior," *Latin American Research Review* 35 (2000): 3–33; Veblen, Thomas T., Kitzberger, Thomas, Villalba, Ricardo, and Joseph Donnegan, "Fire History in Northern Patagonia: The Roles of Humans and Climatic Variation," *Ecological Monographs* 69 (1999): 47–67.

PETER KLEPEIS

Spatial Diffusion.   The movement and dispersion of phenomena across geographic space, and through time, from an initial location to a wider area. The diffusing phenomena can include people (migration), ideas (innovation), goods, services, urban development, species, or diseases. Diffusion reflects human interaction in space and time, and diffusion processes contribute to the development of spatial structure and land-use patterns. There are two basic types of spatial diffusion: relocation diffusion, where an object no longer remains in the originating area, for example, the migration of people; and expansion diffusion, where an object passes between areas, but its influence remains in the originating area, for example, the movement of information. Expansion diffusion further subdivides into contagious diffusion, which depends on direct contact between people and is, therefore, a function of proximity or distance, for example, the transmission of contagious diseases; and hierarchic diffusion, in which movement occurs through sequences of classes or hierarchies, for example, transmission of ideas through social networks.

The diffusion of innovation is of particular interest in land-use studies because it can affect the ways in which land-use decisions are made and the rates of land-use change. The diffusion of innovation often follows a logistic curve, which reflects an initially rapid growth rate in the number of adopters of an innovation, after which the rate of adoption slows with time as the number of adopters reaches saturation. The presence of diffusion processes can often be observed in geographic space through the typical wave patterns they produce. The wave height may, for example, reflect the number of new adopters of an innovation, the number of migrating people, or the incidences of a disease. Diffusion waves change character with distance from the time and point of origin.

Torsten Hägerstrand developed perhaps one of the earliest and best-known models of spatial diffusion. The Hägerstrand model was used to simulate the diffusion of knowledge about a new **pasture** management technique in an agricultural community in Sweden. The model represented the spread of information between farmers through personal contact (contagious diffusion). Distance was treated explicitly as the factor controlling the frequency of contacts between farmers and therefore the rate and location of new adoption. Farmers in close proximity had a higher probability of contact than farmers further apart, although these contacts were also driven by a random element implemented using Monte Carlo simulation techniques. Each farmer could theoretically have his own contact-distance relationship (depending on the range and frequency of his movements); however, in practice a mean information field was used to represent the probability of contact between all farmers. The model was further developed to favor the movement of knowledge through farmers with larger farms (i.e., hierarchic diffusion). The Hägerstrand model was very important in land-use science as it demonstrated how land-use change could be understood as a spatial dynamic process driven by the diffusion of knowledge. The model was the precursor of many present-day land-use models based on **cellular automata** and agent-based approaches.

See also: **Agent-Based Model; Decision-Making; Economic Restructuring; Integrated Model; Modeling.**

Further Reading

Hägerstrand, Torsten, *Innovation Diffusion as a Spatial Process*, Chicago: The University of Chicago Press, 1967.

MARK ROUNSEVELL

Species Extinction. The disappearance of an organism or naturally occurring population that can interbreed only with others of its species (microorganisms, fungi, plants, and animals), as a result of hunting, natural disaster, climate change, overspecialization, unsuccessful competition for resources, introduction of **exotic species**, and/or human-caused stresses associated with the **proximate causes** and underlying **driving forces** of land-use/cover change, including the use of **fire**. Extinction of species has occurred since the beginning of life on earth. In fact, most of the organisms that have ever existed are now extinct: the species currently living amount to 2–3 percent of those that have ever lived; all others have become extinct, typically within about 10 million years of their first appearance. Five significant extinction events occurred, including the extinction of megafauna at the end of the Pleistocene era, but these events collectively seem to have ended no more than 5–10 percent of the species that ever lived. The current estimate for the total number of species possibly lies in the 5–15 million range, with a best guess of about 7 million. However, the species at risk now represent an unusually high proportion of all those that ever lived. It appears that a recent pulse of extinction started during the late Quaternary period. Evidence indicates that a massive (sixth) extinction event, driven by human activities, has been underway for some 40,000 years. Human activities have greatly accelerated the rate of species extinction far beyond the natural rate. There is agreement that the average species has a life span of about 5–10 million years, and that the rate of extinction is 1–0.1 species per million species-years. For the past 300 years, recorded extinctions for a few groups of organisms reveal rates at least several hundred times the rate expected on the basis of geological record. Rodolfo Dirzo and Peter Raven estimate that extinction rates in the next fifty years are some 1,000 times higher than the background rate of 1.0. Historically, overhunting has been the most consistent explanation of the megafaunal extinctions of the past 40,000 years. In addition, "first-contact extinctions" of species over the last 1,000 years are numerous on islands such as Madagascar and New Zealand, that is, extinction correlated with the arrival of humans including the introduction of alien invasive species. Today's major causes of threat are habitat loss and degradation related to agricultural activities (plantations, crops, and livestock farming), extraction activities (logging, harvesting, mining, and fishing), and the development of infrastructure (human settlements, industry, roads, dams, and power plants and lines). Direct exploitation such as hunting (also trading and collecting) is the current driver second in importance.

The rate of extinction had and still has an important influence on the world's **biodiversity**. This is true despite the fact that only a very few groups of organisms are well-enough known to be assessed for extinction. For example, some nineteen of each twenty tropical moist **forest** species are unknown to science at present, so that the effects of ongoing forest burning result in tremendous biodiversity losses. The International Union for Conservation of Nature and Natural Resources (IUCN) lists about 11,200 threatened species facing a high risk of extinction in the near future resulting from human activities. Although this is a small number relative to the total number of species (i.e., less than 1 percent), it includes 24 percent and 12 percent of all mammals and birds, respectively. For plants, the IUCN number represents only 2–3 percent of the known species, but individual country evidence suggests far more extinctions. In the **United States of America**, for example, it is estimated that about 33 percent of the plant species native to the country are threatened with extinction, including 24 percent of the conifers (most seriously threatened, though, are 69 percent of the freshwater mussels and 51 percent of the crayfish). Therefore, Rodolfo Dirzo and Peter Raven propose that a reasonable *interim* estimate would be that, at present, a third of the plant species of the world are threatened. As for birds and mammals, they provide a global picture of extinction as follows: at least 500 (but probably closer to 600) out of 1,192 threatened bird species and some 565 of the 1,137 threatened species of mammals will go extinct in the next fifty years, due to habitat loss and fragmentation, and mainly in the tropical forest zone. They conclude that, at any event, more than a third of the existing species on earth could disappear with the destruction of tropical forests only, and that it is reasonable to envision the loss of two-thirds of the species on earth by the end of the twenty-first century.

See also: **Agrodiversity; Biodiversity Novelties; Carson, Rachel Louise; Conservation; Cumulative Change; Deforestation; Ecological Colonization; Forests; Marsh, George Perkins; Paleo Perspectives; Tropical Dry Forest; Tropical Humid Forest; Turnover.**

Further Reading

Dirzo, Rodolfo, and Peter H. Raven, "Global State of Biodiversity and Loss," *Annual Review of Environment and Resources* 28 (2003): 137–67; International Union for Conservation of Nature and Natural Resources (IUCN), Species Survival Commission [Online, September 2004], Red List of Threatened Species Web Site, www.redlist.org.

HELMUT GEIST

SPOT Vegetation.   A series of satellites originally developed by the French Space Agency (CNES), the first being launched in 1986 and the latest (SPOT-5) launched in 2002. The latest two platforms carry the Vegetation sensor. The SPOT Vegetation system has four spectral bands at visible and infrared wavelengths with a resolution of one kilometer. The Vegetation Programme was developed by a number of partners for the **monitoring** of land-surface parameters with daily frequency at global scale. The Vegetation system will complement the existing high-spatial resolution capabilities of the SPOT series. Data from SPOT Vegetation has been used to map the distribution of burnt areas at the global scale, assess the impact of natural disasters, and map land cover and **land-cover change**.

See also: **Advanced Very High Resolution Radiometer; Global Burnt Area 2000; Global Land Cover Map of the Year 2000; GLOBCARBON; Normalized Difference Vegetation Index; Remote Sensing; Tropical Ecosystem Environment Observations by Satellite (TREES) Project; Vegetation.**

Further Reading

French Space Agency (CNES) [Online, August 2004], The SPOT Vegetation Programme Web Site, www.spot-vegetation.com; Lupo, Fred, Reginster, Isabella, and Éric F. Lambin, "Monitoring Land-Cover Changes in West Africa with SPOT VEGETATION: Impact of Natural Disasters in 1998–1999," *International Journal of Remote Sensing* 22 (2001): 2633–9; Mayaux, Philippe, Bartholomé, Etienne, Fritz, Steffen, and Alan Belward, "A New Land-Cover Map of Africa for the Year 2000," *Journal of Biogeography* 31 (2004): 861–77.

KEVIN J. TANSEY

Subsistence Agriculture. Agricultural production that is oriented toward home **consumption** and characterized by low application of external production factors and low output per unit of land, in contrast to market-oriented or commercial agriculture. The main farm inputs in subsistence agriculture (or subsistence farming) are family labor and land, while capital **investment** is minimal to nonexistent. Subsistence farming relies on the cultivation of **food crops** to fulfill the immediate nutrition demands of the household. The amount of farm output determines the amount a family can consume in pure subsistence agriculture. However, the term subsistence agriculture involves several ambiguities. This stems from the fact that farm households are producers and consumers at the same time, in contrast to conventional economic theory where firms produce goods by maximizing their profits, while households supply labor and consume the products. Consequently, subsistence-oriented farm households are economic entities, which make both production and consumption decisions. The objective of subsistence land users is not profit maximization but the maximization of subjective utility, which includes **tradeoffs** in limited labor allocation between the necessary satisfaction of consumption requirements and the achievement of leisure time. Pure subsistence farmers are completely outside of market exchanges and produce exclusively for home consumption, while a pure commercial farmer produces solely for sale on markets. The term subsistence agriculture is not clearcut and implies a certain threshold along a gradient between zero market integration and a 100 percent share of marketed agricultural production. The proportion of production devoted to a household's home consumption is termed the "degree of subsistence" of a farm household. Often, a 50 percent threshold is used. Historically, most farmers were engaged in subsistence-oriented agricultural production. This is still the case in the least-developed countries (LDCs), but pure subsistence farming is becoming increasingly rare and most farmers today produce some marketable surplus. The degree of subsistence of farming households shows substantial geographic variations, but is especially high in LDCs. Subsistence agriculture also increased after the collapse of socialism in the transition countries of **east-central Europe**.

The use of only family labor in combination with traditional farming tools limits the amount of land a farm household can cultivate. As cultivated fields on sloping land tend to lose fertility under permanent cultivation, the most economic method with the highest return per unit of labor is shifting or **swidden cultivation** to restore the fertility of plots under **fallow**. The expansion of areas devoted to subsistence production in permanent and shifting agricultural cultivation is identified as one of the major causes of tropical **deforestation**. Population growth through natural growth or inmigration often leads to this expansion of cropped areas in traditional subsistence-based communities. Increasing market orientation and intensification of subsistence-based economies brought about by economic development may be a major factor contributing to land-use and **land-cover changes** in the future.

See also: **Agricultural Frontier; Agriculture; Cassava; Decision-Making; Human Immunodeficiency Virus (HIV)/Acquired Immunodeficiency Syndrome (AIDS); Landholding; Land-Use System; Population Dynamics; Proximate Causes.**

Further Reading

Chayanov, Alexander V., *The Theory of Peasant Economy*, Madison, WI: University of Wisconsin Press, 1986; Ellis, Frank, *Peasant Economics*, Cambridge, UK: Cambridge University Press, 1993; Upton, Martin, *The Economics of Tropical Farming Systems*, Cambridge, UK: Cambridge University Press, 1996; Singh, Inderjit, Squire, Lyn, and John Strauss, *Agricultural Household Models: Extensions, Applications, and Policy*, Baltimore, MD: Johns Hopkins University Press, 1986.

DANIEL MÜLLER

Suburbanization.   The decentralization of cities caused by outward movement of households and firms to the urban periphery. The process of suburbanization has been taking place for well over a century and has occurred in cities worldwide. However, it accelerated markedly in the latter half of the twentieth century and has proceeded faster and farther in the **United States of America** than in any other country.

As Alex Anas, Richard Arnott, and Kenneth Small detail, suburbanization is fundamentally tied to changes in transportation and communication systems. At the beginning of the nineteenth century in pre-industrialized cities, intra-urban transport took place mainly by horse and wagon or by foot. People lived where they worked, and there was a clear demarcation line between the dense city and sparsely settled countryside. Within cities, the rich and poor lived together, although the rich, for the most part, outbid the poor for the most central and convenient locations. The industrial age led to massive rural-to-urban migration as newly built factories and mills attracted people from the countryside and spurred urban agglomerations. In the United States, the advent of the railroad in the 1850s and then the electric streetcar in the 1870s brought about the first real suburbs as increasing numbers of upper- and middle-income commuters moved out along railcar and then streetcar lines. The result was a "hub-and-spoke" pattern of development: a compact production-oriented core and radial streetcar lines with residences concentrated around stations located along these lines. The "streetcar suburbs" marked the first real emergence of land-use patterns characterized by a clear separation between residential and commercial/industrial uses. The separation of land uses was further reinforced by local zoning laws that became common in the United States by the 1920s.

By far the most influential force underlying suburbanization has been the automobile. Assembly-line production beginning in 1908 made the automobile increasingly affordable for households, and trucks gradually replaced the horse-drawn vehicles as the preferred intra-city transport for businesses. As a result, businesses and households were able to take advantage of lower land values further away from the city center; residential suburbs developed at the edges of cities while businesses moved further from the central city core as business districts expanded. In the United States, the federal government initiated support for road building in the 1920s that by the 1970s had led to an extensive network of high-speed, interstate highways and secondary roads.

After World War II, the increase in automobile ownership and roads led to unprecedented suburban growth in the United States. The share of metropolitan population living in central cities declined from 64 percent in 1948 to 39 percent in 1990. Spurred by rising household incomes, abundant and accessible rural land, and the relatively lower suburban housing prices, many city residents flocked to the suburbs. Higher crime rates and racial tensions in cities, exacerbated by forced desegregation of schools in the 1950s, led to "white flight" from the cities and the emergence of homogenous, segregated suburban jurisdictions comprising higher-income households that offered higher-quality schools and services. A vicious cycle of central city decline and suburban growth resulted as more and more middle class families left the city for better suburban neighborhoods, the tax base of central cities eroded further, and **poverty** became concentrated in the central cities.

In addition to highway and road-building subsidies, other federal policies have fueled suburbanization in the United States. For example, federal income tax policies allow taxpayers to deduct mortgage interest and **property** taxes from taxable income, which creates a substantial subsidy for homeownership. Because land for new residential development is located at the urban fringe, this subsidy has fostered even greater suburbanization.

The suburbanization of employment in the United States also accelerated markedly after World War II. The percent of manufacturing located in U.S. central cities declined from 67 percent in 1948 to 45 percent in 1990. Corresponding percentages of wholesale and retail trade declined from 92 to 49 percent and from 75 to 48 percent respectively between these years. Intracity and intercity trucking, commuting via automobile, and the switch from multistory to single-story plants that exploited new horizontal production technologies (e.g., assembly-line production) led manufacturing to locate to suburban areas where land was cheaper and circumferential highways provided easy access to interstate highways. Advances in information technologies in the 1980s and 1990s (including fax machines, e-mail, the Internet, and teleconferencing) further eroded the benefits of physical proximity and have allowed firms to be increasingly footloose.

Suburbanization has also occurred outside of the United States during the post-World War II era, although to a lesser degree. Cross-country comparisons of urban **decentralization** indicate that cities in other developed countries, including **Japan**, **Canada**, and Germany, are relatively less suburbanized. Likewise, a number of cities in developing countries have become more decentralized over time. Interestingly, some evidence suggests that, while central-city densities are much higher in non-U.S. cities, the rate of decentralization of some non-U.S. cities has been comparable to that of U.S. cities. It is difficult to isolate the factors that contribute to the differences in suburbanization between the United States and other countries. Scholars have emphasized a variety of factors, including the greater abundance of land in the United States, a greater reliance on automobiles versus public transportation, a more extensive system of intra-metropolitan highways, less government protection of rural land, greater suburban fiscal autonomy, higher crime rates in central cities, and greater ethnic and racial diversity in the United States.

Urban decentralization has grown more complex in recent decades. Rather than uniform decentralization, cities have evolved into polycentric forms characterized by a number of suburban subcenters in addition to the traditional urban center. Agglomeration economies in production, for example, from manufacturers that share the same input supplier or from office firms that are supported by the same nearby hotel and restaurants, have led to clusters of suburban employment and, in some cases, the emergence of "edge cities"—new concentrations of office and retail activities outside core areas of metropolitan areas. The emergence of edge cities, along with continued road building and advances in communications technologies, has pushed the urban periphery out even further into rural areas. In many developed countries, suburbanization has given way to exurbanization—development of peri-urban regions that are beyond the suburbs, but still within the commutershed of an urban area.

See also: **Industrialization; Leapfrogging; Public Policy; Urban Sprawl; Urbanization; Urban-Rural Fringe.**

Further Reading

Anas, Alex, Richard Arnott, and Kenneth A. Small, "Urban Spatial Structure," *Journal of Economic Literature* 36 (1998): 1426–64; Jackson, Kenneth T., *Crabgrass Frontier: The Suburbanization of the United States*, New York: Oxford University Press, 1985; Mieszkowski, Peter, and Edwin S. Mills, "The Causes of Metropolitan Suburbanization," *Journal of Economic Perspectives* 7 (1993): 135–47; O'Sullivan, Arthur, *Urban Economics*, 5th ed., New York: McGraw Hill, 2004.

E L E N A   G .   I R W I N

Sudan-Sahel. Bioclimatic region lying south of the Sahara Desert and stretching from the Atlantic Ocean in the west through Sudan in the east. The Sudan-Sahel is a semiarid tropical region experiencing high temperatures and strongly seasonal rainfall (most all rain falling from May through September). The northern and southern boundaries of the Sahel and the Sudan regions have been generally defined by long-term mean annual rainfall (LMAR). Climate in the region is driven by the movement of the Intertropical Convergence Zone. Annual rainfall increases as one moves south from the desert edge into the Sudan-Sahel with a corresponding increase in the length of the summer rainy season from one to four months. The boundary between the Sahel and the Sahara Desert to the north has been variously defined as 100–200 millimeters LMAR. To the south, the boundary between the Sahel and the Sudan has been variously defined as the 500–600 millimeter LMAR. The Sudan borders the Guinean bioclimatic zone to the south at a variously defined boundary of 800–1000 millimeter LMAR.

Vegetation in the Sudan-Sahel grades from **grassland** steppe in the northern Sahel to open savanna in the southern Sudan. Herbacous vegetation in the Sahel is dominated by short-cycle annual grasses and forbs, the species composition of which changes from year to year based on the magnitude and distribution of summer rains. Perennial grasses increase in prevalence in the southern Sudan. Vegetative productivity in both zones is limited not only by sparse and highly seasonal rainfall but by low soil fertility. Due to a combination of underlying sedimentary rock formations and earlier historical periods of higher humidity and weathering in the region, soils are acidic, low in organic matter content and exchange capacity, and with particle size distributions dominated by fine sands. Plant growth is thus multiply constrained in the region, with the availability of nitrogen, phosphorus, and moisture seen as the major constraints. The relative importance of these limiting factors changes seasonally and year-to-year within any particular location, but the importance of moisture availability is generally higher in locations of the northern Sahel while soil fertility is more likely to be the dominant constraint in the southern Sudan. The Sudan-Sahel is a region with multiple constraints to vegetative productivity but one where the relative importance of these constraints varies along a sharp north-south gradient.

The Sudan-Sahel is where many African precolonial empires developed (e.g., Nubia, Ghana, Mali, Songhay, Segou, Kanem-Bornu, Haussaland, Sokoto, Toukouleur, and Maasina). While these larger polities developed for different reasons and sustained themselves in different ways, one general reason for the succession of empires in the region was the economic surplus gained through the control of regional trade. North-south regional trade networks in cattle, ivory, gold, slaves, cola nuts, salt, etc. crossed through the Sudan-Sahel region. These goods were produced at different points along the steep north-south ecological gradient from the rainforest to the south of the region extending to the desert north. Ecological specialization of production across a relatively short north-south distance was a major stimulus for the development of this trade in the region that played a role in the formation of precolonial states and cultural identities.

Agricultural strategies have developed in response to this environment. Major grain crops in the region are millet and sorghum (with some floodplain **rice**). Millet is grown as a rainfed crop in all of the region except the northern half of the Sahel zone (less than 300 millimeters LMAR). Sorghum, while rare in the Sahel (cultivated on low-lying fine-grained soils), becomes more common in the Sudan zone. Farmers' agricultural strategies reflect the multiple constraints that crop agriculture faces in the region. Risk of crop failure due to insufficient moisture during a portion of the growing season is a major issue particularly as one moves north into the Sahel zone. Where arable land is not scarce, farmers will often manage widely dispersed fields, only committing the necessary labor (for weeding, etc.) to bring a field to harvest for those fields that have received sufficient rainfall. Farmers have historically managed the wide variety of millet and sorghum cultivars that require different-length growing seasons and with different resistances to drought. Farmers will often be required to reseed their fields as crops fail during the beginning of the rainy season—utilizing different cultivars as the remaining growing season shortens.

When moisture is sufficient, soil fertility quickly becomes a major limiting factor in crop **agriculture**. While inorganic fertilizers have become more prevalent in the region over the past several decades, there remain significant economic constraints to their widespread use. Historically, farmers have relied on a combination of legume intercropping (e.g., cowpea, *Faidherbia albida*), long-term **fallow** systems (outfields) and nutrient applications as organic matter amendments (infields). While green manure systems exist, the major form in which nutrients are harvested from a wider area and applied to cropped fields has been through animal grazing and manure application. It is the Sudan-Sahel region where systems of manure contracting between livestock herders and farmers are most well documented.

Livestock producers in the region have long relied on seasonal mobility to reduce their exposure to drought risk and to take advantage of the different forage opportunities across the north-south ecological gradient. Seasonal movements to the northern Sahel during the rainy season and back to agricultural areas in the southern Sahel and Sudan during the dry season are commonly referred to as **transhumance**. Long-distance transhumance has worked to reduce the risk of crop damage in the agricultural zone while presenting regional livestock to the sparser but more nutritious forage in the northern Sahel. These systems of long-distance transhumance have been seriously eroded over the past thirty years as drought-induced decapitalization of the herds' "pastoralists" have caused many to rely on crop agriculture; rebel activity in the north has increased risks; and new livestock owners have preferred to manage their livestock within the agricultural zones during the rainy season. Therefore, **pastoral mobility** has generally become more circumscribed in the region. Notable exceptions are those cases where mobility has increased due to local shortages of **pastures**.

Ethnic and caste identities have long been associated with different occupational specializations. Production strategies of Sudano-Sahelian households are multifaceted, often extending across categories such as pastoralist, farmer, city dweller, etc. Still, there remain dominant economic activities associated with different ethnic identities. For example, a wide range of ethnic groups (e.g., Wolof, Malinke, Bambara, Songhay, Mossi, Haussa, etc.) are seen as dryland farmers of the region. Likewise, there are a number of different ethnic groups whose identities are tied to livestock husbandry (e.g., Fulani, Touareg, Toubou, Dinka). In all cases, actual productive activities are broader and more diverse than those considered prototypic or ideal for a particular ethnic or caste group.

The countries of Senegal, Mali, Burkina Faso, Niger, Chad, and Sudan are countries that contain substantial areas of both Sahelian and Sudanian zones. Except for Sudan, these countries were part of the French West African colonial system from the late nineteenth century to the early 1960s. The French, through their system of taxation, encouraged the production of **cash crops** such as **cotton** and groundnut. With the expansion of cash cropping and an acceleration of population growth, cropped area expanded rapidly in the 1950s and 1960s—a historic period of relatively high rainfall. During this period, cropped area expanded northward onto pastoral rangelands. With the drought of 1968–1973, this period of expansion came to an abrupt halt, leaving those who had invested their labor into cash crop production, as during the early colonial period, particularly vulnerable.

Since the 1968–1973 drought, rainfall in the Sudan-Sahel region has remained below its long-term average. This represents one of the best examples of a seemingly persistent shift in rainfall regime in the world—leading to speculation as to its relationship to global climate change or more regional biogeophysical **feedbacks**. Reduced vegetative productivity and rapidly increasing human populations have raised concerns about anthropogenic environmental degradation in the region. In fact, the Sudan-Sahel has been commonly cited as one of the best examples of widespread **desertification**. In addition, it served as a prominent example for linking **poverty** and **land degradation** and as such contributed to the development of the concept of sustainable development. Contemporary assessments have greatly reduced the prior estimates of anthropogenic environmental degradation in the region.

This dry period has severely affected people living in the Sudan-Sahel. Rural peoples have continued their efforts to reduce their exposure to risk by pursuing non-agricultural occupations (for some family members) and moving their production activities to the southern Sudanian and even Guinean zones where rainfall is more consistent. As a result, there has been a southerly shift of cropping and livestock presence and an acceleration of rural-to-urban migration over the last thirty years. **Conservation** and development funding for the Sahel have declined significantly since the 1980s and are now focused on economic and ecological stabilization through **decentralization** of **decision-making**. The most common forms of rural development activity in the zone include poverty banks, gardening projects, and programs to improve the management of village lands. In the Sudan, there has been greater private **investment** in agriculture with an expansion of cash crop production in some areas, particularly cotton (e.g., the Sikasso region of Mali).

See also: Modification; People at Risk; Sahel Land Cover; Sahelian Land Use (SALU) Model; Vulnerability.

Further Reading

Gritzner, Jeff, *The West African Sahel: Human Agency and Environmental Change*, Geography Research Paper no. 226, Chicago: The University of Chicago, 1988; Le Houérou, Henri N., *The Grazing Land Ecosystems of the African Sahel*, Berlin, New York: Springer Verlag, 1989; Mortimore, Michael, *Adapting to Drought: Farmers, Famines and Desertification in West Africa*, Cambridge, UK: Cambridge University Press, 1989.

MATTHEW D. TURNER

Sustainable Land Use. The management of land for agricultural and other purposes, usually through the integration of ecological with socio-economic and political principles, in order to ensure continuous satisfaction of human needs for present and future generations without comprising the ability of future generations or populations in other locations to meet their needs (intra- and intergenerational equity). According to Euro Beinat and Peter Nijkamp, at least three dimensions of sustainable land use can be distinguished. First, the husbandry dimension relates to the durability, exploitability, and continuity of natural resources over a long time horizon. Related actions, for example, are the application of crop-rotation systems—rather than **monocultures**—the careful use of scarce natural resources, and the rehabilitation of degraded land. Second, the interdependence dimension implies aspects of **land fragmentation**, land segmentation, and relations among different types of land uses. In contrast to **agro-industry**, for example, traditional farming is more often based upon interdependence according to which the farm and the surrounding (semi)natural areas achieve a flowing equilibrium based on interaction and mutual system resilience. Third, with concepts such as option value or existence values, the ethics dimension refers to obligations toward future generations: sustainable land use is given if the use and management of resources do not lead to the loss of future options, with **resilience** as a **land quality** feature that renders land use sustainable.

From a wide array of scientific evidence, Éric Lambin distilled three general conditions or components of human-environment interactions, involving multiple—that is, temporal, historical, social, economic, socio-political, cultural, technological, financial, and institutional—dimensions each. These conditions are shown to control the transition toward sustainable land use, that is, success or failure (including collapse) in land use and management. The first component is information on the state of the environment. It relates to the understanding by decision-makers of **land degradation** and of alternative management practices, as driven by knowledge, information, and communication. The second component is motivation to sustainably manage the environment, and it relates to the sources of the behavior of agents. The third component is the capacity to implement sustainable management, and it relates to the provision of appropriate physical, technical, and institutional infrastructure necessary for sustainable land management.

See also: **Agrodiversity; Alternatives to Slash-and-Burn (ASB) Programme; Community Involvement; Decision-Making; Endogenous Factor; Forest Transition; Indigenous Knowledge; Institutions; Land Rehabilitation; Land-Use Planning; Land-Use Transition; Public Policy; Tradeoffs.**

Further Reading

Beinat, Euro, and Peter Nijkamp, "Land Use Planning," in *Encyclopedia of Global Change: Environmental Change and Human Society*, vol. 2 (J–Z), eds. Andrew S. Goudie and David J. Cuff, Oxford, UK, New York: Oxford University Press: 2002, pp. 27–33; Lambin, Éric F., "Conditions for a Transition Towards a Sustainable Use of Environmental Resources," *Global Environmental Change* 15 (3) (2005): 177–80; Turner, B. L., "The Sustainability Principle in Global Agendas: Implications for Understanding Land-Use/Cover Change," *Geographical Journal* 163 (1997): 133–40.

HELMUT GEIST

Swidden Cultivation.   Also called shifting cultivation and derogatively **slash-and-burn agriculture**, swidden cultivation is among the oldest, most complex, and multifaceted forms of **agriculture** in the world. Estimates of the amount of land cultivated worldwide by swidden cultivators vary from 2.9 billion hectares to roughly half of the land area in the tropics.

Swidden is a system in which **vegetation** felled in patches of **forest** during the dry season is burned before the onset of the rainy season to open the site and release nutrients. The cleared fields are cultivated and harvested for one or more years, and then left to lie **fallow** for varying periods to allow secondary forest to regrow. Indigenous farmers manage the system in ways that integrate production from both cultivated fields and diverse secondary forests, including everything from grass and bushes in the early stages, to young open-canopy trees, to mature closed-canopy tree communities. Clifford Geertz described a swidden as "[a] natural forest . . . transformed into a harvestable forest," and others have suggested that vast areas considered primary or virgin forest are really late-secondary forests on lands previously cleared for swiddens. **Harold Brookfield** and Christine Padoch argue that swidden cultivation is not one system but many hundreds or thousands of systems. Some cultivators rotate some of their fields but also practice permanent cultivation on part of their land. Other swidden farmers create complex agroforests in which every plant has its uses, and they modify the landscape by making small soil-retention terraces on steep slopes.

Opposition to swidden cultivation is based mainly on misconceptions of how the system works, especially misunderstanding of its effects on soil as compared to the effects on soil of permanent agriculture. Swidden cultivation involves only temporary use of forestlands, not permanent loss, as is the case with permanent agriculture, human settlements, mining, and dam-building. In fact, swidden has a significant advantage over permanent agriculture: the fallow periods allow soil to stabilize and give forest vegetation an opportunity to regrow, providing a home for a variety of life forms. Because of its fallow period, swidden agriculture promotes both greater **carbon sequestration** and **biodiversity** conservation than permanent agriculture. Failure to understand the swidden agricultural system and its effects on forest regeneration has led scientists to overestimate the amount of "**deforestation**" that has occurred. Failure to understand successional vegetation has also led to government policies, mostly failures, for settling swidden agriculturists.

See also: **Agricultural Intensification; Agroforestry; Alternatives to Slash-and-Burn (ASB) Programme; Conservation; Degradation Narrative; Pristine Myth; Public Policy; Regrowth; Secondary Vegetation.**

Further Reading

Brookfield, Harold, and Christine Padoch, "Appreciating Agrodiversity: A Look at the Dynamism and Diversity of Indigenous Farming Practices," *Environment* 36 (1994): 7–11; Brown, Sandra, and Ariel E. Lugo, "Tropical Secondary Forests," *Journal of Tropical Ecology* 6 (1990): 1–32; Geertz, Cliford, *Agricultural Involution: The Processes of Ecological Change in Indonesia*, Berkeley, CA: University of California Press, 1963; Thrupp, Lori Ann, Hecht, Susanna B., and John O. Browder, *The Diversity and Dynamics of Shifting Cultivation: Myths, Realities, and Policy Implications*, Washington, DC: World Resources Institute, 1997.

JEFFERSON FOX

Syndrome.   Fundamental high-level cause of land-use/cover change. Summarizing a large number of case studies under the umbrella of the **Land-Use/Cover Change (LUCC) project**, Éric Lambin and colleagues found that land-use change was driven by a combination of the following five syndromes or high-level causes: resource scarcity leading to an increase in the pressure of production on resources; changing opportunities created by markets; outside

Typology of land-change syndromes

	Resource scarcity causing a pressure of production on resources	Changing opportunities created by markets	Outside policy intervention	Loss of adaptive capacity, increased vulnerability	Changes in social organization, in resource access, and in attitudes
Slow	Natural population growth and division of land parcels; Domestic life cycles leading to changes in labor availability; Loss of land productivity on sensitive areas following excessive or inappropriate use; Failure to restore or to maintain protective works of environmental resources; Heavy surplus extraction away from the land manager;	Increase in commercialization and agroindustrialization; Improvement in accessibility through road construction; Changes in market prices for inputs or outputs (e.g., erosion of prices of primary production, unfavorable global or urban-rural terms of trade); Off-farm wages and employment opportunities;	Economic development programs; Perverse subsidies, policy-induced price distortions and fiscal incentives; Frontier development (e.g., for geopolitical reasons or to promote interest groups); Poor governance and corruption; Insecurity in land tenure;	Impoverishment (e.g., creeping household debts, no access to credit, lack of alternative income sources; weak buffering capacity); Breakdown of informal social security networks; Dependence on external resources or on assistance; Social discrimination (ethnic minorities, women, lower classes people, or caste members);	Changes in institutions governing access to resources by different land managers (e.g., shift from communal to private rights, tenure, holdings, and titles); Growth of urban aspirations; Breakdown of extended family; Growth of individualism and materialism; Lack of public education and poor information flow on the environment;
Fast	Spontaneous migration, forced population displacement, refugees; Decrease in land availability due to encroachment by other land uses, e.g., natural reserves (i.e., tragedy of enclosure);	Capital investments; Changes in national or global macro-economic and trade conditions leading to changes in prices (e.g., surge in energy prices, global financial crisis); New technologies for intensification of resource use;	Rapid policy changes (e.g., devaluation); Government instability; War;	Internal conflicts; Illness (e.g., HIV); Risks associated with natural hazards (e.g., leading to a crop failure, loss of resource or loss of productive capacity);	Loss of entitlements to environmental resources (e.g., expropriation for large-scale agriculture, large dams, forestry projects, tourism, and wildlife conservation), leading to an ecological marginalization of poors;

Lambin, Geist, and Lepers (2003).

policy intervention; loss of adaptive capacity and increased **vulnerability**; and changes in social organization, in resource **access**, and in attitudes. Some of the syndromes are experienced as constraints, forcing local land managers into degradation, innovation, or displacement pathways. Other fundamental causes are associated with the seizure of new opportunities by land managers seeking to realize diverse aspirations. Each of these syndromes can apply as a slow evolutionary process that changes incrementally at the time scale of decades or more, or as a fast change that is abrupt and occurs as a perturbation that affects human-environment systems suddenly (see table titled "Typology of land-change syndromes"). Only a combination of several causes, with synergetic interactions, is likely to drive a region into a critical trajectory. Some of the syndromes leading to land-use change are predominantly **endogenous factors**, such as resource scarcity, increased vulnerability, and changes in social organization; while others, such as changing market opportunities and policy intervention, are mostly **exogenous factors**, though mediated by some local factors.

The basis of the syndrome approach is the insight from a large array of case studies that not all causes and all levels of organization are equally important. For any given human-environment system, a limited number of causes are essential to predict the general trend in land use. The syndrome approach dates back to the pioneering work of Gerhard Petschel-Held and colleagues at the Potsdam Institute for Climate Impact Research (PIK) in Germany. PIK has described archetypical, dynamic, coevolutionary patterns of human-environment interactions; and a taxonomy of syndromes links processes of **land degradation** to both changes over time and status of state variables. The approach is applied at the intermediate functional **scales** that reflect processes taking place from the household level up to the international level. For example, the "overexploitation syndrome" represents the natural and social processes governing the extraction of biological resources through unsustainable industrial logging activities or other forms of resource use. The typology of PIK syndromes reflects expert opinion based on local-scale examples, and the approach aims at a high level of generality in the description of mechanisms of environmental degradation.

See also: Cumulative Change; Driving Forces; Human Immunodeficiency Virus (HIV)/ Acquired Immunodeficiency Syndrome (AIDS); Mediating Factor; Proximate Causes; Tragedy of Enclosure.

Further Reading
Lambin, Éric F., Geist, Helmut J., and Erika Lepers, "Dynamics of Land Use and Cover Change in Tropical and Subtropical Regions," *Annual Review of Environment and Resources* 28 (2003): 205–41; Petschel-Held, Gerhard, Lüdeke, Matthias K. B., and Fritz Reusswig, "Actors, Structures and Environments: A Comparative and Transdisciplinary View on Regional Case Studies of Global Environmental Change," in *Coping with Changing Environments: Social Dimensions of Endangered Ecosystems in the Developing World*, eds. Beate Lohnert and Helmut Geist, Aldershot, UK: Ashgate, 1999, pp. 255–93.

HELMUT GEIST

Systemic Change.   A characteristic of biogeochemical cycling in which change in any part or at any location in the earth system can create significant consequences throughout the system. The structure and function of the earth system, including the biosphere or that part of the system in which life can exist, is maintained by the cycling of biogeochemicals such as carbon, nitrogen, and water among the oceans, land, and atmosphere. This cycling, for the most part, is globally fluid; changes in the ocean, land, and atmospheric stocks or disruption in the cycling among these stocks, if of sufficient magnitude, affect the earth system at large. The capacity of humankind to affect directly and sufficiently most biogeochemical cycles to trigger systemic change in the earth system is recent, a product of advances in fossil fuel and synthetics technology over the past 300 years, although the systemic consequences of **deforestation** may have begun much earlier. By the end of the twentieth century, the production and **consumption** activities of humankind matched nature in terms of affecting critical biogeochemical cycles. Perhaps the most notable example is the human impact on the **carbon cycle**. The burning of fossil fuels primarily in highly developed economies and, to a lesser extent, global deforestation, has released terrestrial-based stocks of carbon, generating unprecedented, rapid buildup of carbon dioxide (CO_2) in the atmosphere, raising concerns about changes in global climates.

See also: **Atmosphere-Land Interlinkages; Change Detection; Cumulative Change; Economic Growth; Nutrient Cycle; Water-Land Interlinkages.**

Further Reading

Kasperson, Roger E., Turner, B. L., II, Meyer, William B., Dow, Kirstin, Golding, Dominic, Kasperson, Jeanne X., Mitchell, Robert C., and Samuel J. Ratick, "Two Types of Global Environmental Change: Definitional and Spatial-Scale Issues in Their Human Dimensions," *Global Environmental Change: Human and Policy Dimensions* 1 (1990): 14–22; Steffen, Will, Sanderson, Angelina, Tyson, Peter, Jäger, Jill, Matson, Pamela, Moore, Berrien III, Oldfield, Frank, et al., eds., *Global Change and the Earth System: A Planet under Pressure*, Berlin: Springer Verlag, 2004.

B. L. TURNER II

T

Taiga. The northern **forest** belt in Eurasia and North America, now often used synonymously in both continents for boreal forest (to which *Boreas*, the Greek mythological god of northern winds, gave his name). The Russian term тайга for "forest" is of Altaic origin for "little twigs." The taiga consists of a nearly continuous circumpolar belt of coniferous trees that covers about 17 percent of the earth's land surface, between 50° and 60° northern latitudes (about 1.5 billion hectares or 3.7 billion acres) and one-third of all forested land, thus representing the single largest terrestrial **biome** and wilderness of the planet. The region is of great importance because of its size, **climate impacts**, and **biodiversity**.

The taiga belongs to the cold continental climate zone with long, severe winters (up to six months with mean temperatures below freezing) and short summers (50–100 frost-free days) with high temperatures. Mean annual precipitation is 400–500 millimeters, mostly as summer rains, but low evaporation rates make this a humid climate.

The boreal forest communities of North America and Eurasia display a number of ecological similarities. For this reason, Taylor Ricket and colleagues assign them into their only circumpolar (of fourteen) biome, the boreal forest/taiga biome. Overlying formerly glaciated areas and partially permafrost locations on both continents, the forest is a mosaic of successional and subclimax plant communities sensitive to varying environmental conditions. The forests are dominated by dark evergreen conifers including *Pinus* (pine), *Abies* (fir), *Picea* (spruce), and *Larix* (larch or tamarack), and *Thuja* (arborvitae) in North America. Hardwoods include paper birch (*Betula papyrifera*) and quaking aspen (*Populus tremuloides*). Shrubs include willows and members of the heath family (*Ericaceae*). Ground cover is dominated by mosses and lichens. Large tracts are formed by recurring disturbances, such as forest fires and insect infestations. **Fires** have occurred more frequently in the past decades than they would if lightning had been the only cause. The Russian Forest Service estimates that 90 percent of fire events are caused by people.

Northward, the boreal forest merges into the circumpolar tundra biome. This transition zone represents the climatically sensitive northern timberline. Further latitudinal differentiation is undertaken into "sparse northern taiga" in northern Eurasia and "lichen woodland" in **Canada**, respectively. This open coniferous forest is followed southward by "middle and southern taiga" and "closed forest," respectively. In the **United States of America**, the southern border—temperate broadleaf deciduous forest—is dominated by white pine (*Pinus strobus*), sugar maple (*Acer saccharum*), and American beech (*Fagus americanus*). Only in northern Eurasia does there exist a strong longitudinal distinction: the western taiga, with less severe climate conditions, and the eastern taiga beyond river Yenisey (at 90° E) with extreme continentality. In the western section, dense forests of spruce and fir ("dark taiga") on moister locations alternate with pine, shrubs, and grasses on drier soils. Extreme frost resistance enables the deciduous larch to be dominant in the east ("light taiga").

Fur-bearing predators like the lynx (*Felis lynx*) and various members of the weasel family (e.g., wolverine, fisher, pine marten, mink, ermine, and sable) belong to the characteristic fauna. Large herbivores are more closely associated with successional stages where there is more nutritious browse available and include elk or wapiti (*Cervus elaphus*, known as red deer in **Europe**) and moose (*Alces alces*, known as elk in Europe). Among birds, insect-eaters like wood warblers are migratory and leave after the breeding season.

Many indigenous people live in boreal forests, including the Dene, Inuit, Cree, and Athabaskans of North America; the Saami of Scandinavia; the Ainu of northern **Japan**; and the Nenets, Yakut, Udege, and Altaisk of Siberia. In recent times, settlement of peoples from outside the **boreal zone** has taken place, and the exploitation of the resources is now posing a threat to the fragile ecosystem and its native peoples.

Agricultural land use occurs in Canada and mainly in the European part of northern Eurasia. The taiga remains the world's largest timber reserve and carbon stock despite huge stretches of swamp and peat bog. Large-scale industrial **forestry** is by far the most important threat affecting boreal forests today. In Scandinavia, large-scale exploitation of the forests has transformed virtually all forestland into intensively managed **secondary vegetation**. In Siberia and parts of Alaska and Canada, primary forest still exists, but forest demand from **China**, Japan, and Mongolia transforms the taiga into a "pulp factory." Other pressures include oil and gas exploration, road building, and mining.

See also: Hot Spots of Land-Cover Change; Peatland; Russia; Wetlands.

Further Reading

Ricketts, Taylor, Dinerstein, Eric, Olson, David, Loucks, Colby, Eichbaum, William, DellaSala, Dominick, Kavanagh, Kevin et al., *Terrestrial Ecoregions of North America: A Conservation Assessment*, Washington, DC: Island Press, 1999; Taiga Rescue Network, The Boreal Forest [Online, December 2004], The Taiga Web Site, www.taigarescue.org; Yaroshenko, Alexey, Potapov, Peter, and Svetlana Turubanova, *The Last Intact Forest Landscapes of Northern European Russia*, Moscow: Greenpeace Russia, 2001.

CHRISTIANE SCHMULLIUS

Technological-Scientific Revolution.   According to Jaroslav Purš, the technological-scientific revolution (TSR) is characterized by the use of electric power to drive machines, by combustion engines, the development of heavy chemistry, the introduction of improved machines and technological-chemical processes in a number of major industries, the beginning of automation, production belt systems, and a more extensive use of scientific knowledge in production practice, including **agriculture**. The TSR, following the **Industrial Revolution**, had three phases: the first from the 1870s to the turn of the century, the second from then until the 1930s, and the third lasting until the end of the 1940s. In parts of the literature, the term "Second Industrial Revolution" is used as a synonymous term. TSR also indicates a period in which industrial hegemony moved from Britain to Germany and the **United States of America**.

After culmination of the **Agricultural Revolution**, the feeding of the rapidly increasing non-agricultural population in the growing cities required an increase of agricultural production efficiency instead of the expansion of cultivated land area. The only solution was a more intensive exploitation of inherited land-use structure. This has been carried out through capital **investments** into the most fertile plots by mechanization, deep tillage, use of chemicals, crop rotation, etc.

Not all elements of TSR (in industry) could be directly related to agriculture, because of the special conditions for circuit of capital depending on natural processes of organic growth, physico-geographical conditions, and limited area of land suitable for cultivation. The TSR in agriculture then meant the complete introduction of free crop rotation systems, mechanization by extensive use of combustion engines, electricity as a new means of energy, use of artificial fertilizers produced in industry, land reclamations (ameliorations), and the use of the tractor as the "universal machine in agriculture."

Its first phase was typical for the rapid introduction of more sophisticated agricultural machines and tools (yet driven by animal power and only at big estates, very rarely also by steam engines); by slow increase of synthetic fertilizer use, the acceleration of land improvements (including rejoining of field patterns); application of scientific methods in plant and animal production; and by the founding of agricultural high schools, research stations, and farmer cooperatives (credit, storage, purchasing, marketing, etc.). Such innovations are named the **Green Revolution** in parts of the literature.

The second phase meant the quantitative and qualitative boom of all features of the first phase, and mainly their growing penetration into small farming. The larger use of chemicals started together with deep tilling after World War I to play the decisive role in reviving and improving the natural fertility of land and in protection of plants against insects and diseases. The principal role was played by the internal combustion engine and electricity, which brought fundamental change in the energy balance in agriculture: the move from solar energy to fossil fuel energy for pulling or driving the drilling, sewing, grain threshing, and cleaning machines; and in animal production for preparing of feed (cutters, grinders, pumping water, centrifuging milk, etc.). Safe electric lighting enabled the night labor to work in peak seasons and spread the indoor breeding of livestock. The tractor as a universal tractive and motive machine, comparable to the steam engine in the Industrial Revolution, was suitable mainly in plant production because of its "areal" use character, and at large farms replaced horse power and cattle. It also made it possible for machines of greater weight and higher performance to be used in the fields, for deeper and faster tilling, and for faster transportation in agriculture.

These processes made a large part of the human labor force available for industry. The number of horses used could be reduced, and permanent

grassland that produced fodder could now be used for production of other plants. The TSR in agriculture needed a much larger amount of capital, because it was re-oriented toward **agricultural intensification**. The land-use trends at the beginning of the twentieth century changed dramatically: generally, the acreage of arable land decreased from this time until today; and permanent grasslands decreased at even higher rates, all to the detriment of **forest** and other areas or land uses such as roads, mining, and built-up areas.

See also: **Agro-Industry; Boserup, Ester; Carson, Rachel Louise; Driving Forces; Economic Restructuring; Extensification; Land Reform; Land Rehabilitation; Land Rent; Proximate Causes; Urbanization.**

Further Reading

Jeleček, Leoš, "Changes in the Production and Techniques in the Agriculture of Bohemia 1870–1945," in *Agriculture in the Industrial State*, eds. Michael A. Havinden and Edward J. T. Collins, Reading, UK: University of Reading, Rural History Centre, 1995, pp. 126–45; O'Brien, Patrick, and Prados Leandro de la Escosura, "Agricultural Productivity and European Industrialization, 1890–1980," *Economic History Review* XLV (1992): 60–92; Purš, Jaroslav, "Complex Revolution of the Modern Age and Industrial Revolution," *Historica* XIX (1980): 135–70; Thompson, Francis M. L., "The Second Agricultural Revolution, 1815–1880," *Economic History Review* 21 (1968): 62–73; Zanden, J. L. van, "The First Green Revolution: The Growth of Production and Productivity in European Agriculture, 1870–1914," *Economic History Review* XLIV (1991): 215–39.

LEOŠ JELEČEK

TERRA-ASTER. An Advanced Spaceborne Thermal Emission and Reflection Radiometer (ASTER) on-board the Earth Observing System (EOS) TERRA satellite of the National Aeronautics and Space Administration (NASA), launched in the year 2000. ASTER acquires spectral information at fourteen different wavelengths from 0.52 to 11.65 µm (visible and near-infrared to thermal infrared). The spatial resolution of the data set varies between fifteen and ninety meters. The ASTER detectors are pointable in the across-track direction, which enables detailed terrain models to be produced. ASTER data has been used to investigate the spatial and temporal distribution of evapotranspiration rates, for **monitoring** glaciers, and to study thermal properties of the urban environment.

See also: **Remote Sensing.**

Further Reading

National Aerospace Agency, Earth Observing System TERRA [Online, August 2004], The Advanced Spaceborne Thermal Emission and Reflection Radiometer (ASTER) Web Site, terra.nasa.gov/About/ASTER.

KEVIN J. TANSEY

Thermal Band Analysis. All objects with a temperature above absolute zero emit energy. Within land-use/cover-change analyses, thermal band (3–14 µm) satellite data measure the emission of energy from the earth's surface. This information, when converted into temperatures, can be used to link directly to other processes (e.g., micrometeorological). Satellite-based thermal analysis offers an accurate and viable means of **monitoring** different environments and estimating spatial aspects of their energy budget. Land surface processes are of paramount importance for the redistribution of energy and moisture between the land and the atmosphere. These exchanges of radiative, moisture, and heat fluxes affect biosphere development and physical living conditions at the earth's surface. The thermal band allows for a continuous representation of land cover and may be as important as reflectance-based **vegetation** indices used in much of today's research using **remote sensing**.

See also: **Advanced Very High Resolution Radiometer; Along Track Scanning Radiometers (ATSR) World Fire Atlas; Atmosphere-Land Interlinkages; Continuous Data; Pixel.**

Further Reading

Southworth, Jane, "An Assessment of Landsat TM Band 6 Thermal Data for Analyzing Land Cover in Tropical Dry Forests," *International Journal of Remote Sensing* 25 (2004): 689–706.

JANE SOUTHWORTH

Threshold. A point where a system is unstable and may switch abruptly between alternative stable states or equilibria. Thresholds may occur in all kinds of systems: ecological, social, or both. When a system moves from one equilibrium to another, this is often denoted as transition. Stable systems respond smoothly to gradual changes; near a threshold rapid, often unpredictable changes in the system can occur. Small changes in conditions may then result in qualitatively different trajectories. Consider the following example of a socio-ecological system. Woodlands and grassy open landscapes can under certain climatic/soil conditions be alternative stable states. Landscapes can be kept open by herbivores (domesticated or not) or **fires** even under conditions where woodlands, once established, are stable, because herbivores cannot destroy adult trees and shading reduces grass cover so that fires cannot spread. Increasing human use can result in a loss of **resilience** of woodlands until the threshold is reached—then the system flips into the **grassland** state.

See also: **Biomass; Canada; Change Detection; Disequilibrium Dynamics; Environmental Change; Land-Use Transition; People at Risk; Tradeoffs; Vulnerability.**

Further Reading

Berkes, Fikret, and Carl Folke, eds., *Linking Social and Ecological Systems*, Cambridge, UK: Cambridge University Press, 1998; Muradian, Roldan, "Ecological Thresholds: A Survey," *Ecological Economics* 38 (2001): 7–24; Scheffer, Marten, Carpenter, Steve, Foley, Jonathan A., Folke, Carl, and Brian Walker, "Catastrophic Shifts in Ecosystems," *Nature* 413 (2001): 591–6.

HELMUT HABERL

Thünen, Johann Heinrich von (1783–1850). Landowner near Rostock, Germany, who is noted for his work on the relationship between location and land use. Interested in how the economics of land use were influenced by distance to market, he published his method of analysis in *The Isolated State*, in 1826. His analysis focused on transport costs. Net profit depended on the market price of a commodity and its production cost plus transport cost. Transport cost for any one commodity, and hence profitability, depended on distance to market. Different commodities had different characteristics in relation to transport: the profitability of bulky, low-value commodities would decline more steeply with increasing distance than would that of compact, high-value commodities. The relative profitability of different commodities would therefore vary with distance from market. This, in turn, would translate into concentric zones or rings of different commodity production (i.e., land uses) arranged around the central market, sometimes referred to as a von Thünen pattern. Thünen's analysis was based on several assumptions—that profit maximization was the objective, that land quality was uniform, that a single market existed, and that transport costs varied with distance. When these assumptions are relaxed, the simplicity and regularity of the model break down, although the logic of the method of analysis survives. Thünian approaches have subsequently been applied at a variety of **scales**, including the global scale (with a central market in northwest **Europe**) as well as that of the village and farm. They have also been applied to study **deforestation** processes and to intra-urban land uses. Revisionists have attempted to adapt the static model to dynamic situations and in particular to regions with expanding cities.

See also: **Agent-Based Model; Cash Crops; Decision-Making; Economic Growth; Food Crops; Land Rent; Land-Use System; Modeling.**

Further Reading

Grotewold, Andreas A., "Von Thünen in Retrospect," *Economic Geography* 35 (1959): 346–55; New School University, The History of Economic Thought [Online, December 2004], The Johann Heinrich von Thünen Web Site, http://cepa.newschool.edu/het/profiles/thunen.htm; Thünen, Johann Heinrich von, *Von Thünen's Isolated State: An English Translation of Der Isolierte Staat* (trans. C. M. Wartenburg, ed. P. Hall), Oxford and New York: Pergamon, 1966.

ALEXANDER S. MATHER

Tobacco. The world's most widely grown non-food crop (in more than 120 countries) with a fairly low share of less than 1 percent of all arable land, though, and a globally widespread legal drug the use of which is one of the largest causes of premature death and the most common preventable cause of death in the world. Since about the mid-1960s, a global shift of tobacco production has occurred. Compared with 1700, when nearly the entire world production of tobacco was concentrated in Brazil, parts of the Caribbean, and in the Chesapeake colonies of New England (where most tobacco was grown and harvested by slave labor), with the breakdown of colonial rule from the mid-nineteenth century onward, tobacco farming spread nearly all over the world. Farming is extremely labor-intensive (100 times that of wheat), and the number of women and child laborers has increased with a production shift to developing countries. There, contract farming under the

Tobacco field in Windsor Locks, Connecticut, September 1937.
PHOTO: Edwin Locke. [Courtesy Library of Congress]

Flue-cured (Virginia) tobacco barn on Irwinville Farms in the Rural Resettlement Administration Project at Irwinville in Irwin County, Georgia, June 1936.
PHOTO: Carl Mydans. [Courtesy Library of Congress]

supervision of multinational cigarette or leaf buyer corporations is the current major trend. It has been described by Jason Clay as follows: tobacco companies act as banks, extending credit to farmers at the beginning of the season (for seed, fertilizer, pesticides, and technical support); in return, farmers agree to sell their crop to the company at harvest; company inspectors make regular visits to the farms to make sure their guidelines are followed; when the harvest is in and the time comes to pay the farmer, the leaf buyers determine the leaf grade and the price; as a common result, they often end up paying producers less than the original loans.

At present, most of the quantities entering world trade originate from zones of the developing world where frost-free days, a sufficient long dry season (allowing for harvesting and curing the crop), and low-cost conditions of production are optimal. By 2000, around four-fifths of the more than 120 growing countries, or the equivalent of around 90 percent of all global land under tobacco, had become located in developing countries. **China**, India, and Brazil produce more than half of the world's tobacco, with China and India being the largest producers as well as consumers (in both instances, most of the tobacco consumed is grown locally). The principal importing countries are **Russia**, followed closely by Germany, the **United States of America**, the Netherlands, the United Kingdom, and **Japan**. In developed market economies, tobacco production is increasingly being phased out, and tobacco use has stabilized (actually expected over the next century in **Europe** and the United States). Driven by these trends, mainly (and large-scale anti-smoking campaigns as well), multinational cigarette companies have applied **corporate strategies** to increasingly shift their manufacturing capacities into low- and middle-income countries where most of today's tobacco is grown and where new and large consumer markets are expected to develop (due to large population size, low rates of smoking among women, increasing disposable incomes available to large numbers of people for **consumption**, and poor law enforcement in terms of tobacco control policies). As corporations such as Altria, Japan Tobacco, Philip Morris, British-American Tobacco, and Imperial Tobacco build new factories; enter into joint ventures; and buy formerly state-owned factories; at the same time international leaf dealers have followed. These are companies that source tobacco around the world, ensure quality control, and then sell to manufacturers. These buyers set up a closed chain of leaf procurement and processing facilities near the factories. The world leaf market is currently dominated by three U.S.-based companies: Dimon, Standard Commercial, and Universal, working with the major tobacco manufacturers to determine which countries will produce how much tobacco leaf and what kind. In a constant drive to increase profits and reap benefits from **economic liberalization**, the companies regularly shift production from one country to another (in the **miombo** zone, for example, from Zimbabwe to Uganda, Tanzania, and Zambia, currently), based on the concessions that they can negotiate with the local governments. This happens irrespective of the impact this may have on local growers, local ecosystems, or national economies.

It has been estimated that tobacco use currently kills 4 million people per year, and that by 2030 it will kill 10 million people annually, with 7 million of these deaths in low- and middle-income countries of the developing world. The global shift of tobacco production coincides with a global shift in tobacco use, and two patterns overlap as follows: first, educated and prosperous people increasingly abandon smoking, while the practice becomes more concentrated among the poor in most societies; and, second, smoke products have undergone a change from dark, naturally cured tobaccos (used predominantly by men) to bright, artificially cured, and highly addictive tobaccos (increasingly used by women and teenagers). In particular, American-blend type cigarettes such as Marlboro bear a globally high prestige, and Virginia or flue-cured tobacco is the major component of these highly addictive cigarettes.

In growing areas of the developing world, tobacco is commonly grown more in highlands than lowlands, and more in drylands (under semiarid to semihumid climates) than in humid ecosystems. The share of tobacco in all arable land is highest in producer countries of subtropical East Asia and in **southern Africa**. Less than half of the global land under tobacco is grown with naturally cured varieties, that is, using natural variations in temperature and humidity to dry up the leaves (air- and sun-curing). The majority of global land under tobacco is cropped with artificially cured varieties, that is, those using heat from external sources such as wood and coal (fire- and flue-curing). While green tobacco leaves are directly exposed to heat and smoke in fire-curing, flue-curing requires the leaves to be dried by radiant heat from flues or pipes connected to a furnace (see photo titled "Flue-cured (Virginia) "tobacco" barn). In Tanzania, for example, the use of wood as firewood and polewood (for curing barn construction) is widespread among flue-cured Virginia (about 80 percent) and dark fire-cured tobacco (about 20 percent of total national production in the 1990s). Experiments with coal in the artificial curing of flue tobacco failed, and no cost-effective alternatives to wood have so far been introduced on a large scale (which is also valid for most African producer countries, at least). Less than 10 percent of the Virginia farmers in the country are known to use wood from their own woodfuel plantings, and most of the wood has been taken from open, accessible forests and woodlands, thus contributing to **deforestation** at an estimated 11,000 hectares annually. In Tanzania, curing of tobacco is the second largest consumer of wood after the domestic sector, producing 4.36 billion cubic meters of carbon dioxide and 238 million cubic meters of carbon monoxide per season. Local case studies point to the fact that in areas where tobacco is grown, the tobacco sector is the leading commercial consumer of wood and forest products, accounting for up to 20 percent of all regional consumption. Globally in the 1990s, an estimated twenty (stacked) cubic meters of wood were used to produce one ton of tobacco in those areas where the energy came from wood. This translates into 200,000 hectares of forests and woodlands that were removed by tobacco farming each year during the 1990s, accounting for nearly 5 percent of total net deforestation in the respective producer countries of the developing world. Tobacco-related deforestation happens in about seventy producer countries and, in particular, serious, high, and medium degrees have been estimated for half of these countries.

Tobacco is grown in **monoculture** and contributes to the simplification of rural landscapes and ecosystems through **agricultural intensification**. When tobacco is grown on the same land repeatedly with minimal rotation with other crops, there is a tendency for the soil to become exhausted and for crop pests to become endemic. This is why monocultural tobacco cropping requires ever-increasing inputs of pesticides and chemical fertilizer. Furthermore, the tobacco plant is unique in that it depletes major soil nutrients such as nitrogen, phosphorus, and potassium at five times the rate of **food**

crops, with especially potassium needed and absorbed in uniquely large amounts. This is due to the design of the crop as a smoke product reaching out for high levels of nicotine enrichment. The uptake of nitrogen is vital for growth, especially the green color and nicotine content; phosphorus is essential for nutrition, especially the promotion of ripening; and potassium has a positive influence on leaf combustibility and the (bright) color of flue-cured leaf. Two agricultural practices, "topping" and "desuckering," are carried out to manipulate natural plant growth before ripening and harvest in order to deliver higher (nicotine) yields (by about 10–15 percent). Topping is the manual removal of inflorescences, done with most cigarette and cigar tobaccos. It forces all nutrients not to go into seed but into leaf production instead, making the upper leaves grow longer, wider, thicker, and darker in color. It further stimulates root growth and slows down ripening rate, thus draining more nutrients from the soil. However, as an unintended consequence, topping stimulates the growth of suckers, which—if not removed at least once a week—would cause more harm than the flower would have done if left. Desuckering is the manual removal of suckers, occasionally supported by chemicals, and needs to be carried out several times, because as soon as suckers are removed, yet more suckers are stimulated to grow. It has been estimated that in air-cured burley tobacco in Malawi, about 1 percent of yield is lost for every day suckers are left unremoved on the plant. Thousands of additives and pesticide residues in a cigarette (an estimated 4,000 in total) not only directly impact the health of smokers, but their application also exacts a toll on agro-ecosystems and the livelihood system of tobacco farming societies. Apart from excessive pesticide usage, one environmental health problem peculiar to tobacco work is green tobacco sickness, with nearly half of all tobacco workers experiencing it at least one time during the tobacco harvest. Green tobacco sickness is a form of nicotine poisoning through dermal absorption (resulting from workers brushing against wet tobacco leaves), and the symptoms include nausea and vomiting, headache, dizziness, blood pressure fluctuation, and abdominal cramping. Because of the toxic chemicals in tobacco, any waste that cannot be used to make other, more valuable tobacco products is used to manufacture insecticides.

The potential to improve tobacco growing is poor for three major reasons: most production has shifted to growing countries where there is less ability to influence production practices; multinational leaf buyers usually have little interest in sustainable production; and few alternatives exist in many developing countries to wood-based energy, especially in Africa. The recommended focus of **public policy** is therefore on curbing the global tobacco epidemic by stopping smoking, but not making production sustainable. Indeed, global regulatory actions were triggered in the late 1990s, and as per February 27, 2005, an international tobacco control convention came into force, with the international standard now to be translated into national law by those countries that signed and ratified.

See also: **Agro-Industry; Cash Crops; Economic Restructuring; Human Health.**

Further Reading

Brown, Valerie J., "Tobacco's Profit, Workers' Loss?" *Environmental Health Perspectives* 111 (2003): 284–7; Clay, Jason, *World Agriculture and the Environment: A*

Commodity-by-Commodity Guide to Impacts and Practices, Washington, DC: Island Press, 2004; Framework Convention on Tobacco Control [Online, March 2005], The FCTC Web Site, www.fctc.org; Geist, Helmut, "Global Assessment of Deforestation Related to Tobacco Farming," *Tobacco Control* 8 (1999): 18–28; Geist, Helmut, "Soil Mining and Societal Responses: The Case of Tobacco in Eastern Miombo Highlands," in *Coping with Changing Environments: Social Dimensions of Endangered Ecosystems in the Developing World,* eds. Beate Lohnert and Helmut Geist, Aldershot, UK, Brookfield, VA: Ashgate, 1999, pp. 119–48; Geist, Helmut, "Transforming the Fringe: Tobacco-Related Wood Usage and Its Environmental Implications," in *Environment and Marginality in Geographical Space: Issues of Land Use, Territorial Marginalization and Development in the New Millennium,* eds. Roser Majoral, Heiko Jussila, and Fernanda Delgado-Cravidão, Aldershot, UK, Burlington, VA: Ashgate, 2000, pp. 87–118; Jacobs, Rowena, Gale, H. Frederick, Capehart, Thomas C., Zhang, Ping, and Prabhat Jha, "The Supply-Side Effects of Tobacco-Control Policies," in *Tobacco Control in Developing Countries,* eds. Prabhat Jha and Frank J. Chaloupka, New York: Oxford University Press, 2000, pp. 311–41.

HELMUT GEIST

Tradeoffs.   A central feature of rational choice among alternatives when resources are scarce, which almost always is the case. Indeed, "making tradeoffs" is one way of referring to inevitable choices in a world of binding resource constraints. Even private choices (within the realm of a single household, farm, or firm) involve problematic tradeoffs. These are not just hard choices, but choices that can be difficult to analyze formally and to quantify; such as those involving natural resources, risk, and uncertainty. One important example is the impact of current management decisions for a particular plot of land on its future productivity.

There also is the prospect—indeed the likelihood—of significant offsite effects from decisions about land-use and **land-cover change** and, in turn, of landscape, larger **scale**, and cross-scale interactions of these effects in space and time. These offsite effects are called environmental or production externalities by economists and are an important class of what others refer to as **ecosystem services**. Pollination and water pollution are two classic examples, respectively, of positive and negative externalities.

Private and social rates of tradeoff differ whenever there are externalities. In this situation, society could do better than the sum total of independent private decisions. Indeed, private decisions alone may produce unsustainable or even disastrous collective results. Because externalities so often are associated with land use and cover change, the focus of analysis appropriately has been on broader challenges of social choice and public policy concerns, including issues of efficiency of economic incentives, human welfare (including health), and environmental impacts of different land-use alternatives, as well as equity of outcomes across people (winners and losers) at a single point in time and across generations.

The potential number of dimensions that could be considered in this social choice problem is vast. Practical attempts at policy analysis, by necessity, have to make pragmatic choices about what to include in the assessment of tradeoffs. For the important case of land-use tradeoffs in developing countries, Steve Vosti and Thomas Reardon have proposed that analyses be organized around a "critical triangle" of public policy goals: productivity growth,

poverty alleviation, and sustainable resource use. Charles Crissman and colleagues extended this policy analysis approach to include **human health**. A collection edited by David Lee and Christopher Barrett includes a large number of examples of approaches to quantitative policy analysis of tradeoffs among the "critical triangle" of **public policy** goals for the case of **agricultural intensification**.

Quantifying tradeoffs starts with assessment of what really is possible. The dimensions of the "possibilities set" to be considered must span multiple policy goals in terms of the attributes of the various goods and services involved. Of particular interest is the "possibilities frontier," which comprises the envelope of efficient outcomes from the set of technically feasible combinations of factors of production (natural, manufactured, financial, social, and human). The shape of this possibilities frontier is the rate of transformation between efficient outcomes or, in more general terms, the rate of tradeoff. "Efficient outcomes" is used here in Pareto's sense: these are states of the world in which it is not possible to produce additional amounts of one good or service without reducing production of another. However, particularly in the case of many developing countries, outcomes may fall far short of possibilities. For such apparent "win-win" opportunities, though, it is important to realize that the necessary collective action or policy reform to achieve a better result for society also entails real costs. (Moreover, what is "good" or "bad" from a broad social perspective is a non-trivial valuation problem that will be taken up below).

Despite a growing body of empirical work, much remains to be done to establish empirical patterns and regularities in the shape of tradeoffs curves. While some relationships among policy goals may be represented by convex possibilities sets with tradeoff curves that are smooth and continuous, there is persuasive qualitative reasoning and a growing body of quantitative results suggesting that non-convexities (discontinuities and **thresholds**) and hysteresis (path dependence) are the norm, particularly for natural resources and environmental aspects of the social choice problem. Indeed, whether or not these tradeoff relationships are smooth is a central feature distinguishing neoclassical and ecological approaches to environmental economics. Others argue that non-convexities and thresholds also should be taken more seriously in development economics. Moreover, in a dynamic context, the rates and patterns of technological change that influence the shape of the tradeoffs curves are mediated by **institutions** and policies, not just biophysical or engineering relationships. The specific shape of these tradeoffs is a fundamental determining factor regarding policy options and the opportunity costs, efficiency, and equity of possible outcomes for society. Thus, approaches that can help refine our empirical understanding have a great deal of practical importance as well as scientific interest.

Two ongoing initiatives for integrated assessment of land-use alternatives illustrate the range of fruitful (yet contrasting) empirical strategies as well as some important common elements. The tradeoff analysis (TOA) method is an example of a more formal approach that combines spatially referenced biophysical and economic data with crop growth models and econometric-process simulation models and biophysical models. By avoiding the "representative farm" concept, a major advantage of this econometric and spatially explicit approach it that the TOA method can represent the heterogeneity of individual land-use decisions made on farms and aggregate their consequences within a landscape context. These advantages, and the additional rigor of the approach, do come at a cost, however, in terms of required data and specialized statistical and modeling skills. By way of comparison, the matrix approach of the **Alternatives to Slash-and-Burn (ASB) Programme** is much simpler in its data requirements and analytical setup, emphasizing replicability and comparability across sites in the search for empirical patterns. The ASB matrix approach uses chronosequences to identify representative land uses that then become the focus of measurement and analysis at benchmark sites across the humid tropics. The rows of the ASB matrix are specific land uses, ranging from natural **forests** and forest extraction activities to agroforests, simple tree-based systems, rotational bush **fallows**, continuous annual cropping, and **pastures/grasslands**. The columns are indicators of specific goals from global perspectives (**biodiversity** and carbon stocks), national policymakers' perspectives (**economic growth**, employment) and local perspectives (agronomic sustainability, adoptability by smallholders). This approach has made it possible to undertake cross-continental comparisons of patterns in tradeoffs from data collected in the **tropical humid forest** margins in the Amazon, Congo basin, and Southeast Asia. The ASB matrix approach is useful for identifying big (order of magnitude) differences and is a useful framework for testing a wide range of indicators for specific policy goals, but is not as sophisticated spatially or statistically as TOA. An additional practical advantage of the ASB matrix is the relative ease of expanding the scope of the comparative analysis of land-use alternatives. Data for additional rows (land-use alternatives) or columns (indicators) can be incorporated without affecting the other elements of the matrix.

Similarities in the TOA and ASB approaches to measuring tradeoffs are at least as important as their methodological differences. Most importantly, each provides a proven framework for planning, coordinating, and reporting multidisciplinary, integrated assessments of policy-relevant tradeoffs associated with land use and cover change. Each also is valuable as a tool for *ex ante* impact assessment. Both engage policymakers and other key stakeholders and likely users of the analysis at the outset in order to identify policy-relevant criteria and indicators for empirical assessment. Each takes a modular multidisciplinary approach to measurement of specific indicators, so that the necessary depth of expertise can be incorporated into the **integrated assessment**. Both also take a multiscale approach, building up from land-use decisions at the farm level as units of analysis but placing these plot-level activities within a broader environmental context.

Perhaps the most interesting similarity between the ASB and TOA approaches concerns the fundamental question of valuation of alternatives. While each approach uses indicators derived from market prices when these are available, neither employs contingent valuation techniques for non-market services, which predominate for environmental issues. Why not use non-market techniques in an effort to aggregate across alternatives and to identify a "socially optimum" point on the tradeoff curve? First, there is no consensus about the utility or reliability of these non-market valuation techniques for the really big issues (e.g., **species extinction**) resulting from land-cover change, particularly in a developing country context. Second, land-use and land-cover change often involve conflicts of interest among diverse stakeholder groups (ranging, for example, from farmers to upstream and downstream communities, national policymakers, and international **conservation** organizations). These groups and others each may well have their own distinct values for alternative outcomes. In a static sense, then, the social optimum probably does not exist. Both ASB and TOA engage with policymakers and other elements of society as a means of discovering social valuation through participation in complex, dynamic political processes. Linking science directly with policy and political processes to support negotiation among different groups is likely to be especially important for complex, interlinked problems where there are conflicting interest groups and uncertainties about alternatives.

See also: **Arena of Land Conflict; Carbon Cycle; Deforestation; Driving Forces; Forest Degradation; Hot Spot Identification; Land-Use Policies; Mediating Factor; Millennium Ecosystem Assessment; Regrowth; Resilience; Sustainable Land Use; Threshold; Vulnerability; Watershed.**

Further Reading

Antle, John M., and Susan M. Capalbo, "Econometric-Process Models for Integrated Assessment of Agricultural Production Systems," *American Journal of Agricultural Economics* 83 (2001): 389–401; Crissman, Charles C., Antle, John M., and Susan M. Capalbo, eds., *Economic, Environmental, and Health Tradeoffs in Agriculture: Pesticides and the Sustainability of Andean Potato Production*, Boston: Kluwer Academic Publishers, 1998; Dasgupta, Partha, and Karl-Goran Mahler, eds., *The Economics of Non-Convex Ecosystems*, Dordrecht, NL: Kluwer Academic Publishers, 2004; Lee, David R., and Barrett, Chris B., eds., *Tradeoffs or Synergies? Agricultural Intensification, Economic Development and the Environment*, Wallingford, UK: CAB International, 2001; Montana State University [Online, November 2004], The Tradeoff Analysis Project Web Site, www.tradeoffs.montana.edu; Stoorvogel, J., Antle, John M., Crissman, Charles C., and W. Bowen, "The Tradeoff Analysis Model: Integrated Bio-Physical and Economic Modelling of Agricultural Production Systems," *Agricultural Systems* 80 (2004): 43–66; Tomich, Thomas P., van Noordwijk, Meine, Vosti, Steve A., and J. Witcover, "Agricultural Development *with* Rainforest Conservation: Methods for Seeking *Best Bet* Alternatives to Slash-and-Burn, with Applications to Brazil and Indonesia," *Agricultural Economics* 19 (1998): 159–74; Tomich, Thomas P., Cattaneo, A., Chater, S., Geist, Helmut J., Gockowski, J., Kaimowitz, David, Lambin, Éric F., et al., "Balancing Agricultural Development and Environmental Objectives: Assessing Tradeoffs in the Humid Tropics," in *Slash and Burn: The Search for Alternatives*, eds. Cheryl A. Palm, Steve A. Vosti, P. A. Sanchez, and P. J. Ericksen, Columbia University Press, 2005; Van Noordwijk, Meine, Tomich, Thomas P., de Foresta, H., and G. Michon, "To Segregate or to Integrate? The Question of Balance between Production and Biodiversity Conservation in Complex Agroforestry Systems," *Agroforestry Today* 9 (1997): 6–9; Van Noordwijk, Meine, Tomich, Thomas P., and Bruno Verbist, "Negotiation Support Models for Integrated Natural Resource Management in Tropical Forest Margins [Online, November 2004], The Conservation Ecology Web Site, www.consecol.org/vol5/iss2/art21; Vosti, Steve A., and Thomas Reardon, T., eds., *Sustainability, Growth, and Poverty Alleviation: A Policy and Agroecological Perspective*, Baltimore, MD: Johns Hopkins University Press, 1997.

THOMAS P. TOMICH

Tragedy of the Commons. First articulated by biologist Garrett Hardin in 1968 as a situation in which a group of self-interested individuals exploit a commonly owned resource to a level beyond that which is socially optimal.

The logic of the tragedy of the commons potentially applies to a range of open access/**common-pool resources**, including fisheries, pasture, unpolluted waterways, and communal **forests**. Such resources are rivalrous (one individual's use of the resources diminishes the benefits available to another user) but non-excludable (**access** to the resource by specific individuals cannot be limited).

Garrett Hardin uses the example of a commonly owned **pasture** in which individuals decide how many of their own cattle to graze on the pasture. Individuals capture the benefits from cattle that they graze, but the costs of this grazing—degradation of the pasture—are shared among all users. When deciding how many cattle to graze, Garrett Hardin assumes that individuals account only for their private benefits and costs. (The difference between the private and social costs of an individual's action is often referred to as an external cost.) Because external costs are not accounted for by individuals, at the individual level the benefits of adding an additional cow to the pasture may exceed the individual's cost, although the social benefits of the additional cow will be below its social cost. The result is that too many cattle are grazed, resulting in **land degradation** beyond a socially optimal level. Garrett Hardin's theoretical arguments were formalized by economist Scott Gordon to demonstrate socially sub-optimal exploitation of an open access fishery by profit-maximizing harvesters. Both authors' models are based on the assumption that individuals act independently to maximize private benefits. Under this theoretical model, privatization of access rights to the common-pool resource can create incentives for socially optimal levels of resource exploitation.

Political scientist Elinor Ostrom has demonstrated that in many common-pool resource regimes, individuals may coordinate to develop a set of institutional rules that lead to lower levels of resource degradation. Case study and experimental research have identified conditions under which successful rules are more likely to develop. These include the ability to establish and enforce clear rules that restrict conditions of resource use, development of rules that allocate access in proportion to access costs, participation in governance by those affected by the rules, and the ability to monitor resource use and impose graduated sanctions for rule violations. If these conditions hold, common-pool resource regimes may be as successful as private-**property** regimes in maintaining a socially optimal level of the natural resource.

See also: **Community Involvement; Decision-Making; Institutions; Land Privatization; Land Rights; Participatory Land Management; Public Policy.**

Further Reading

Gordon, H. Scott, "The Economic Theory of a Common-Property Resource," *The Journal of Political Economy* 62 (1954): 124–42; Hardin, Garrett, "The Tragedy of the Commons," *Science* 162 (1968): 1243–8; Ostrom, Elinor, *Governing the Commons: The Evolution of Institutions for Collective Action*, New York: Cambridge University Press, 1990.

DAWN CASSANDRA PARKER

Tragedy of Enclosure.   Coupled social and environmental consequences of land encroachment and demarcation using boundaries, usually associated with changes in **land rights** and **land-use systems**. In developing nations, the state often acts in conjunction with businesses and multilateral institutions to deny small farmers or grassroots actors access to **common-pool resources** hitherto managed by them through local **institutions**. In effect, common property regimes are taken over by the state for large-scale commercial exploitation, either by its own agencies or by allied business interests using the legal-political powers of the state. Raymond Bryant and Sinéad Bailey point to notable cases or **proximate causes** such as the creation of extensive networks of reserved forests, national parks and "government lands" or "nationalized" territories in many parts of Asia, Africa, and **Latin America**, including areas of large-scale logging, **mineral extraction**, **cattle ranching**, production of **cash crops**, and dam construction. They point out that political ecology has been instrumental in stressing that post-colonial "development" was typically associated with the dissolution of many grassroots or local institutional arrangements that had so far governed the commons. The second point noted by Raymond Bryant and Sinéad Bailey is that the tragedy of enclosure served to further marginalize poor local land managers in the measure that their access to environmental resources essential for their **economic livelihoods** was restricted or denied. Not only was **access** to common resources ended, but these actors were often forced into a situation whereby they had to work ecologically marginal lands elsewhere in order to survive. The end result was that marginalization became a defining trait for most impoverished local groups and individuals as they were displaced from newly created **forest** reserves, national parks, or "development" projects. Within the broad category of impoverished and marginalized people, some actors are usually worse affected than others, notably women and indigenous minorities. The term **economic restructuring**, though unspecific with regard to geographic and historical context, insists that multiple mechanisms or processes of corporate, technological, and institutional change be considered together as a totality (rather than as separate variables) to drive local and regional environmental change associated with the enclosure of the commons. For example, the restructuring of developing economies following the adoption of **economic liberalization** policies includes a "property-regime effect" whereby the institutions of **property** through which land and resources are managed are progressively privatized and the social and environmental "commons" enclosed. The so-called Washington Consensus of liberalization and privatization continues to produce **land change as a forcing function in global environmental change** through the creation of private, exchangeable rights to resources such as land, water, or plant and animal life, which embeds resources in a distinctive management regime, one that prioritizes land-use practices that maximize the production of exchange values rather than use value. Éric Lambin and colleagues found that forced population displacement in conjunction with a decrease in land availability due to encroachment by other land uses (such as natural reserves) features "resource scarcity causing pressure of production on resources" as a fundamental high-level cause or **syndrome** of land change. It is associated here with rapid **land-cover change**, while the overall syndrome also includes slow rates of land change associated with heavy surplus extraction away from the land manager, for example. Summarizing a large number of case studies, the resource scarcity and production pressure syndrome was found to be just one out of five syndromes, and it appears as if only a combination of several causes, with synergetic interactions, is likely to drive a region into a critical trajectory of population displacement and **land degradation**.

See also: Agro-Industry; Conservation; Corporate Strategies; Driving Forces; Land Concentration; Land Rent; People at Risk; Poverty; Private Property; Public Policy; Yellowstone Model.

Further Reading

Bryant, Raymond L., and Sinéad Bailey, *Third World Political Ecology*, London, New York: Routledge, 1997; Geist, Helmut, "Exploring the Entry Points for Political Ecology in the International Research Agenda on Global Environmental Change," *Zeitschrift für Wirtschaftsgeographie* 43 (1999): 158–68; Glennie, Paul, "Enclosure," in *The Dictionary of Human Geography*, 4th ed. (reprinted), eds. Ron J. Johnston, Derek Gregory, Geraldine Pratt, and Michael Watts, Oxford, UK, Malden, MA: Blackwell Publishers Ltd., 2001, p. 206; Kates, Robert W., and Viola Haarman, *Poor People and Threatened Environments: Global Overviews, Country Comparisons, and Local Studies* (Research Report 91–20), Providence, RI: Brown University, The Alan Shawn Feinstein World Hunger Program, 1991.

HELMUT GEIST

Transhumance.   Transhumance involves the seasonal movement of livestock herds in order to make the best use of resources and to minimize risk. All pastoralist systems juggle the basic production constraints of forage, water, minerals, and disease in a wider context of alternative possible economic activities, changing markets, and shifting security. Many operate in arid or semiarid environments, characterized by variability and unpredictability of both rainfall and primary production, and most pastoralist strategies involve movement and patch use. Transhumance is found in most pastoralist systems, such as sub-Saharan Africa; the Mediterranean including southern **Europe**, the Near East, and the Maghreb; the European, for example, Swiss Alps, Pyrenees; Middle East (Kirghizstan, Turkmenistan, and Afghanistan); Central Asia (Mongolia and Siberia); and New World systems, e.g., Andean and Australian.

Where key resources are predictable, transhumance involves regular movements between wet/dry seasons or winter/summer **pastures**. Arid and variable systems involve more frequent, less predictable movements in response to patchy rainfall and forage. Movements may be influenced by raiding and violent conflict, making certain areas too hazardous to use. Depending on the livestock species involved and on local conditions, transhumance may cover anything from a few kilometers (e.g., European Alps) to hundreds (e.g., Fulani in the inland delta of the Niger) or even thousands of kilometers (Saharan Tuareg camel pastoralists). Movements may be altitudinal (e.g., Swiss alpine transhumance), exploit relatively subtle differences between floodplain and higher land (e.g., the Sudd in southern Sudan), or use a latitudinal system (Fulani groups moving among Sahelian, Sudanian, and Guinean zones).

Transhumant movements may be undertaken primarily to manage herd conditions through making best use of forage, water, and mineral availability; and transhumant animals often have significantly better health and

Winter feeding on cattle ranch in Madison County, Montana, March 1942.
PHOTO: John Vachon. [Courtesy Library of Congress]

production. Different livestock species and/or herds may follow different transhumant patterns (as among the Turkana) or exploit the same area in sequence (as for the inland delta of the Niger) as quality and quantity of forage varies with growth stage and with availability of crop residues.

Many pastoralist transhumance systems have developed to minimize exposure to disease risk while optimizing forage use. Fulani cattle transhumance is often timed to leave higher-risk parts of the range as the rains come and vector populations increase, moving into drier rangelands with little vector-borne disease, to exploit the temporary wet season flush of annual grasses. As the rains pass and the seasonal pastures are used up, herds return to wetter areas with more woody **vegetation**, perennial pastures, and crop residues. These areas have potentially high vector populations and disease transmission, but the risk decreases as vector populations decline in the dry season.

Animals in good condition can withstand a degree of disease challenge, and low levels of exposure stimulate a degree of immunity.

Other considerations enter into transhumance decisions. Fulani families in northern Burkina Faso minimize crop damage by keeping the herds away from growing crops after the harvest, when they may move animals to pasture on crop residues as part of a manuring contract with settled farmers. Other Fulani groups move near markets to sell pastoral produce. Transhumance may facilitate participation in social gatherings.

Among systems and households concentrating on pastoralism, the whole family may move with the animals. Families may join together to travel and camp (Turkana); seasonally, large groups of families may congregate (e.g., Tuareg at the *cure salée*). In more agropastoralist groups, one or more young adult males may take the family herd(s) on transhumance, depending on household demography and the broader network of livelihoods strategies, or cattle may be entrusted to a transhumant herder while the household stays put.

Transhumant systems depend on rights of tenure, **access**, and through-passage between point-centered, key limiting common pool resources (water, pasture, minerals) around which the radius of control fluctuates according to conditions. Access has customarily been negotiated by others in need, through token gifts, labor provision, trade, relations of intermarriage, fostering and adoption, and substantive payments in cash or in kind. Negotiated access often operates customarily in a graded way through orders of precedence. Access by users with successively lower priority claims becomes decreasingly likely as conditions deteriorate. The social **institutions** necessary to manage these transhumant systems are inherently flexible, with fuzzy social and spatial boundaries, unlike "modern" westernized **land tenure** systems based on imported European legal codes and imposing total exclusion. Currently, in addition to the restrictions imposed by land being set aside as protected areas, **land privatization** is increasingly threatening transhumant mobility, disrupting through-routes and barring pastoralists from key resources given over to cultivation.

See also: **Andes; Common-Pool Resources; Degradation Narrative; Disequilibrium Dynamics; Grassland; Land Rights; Pastoral Mobility; Property; Sudan-Sahel.**

Further Reading
Adamu, X., and Kirk-Greene, B., eds., *Pastoralists of the West African Savanna*, Manchester, UK: Manchester University Press, International African Institute, 1986; McCabe, J. T., *Cattle Bring Us to Our Enemies: Turkana Ecology, History, and Raiding in a Disequilibrium System*, Ann Arbor, MI: University of Michigan Press, 2004; Niamir-Fuller, M., ed., *Managing Mobility in African Rangelands: The Legitimization of Transhumance*, London: IT Publications, Rome: Food and Agricultural Organization of the United Nations, 1999.

KATHERINE HOMEWOOD

Transition Matrix. A table of numbers arranged in rows and columns that shows the amount of land that transitions from one category to another category between two points in time. Typically, the rows of the matrix show the categories at the initial time and the columns show the categories at the subsequent time. The entry in row *i* and column *j* of the matrix gives the area of land that transitions from category *i* at the initial time to category *j* at the subsequent time. The entries on the diagonal give the area of the map that persists between the two points in time. It is standard practice to append a "totals column" on the right that gives the total areas for each category at the initial time, and a "totals row" on the bottom that gives the total areas for each category at the subsequent time.

When the categories at the initial time are the same as the categories at the subsequent time, it is helpful to create an additional "losses column" on the right that gives the total area lost for each category, and a "gains row" on the bottom that gives the total areas gained for each category. It can be advantageous to express the areas in the matrix as a proportion of the study area by dividing all the entries in the matrix by the total area in the study. The creation of the transition matrix is one of the most common first steps in analyzing change among land categories between two points in time.

One can analyze the numbers in the transition matrix at various levels of detail, depending on the particular question. At the simplest level of analysis, one compares the numbers in the totals column to the corresponding numbers in the totals row in order to compute the net change for each category, which is the change in quantity of area. At a slightly more sophisticated level of analysis, one computes the change in location for each category (i.e., swap), which is formed by pairing as many gaining areas as possible with losing areas. In this manner, a scientist can budget the total change on a map in terms of two pairs of components: gains and losses, and net and swap. In order to examine transitions among individual categories, one must examine the entries within the matrix. The off-diagonal entries can be analyzed at a variety of levels of detail. At the simplest level, scientists consider the size of the entries, in which case the larger numbers are usually more important. In order to detect a systematic transition, each off-diagonal entry should be analyzed with respect to the relative size of all the categories.

The transition matrix is the foundation of Markov chain analysis, which is a common statistical tool. Scientists can use Markov chain analysis to extrapolate land change into the future, where the category of a **pixel** has a certain probability of transitioning to a different category at each time step based on the information in a probability matrix. The probability matrix derives directly form the transition matrix. If the probabilities from one time interval are the same as the probabilities of other intervals, then the probabilities are said to be stationary, in which case it is reasonable to use them to extrapolate over various ranges of time.

There are now methods to compute the transition matrix for cases where each unit of observation (e.g., pixel) is soft-classified, which means that it has simultaneous partial membership in multiple categories. Therefore, the entire transition matrix can be computed and analyzed at multiple **scales**.

See also: Change Detection; Land-Use Transitions; Pattern to Process; Probit/Logit Model.

Further Reading

Baltzer, Heiko, "Markov Chain Models for Vegetation Dynamics," *Ecological Modeling* 126 (2000): 139–54; Pontius, Robert Gilmore, Jr., Shusas, Emily, and Menzie

McEachern, "Detecting Important Categorical Land Changes while Accounting for Persistence," *Agriculture, Ecosystems and Environment* 101 (2004): 251–68; Pontius, Robert Gilmore, Jr., "Statistical Methods to Partition Effects of Quantity and Location during Comparison of Categorical Maps at Multiple Resolutions," *Photogrammetric Engineering and Remote Sensing* 68 (2002): 1041–9.

PONTIUS, ROBERT GILMORE, JR.

Tropical Dry Forest. The preferred zone for **agriculture** and human settlement worldwide, because environmental and biotic stresses are the lowest compared to other tropical life zones, and other life zones are too cold, hot, dry, or wet. When contrasted against wet or **tropical humid forests**, tropical dry forests (TDFs) cannot be defined as the last frontier but the first frontier of economic development. In fact, early farming communities and the earliest civilizations in Mesoamerica, for example, developed in the tropical dry forest regions, and there is a strong possibility that today's major tropical crops and food animals originated from tropical dry forest ecosystems. Worldwide, 42 percent of tropical forests are tropical dry forests.

Various political and economic factors have increased anthropogenic stresses in these ecosystems, leading to severe disturbances and widespread clearing. Currently, a significant gap exists in terms of TDF research when compared with tropical moist or rainforests where, for many complex political and institutional reasons, international funding has been more prominent. Since 1945, a total of 2,300 research articles have been produced from tropical environments, of which (in 2004) only 10 percent related in some form to tropical dry forests, according to the Science Citation Index. Efforts aimed at generating information regarding TDF are scattered and limited to a few sites in the Americas and elsewhere such as in the tropical dry forests of India. No TDF multidisciplinary and comparative studies exist for the Americas and other regions in Africa and Asia.

TDFs have less species diversity than the wet forests, though the former one has more endemic species that the latter one. It is estimated that the Chamela's tropical dry forest region on the Pacific coast of Mexico is probably one of the sites with the highest number of endemic tree species in Mesoamerica. Tropical dry forests serve as home to rare and charismatic creatures as jaguars, howler monkeys, scarlet macaws, toucans, and tapirs. It is estimated that when the Spaniards arrived in Mesoamerica (the area covered by southern Mexico and Central America), TDF covered as much extension as the tropical wet forest, yet today it is estimated that only about 2 percent of all the tropical dry forest in the region may be considered intact. For the Guanacaste Conservation Area in Costa Rica alone, new biodiversity inventories have found in the order of 120,000 species of insects, 60,000 species of fungi, and 9,000 species of plants. Therefore, when **conversion** to agriculture/**pasture** land takes place in tropical dry forest regions, the risk of losing species is high, even in small/localized areas.

Our understanding of the human and biophysical dimensions of TDF change and their cumulative effects is still in its early stages of academic discovery, since more emphasis has been placed on tropical evergreen forests, especially the Amazon basin. Therefore, there is a need for continuous and systematic efforts to understand and integrate TDF phenological patterns and social/anthropogenic mechanisms at three basic levels: in the context of conservation biology, the context of land-use and **land-cover change** that are taking place on this rich **agricultural frontier**, and in the context of local and national development policies that contribute to degradation of tropical dry forests.

Accurate forest-cover assessments are important to define natural resource management strategies and policies for **conservation**. It is through information about the location, the extent, and the status of forest areas that threats to **biodiversity** from **deforestation** fronts ("hot spots") can be identified and solutions can be evaluated and implemented. The understanding of the causes and consequences of land-cover change in dry forest environments and its impacts on the environment through time is key to predicting possible negative effects on biological resources. In addition, their cascading effects across the many components of a functional ecosystem have negative impacts on the services they provide to mankind. In fact, our current understanding of the cumulative effects of human-dominated landscapes in tropical dry forest is strongly linked to a clear understanding of the **proximate causes** and underlying socio-economic, biophysical, and **driving forces** of land-use and land-cover change. These forces acting at different **scales** (from the international sphere to the unit of the production level) contribute to regional environmental deterioration trends that in many cases cannot be fully understood because of the lack of sound spatial, socio-economic, and ecological databases that have been built in the context of an interdisciplinary approach. It is through the combination of these data sets that the cumulative effects of land-use/cover change in a given environment can be soundly evaluated and remediation/control policies can be implemented.

See also: **Hot Spots of Land-Cover Change; Forest; Species Extinction.**

Further Reading

Kalacska, Margaret, Sánchez-Azofeifa, G. Arturo, Calvo, Julio, Rivard, Benoît, Quesada, Mauricio, and Daniel Janzen, "Species Composition, Similarity and Diversity in Three Successional Stages of a Tropical Dry Forest," *Forest Ecology and Management* 200 (2004): 227–47; Sagar, R., Raghubanshi, A. S., and J. S. Singh, "Tree Species Composition, Dispersion and Diversity along a Disturbance Gradient in a Dry Tropical Forest Region of India," *Forest Ecology and Management* 183 (2003): 61–71; Janzen, Daniel H., "Tropical Dry Forests: The Most Endangered Major Tropical Ecosystem," in *Biodiversity*, ed. Edward O. Wilson, Washington, DC: National Academy Press, 1988, pp. 130–7.

ARTURO SÁNCHEZ-AZOFEIFA

Tropical Ecosystem Environment Observations by Satellite (TREES) Project. A research program managed by the **Joint Research Centre (JRC) of the European Commission**, in close cooperation with the Commission's Environment Directorate-General, and local partners in developing countries, and dedicated to the development of **forest**-cover assessment throughout the humid tropics. Initiated in the early 1990s, the project used the global imaging capabilities of a number of earth-observing satellites, including Europe's SPOT 4 and the **European remote sensing (ERS-1/-2) satellites** to provide the latest information on the state of the world's **tropical humid forests**. Indeed, in spite of the importance of the tropical forests, our knowledge concerning their rates of change remains limited. The project made use of an extensive set of **remote sensing** satellite imagery. The initial objectives of the TREES project were to develop techniques for global tropical forest mapping, to develop techniques for **monitoring** active deforestation areas, and to set up a comprehensive tropical forest information system. The ultimate goal is to establish an operational observing system that can detect and identify changes in the tropical forest cover of the world.

The first phase of this research program (TREES-I) has allowed the creation of the most complete, up-to-date set of maps available that document the distribution of the world's remaining humid tropical forests. These maps provide an unprecedented view of one of the most important **biomes** on the planet.

The primary objectives of the second phase of this research program (TREES-II) were to produce relevant information, more accurate than currently available, on the state of the humid tropical forest ecosystems from new remote sensing-based approaches and to analyze this information in terms of **deforestation** and **forest degradation** trends. TREES-II was completed in 2002.

TREES-II's results clearly show that deforestation in the humid tropics is still a major global environmental issue. The project provides the most accurate, consistent figures on rates of deforestation throughout the humid tropics currently available. Between 1990 and 1997 a staggering 5.8 million hectares of humid tropical forest were lost each year. This is an area twice the size of Maryland or Belgium. A further 2.3 million hectares per year of forest are detected as highly degraded—becoming increasing fragmented, heavily logged, and/or burnt. Although the statistics document the trends up to 1997, the most recent maps from the project (around year 2000) provide no grounds to believe that this situation was improving in the recent years.

The maps, information on forest-cover status, and rates of change are based on uniform, independent, and repeatable methods. These new data have already reduced uncertainties in dealing with carbon sink issues, and they provide accurate baseline views of this hugely valuable global resource and help in planning strategies for effective **conservation** of its biological diversity.

TREES, endorsed by the international **Land-Use/Cover Change (LUCC) project**, clearly demonstrates the important role of sound scientific evidence to support policy. The close collaboration with local partners in developing countries and international governmental or non-governmental organizations combined with state-of-the-art analysis of satellite imagery has proved a powerful combination.

See also: **Biodiversity; Carbon Cycle; Forests; SPOT Vegetation.**

Further Reading

Achard, Frédéric, Eva, Hugh, Stibig, Hans-Jürgen, Mayaux, Philippe, Galego, Francisco, Richards, Timothy, and Jean-Paul Malingreau, "Determination of Deforestation Rates of the World's Humid Tropical Forests," *Science* 297 (2002): 999–1002;

Joint Research Centre of the European Commission, Institute for Environment and Sustainability, Global Vegetation Monitoring Unit [Online, November 2004], The Tropical Ecosystem Environment Observations by Satellite Web Site, www-gvm.jrc.it/tem/ProductsArchive.htm.

FRÉDÉRIC ACHARD

Tropical Humid Forest. Major forest type of the world that has year-round abundant precipitation (typically more than 1,600 millimeters annually), occurs either near the equator or on low-latitude windward coasts, and which is currently undergoing rapid changes as humans increasingly extract timber resources, expand and intensify agricultural areas, and migrate into previously unsettled areas. The evergreen and seasonal forests of the tropical humid bioclimatic zone, also called rainforests, differ from **tropical dry forests** (and woodlands) in that the latter occur in drier areas farther from the equator, holding broad-leaved deciduous trees. The domain of the dry seasonal forests of continental Southeast Asia is intermixed with the humid forests domain, while the separation between these two domains is clearer in Africa and Latin America. Yet in spite of the importance of tropical rainforests, our knowledge concerning their distribution and rates of change remains surprisingly limited.

Monitoring systems based on **remote sensing** provide information on **land-cover change** (including cover type and biophysical parameters) and disturbance due to **fire**. By using the repeat coverage of the earth observation (EO) satellites, it has been possible to monitor forest-cover change in an operational mode since the early 1980s. In the following, summary findings are presented originating from the Forest Resources Assessment program of the **United Nations Food and Agriculture Organization** (FAO), initiated in the early 1980s, and of the **Tropical Ecosystem Environment Observations by Satellite (TREES) Project**, estimating tropical forest remaining cover and its rate of change for the 1990s.

FAO Forest Resources Assessment Programme

The FAO Forest Resources Assessment (FRA) program provides, in principle, two estimates for the full tropics separately through two methodological approaches: the country survey (CS), which is based on the compilation and standardization of national data and the estimates of which are generally considered as reference figures for the tropics; and the remote sensing survey (RSS), which provides statistical estimates at a continental level, derived from forest-cover change maps (interpretation of thirty-meter resolution satellite imagery). Internal inconsistencies have been highlighted for the FRA CS estimates, which might be due to the difficulties of standardizing country level data obtained from official intergovernmental processes. On the other hand, it is well accepted that a sample can provide reliable estimates of **deforestation** if the sample size is sufficiently large. The FRA RSS uses a random sample covering 10 percent of the whole tropical belt, designed for forest cover estimation.

For both surveys (CS and RSS), a 10 percent canopy cover threshold is used as the minimum definition of forest in tropical countries. The FRA RSS findings indicate that the world's tropical forests covered 1.57 billion hectares in 2000, about 12 percent of the world's land area. The net change in tropical forest area was −8.6 million hectares per year, representing the difference between a deforestation rate of 9.2 million hectares per year of natural forests and an expansion of 0.6 million hectares per year of natural forests and forest plantations. The reported FAO CS rates of change in forest area appear larger than the estimates from satellite-based studies. Most of the world's forest losses were in the tropics. The rate of net change was slightly lower in the 1990s compared to the 1980s, due to a higher estimated rate of forest expansion in the 1990s. Country studies point at **land rights** as a

Estimates of tropical forest area and forest-cover change during the 1990's from FAO and TREES surveys

Continent	Forest area 2000 (10^6 ha)		Annual forest area change 1990–2000	
	TREES Domain	Non-TREES Domain	TREES Domain	Non-TREES Domain
Southeast Asia	260		−2.3	
Africa	191	292	−0.7	−1.5
Latin America	646	121	−2.3	−1.9
Tropics		1,510		−8.7

Note: "TREES domain" refers to: (a) the humid tropical forest biome of Latin America excluding both Mexico and the Atlantic forests of Brazil, (b) the humid tropical forest biome of Africa, and (c) the tropical forest biome of Southeast Asia and India, excluding only the dry biome of India. "Non-TREES domain" is defined as all remaining tropical forests, i.e., dry forests of Latin America, Africa and India, and all Mexico forests and Atlantic forests of Brazil.

common main determinant behind deforestation, and the RSS indicates that direct **conversions** of **forests** to permanent **agriculture** were more prominent than shifting agriculture in forest change processes.

TREES' Monitoring of Humid Tropical Forests

The TREES project examines **hot spots of land-cover change**, namely deforestation with fine spatial resolution (thirty meters) satellite imagery. TREES uses a stratified systematic sample covering 6.5 percent of the humid tropical domain, designed for forest cover change estimation. The project provides the most accurate, consistent figures on rates of deforestation throughout the humid tropics currently available. The TREES and FRA RSS estimates can be considered complementary, as they are both based on a statistical sampling strategy using satellite imagery and have been designed to give continental estimates. Both methods provide measurement of tropical forest-cover change in a uniform, independent, and repeatable manner, and moreover are consistent with each other (see table titled "Estimates of tropical forest").

The results of the TREES project clearly show that deforestation in the humid tropics is still a major global environmental issue. Analysis is now complete and shows that in 1990 there were some 1.15 billion hectares of humid tropical forest. Between 1990 and 1997 a staggering 5.8 million hectares of humid tropical forest were lost each year. This is an area twice the size of Belgium or Maryland. A further 2.3 million hectares per year of forest are detected as highly degraded—becoming increasing fragmented, heavily logged, and/or burnt. Although the statistics document the trends up to 1997, the most recent maps from the project (from 1999 and 2000) provide no grounds to believe that this situation is improving.

Although a global phenomenon, the spatial detail and ability to compare different regions of the world provided by TREES reveals considerable variation around the world. An annual global deforestation rate of 0.5 percent is alarming enough, but the project shows that Southeast Asia is losing its tropical forests twice as fast as anywhere else. While the regional forest change rate for **Latin America** is relatively low, the overall gross deforestation is a dramatic 2.5 million hectares per year, the same as that of Southeast Asia. Furthermore, this change is confined to several "hot spots," where rates can reach at least 3 percent loss per year.

See also: **Agricultural Frontier; Alternatives to Slash-and-Burn (ASB) Programme; Biome; Forest Degradation.**

Further Reading

Achard, Frédéric, Eva, Hugh, Stibig, Hans-Jürgen, Mayaux, Philippe, Galego, Francisco, Richards, Timothy, and Jean-Paul Malingreau, "Determination of Deforestation Rates of the World's Humid Tropical Forests," *Science* 297 (2002): 999–1002; DeFries, Ruth, Houghton, Richard A., Hansen, Matthews, Field, Chris, Skole, David, and John Townshend, "Carbon Emissions from Tropical Deforestation and Regrowth based on Satellite Observations from 1980s and 90s," *Proceedings of the National Academy of Science* 99 (2002): 14,256–61; Food and Agriculture Organization of the United Nations, *The Global Forest Resources Assessment 2000* (FRA 2000): Main Report, Rome: FAO, 2001; Harcourt, Caroline S., and Jeffrey A. Sayer, *The Conservation Atlas of Tropical Forests: The Americas*, London: Macmillan, 1996.

FRÉDÉRIC ACHARD, HUGH EVA, PHILIPPE MAYAUX, HANS-JÜRGEN STIBIG

Turnover. Amount of time it takes to replace the **biomass** of a population or ecosystem, measured as the ratio of stock (mass, e.g., standing crop of biomass) and flow (mass per time unit, e.g., net primary production) and expressed in time units (e.g., per year) or a synonym: for example, transit time or residence time. Habitat ecologists sometimes use the term in the sense of turnover of species, indicating the time required for a complete change in species composition by means of **ecological colonization** and **species extinction**. Ecosystems are characterized by typical rates of biomass turnover. Turnover time is probably related to the resistance and resilience of ecosystems. Systems with fast turnover may be displaced from their state easily, but recover quickly, that is, their resistance is low but their **resilience** high. For example, an annual **grassland** with a turnover of one year recovers quickly after a **fire**. By contrast, a system with a slow turnover is not easily changed completely, but if it is, it recovers very slowly if at all, that is, its resistance is high and its resilience low (e.g., a **forest**). Land use fundamentally alters ecosystem patterns and processes. As a consequence, the turnover of biomass is usually accelerated considerably. Societies show preferences for annual plants such as cereals, and very often replace systems with larger biomass stocks such as forests and savannas. Moreover, as **colonization** aims at increasing useful biomass output (e.g., net primary productivity of harvestable goods), stocks are reduced due to juvenilization effects (e.g., reduction of stand age in forests). The consequences of this acceleration of turnover for ecosystem functioning and patterns have received little, if any, scrutiny and are therefore at present largely unknown.

See also: **Agriculture; Agrodiversity; Biodiversity; Disequilibrium Dynamics; Domestication; Forestry; Human Appropriation of Net Primary Production; Proximate Causes.**

Further Reading

Haberl, Helmut, Erb, Karlheinz, Krausmann, Fridolin, Loibl, Wolfgang, Schulz, Niels B., and Helga Weisz, "Changes in Ecosystem Processes Induced by Land Use: Human Appropriation of Net Primary Production and Its Influence on Standing Crop in Austria," *Global Biogeochemical Cycles* 15 (2001): 929–42; Ricklefs, Robert E., and Gray L Miller, *Ecology*, 4th ed., New York: Freeman and Company, 1999.

KARLHEINZ ERB

U

United Nations Convention on Biological Diversity. Presented at the United Nations Conference on Environment and Development (UNCED) in Rio de Janeiro in 1992 (Earth Summit), the convention defines biological diversity as the variability among living organisms from all sources, including terrestrial, marine, and other aquatic ecosystems and the ecological complexes of which they are a part. Thus, it encompasses diversity within species, between species, and of ecosystems. The Convention on Biological Diversity (CBD) came into effect in 1993, and a summary of the key elements that highlight and characterize the **biodiversity** problem is provided in the preamble:

1. The importance of biological diversity for maintaining life-sustaining systems of the biosphere
2. The concern that biodiversity is being significantly affected by certain human activities
3. The awareness that **conservation** and sustainable use of biodiversity are critical in meeting the food, health, and other needs of the growing world population
4. The awareness that **access** to and sharing of both genetic resources and technologies are essential.

However, the CBD failed to address the "theft" of biological resources that developing countries claim has occurred since colonialism. In contrast, developed countries have argued that biodiversity is part of the "global commons" for the benefit of humankind, regardless of where it is located.

See also: **Agenda 21; Agrodiversity; Biodiversity Novelties; DIVERSITAS; Domestication; Exotic Species; Millennium Ecosystem Assessment; Proximate Causes; Species Extinction; Tragedy of the Commons.**

Further Reading

McConnell, F., "The Convention on Biodiversity," in *The Way Forward: Beyond Agenda 21*, ed. Felix Dodds, London: Earthscan, 1997, pp. 47–54; United Nations Environment Programme, Secretariat of the Convention on Biological Diversity [Online, September 2004], The Convention on Biological Diversity Web Site, www.biodiv.org.

HELMUT GEIST

United Nations Convention to Combat Desertification. An international agreement negotiated by the **United Nations Environment Programme** that addresses **desertification**. Through today, the United Nations

Convention to Combat Desertification (UNCCD) provides the most authoritative definition of the issue at stake, stating that desertification means **land degradation** in arid, semiarid, and dry sub-humid areas resulting from various factors, including climatic variations and human activities. In 1977, the United Nations Conference on Desertification (UNCOD) adopted the Plan of Action to Combat Desertification (PACD), and the question of how to tackle desertification was still a major concern for the United Nations Conference on Environment and Development (UNCED) in Rio de Janeiro in 1992. UNCED supported a new, integrated approach to the problem, emphasizing actions to promote sustainable development at the community level. It also called on the United Nations (UN) General Assembly to establish a procedure for the preparation of an agreement to combat desertification, particularly in Africa. In December 1992, the UN General Assembly agreed and adopted resolution 47/188. The convention was adopted in June 1994, and came into force on December 26, 1996, that is, ninety days after the fiftieth ratification was received. The Conference of the Parties (COP) is the convention's supreme governing body, and in 2003 over 180 countries were parties. Not denying the fact that many people are failing to thrive and, indeed, survive in drylands, the realization of UNCCD is a major causal element of why desertification continues to be perceived as a global environmental change issue despite weak empirical data on the global **scale**. Following UNCOD from the 1970s onward, many governments of countries with large dryland zones, but also international development and relief agencies, have been using the coupled notion of desiccation and **poverty** to rescue an ideology of authoritarian intervention in rural **land-use systems** and to claim rights to stewardship over resources previously outside their control. Despite the fact, for example, that the landscape in Ivory Coast is becoming more wooded and not desertified, the desertification and **degradation narrative** is present in the National Environmental Action Plan. Still, there is an explicit focus of intervention measures on Africa, and the full text of UNCCD consequently reads as the "United Nations Convention to Combat Desertification in Those Countries Experiencing Serious Drought and/or Desertification, Particularly in Africa." This is directly tied to the political and managerial outcomes of the Rio conference rather than based on empirical evidence, which shows current and rapid **hot spots of land-cover change** in the drylands of Central Asia rather than in Africa. Therefore, political ecologists like Neil Adger and colleagues recognize a "desertification crisis rhetoric," since the theme was added to the global agenda at UNCED due to persistence from African countries and because the **United States of America** unexpectedly supported the African position. It is believed that the support came in response to the environmental criticism the United States had received for failing to support the **United Nations Convention on Biological Diversity** and for dragging its feet in the preparations for the UN Framework Convention on Climate Change. Others interpret the concession of U.S. support for UNCCD as part of a **tradeoff** between developing and developed countries insofar as the European Union together with the United States expected African states to be supportive regarding the remaining issues in the Rio declaration.

See also: **Community Involvement; Institutions; Land Degradation Assessment in Drylands (LADA) Project; Millennium Ecosystem Assessment; Public Policy; Tragedy of Enclosure.**

Further Reading

Adger, Neil W., Benjaminsen, Tor A., Brown, Katrine, and Hanne Svarstad, "Advancing a Political Ecology of Global Environmental Discourses," *Development and Change 32* (2001): 681–715; Geist, Helmut, *The Causes and Progression of Desertification* (Ashgate Studies in Environmental Policy and Practice), Aldershot, UK, Burlington, VT: Ashgate, 2005; United Nations Convention to Combat Desertification [Online, February 2005], The UNCCD Web Site, www.unccd.int.

HELMUT GEIST

United Nations Environment Programme. A specialized agency of the United Nations that was created in 1972 in response to a recommendation from the United Nations (UN) Conference on the Human Environment in Stockholm, Sweden. Located in Nairobi, Kenya, the United Nations Environment Programme (UNEP) coordinates the environmental activities of UN agencies, works to increase national capacity for **conservation** or environmental protection in member states, and oversees environmental **monitoring** and assessment activities under the Earthwatch program (such as the Global Environmental Monitoring System, GEMS) and the gathering of scientific information (such as through the **Millennium Ecosystem Assessment**). In 2004, UNEP and the **United Nations Food and Agriculture Organization (FAO)**, based on the recommendations of **Agenda 21**, responded to the needs for standardized land-cover data by initiating the **Global Land Cover Network**. For example, the global data set of the **Global Land Cover Map of the Year 2000** has been published by UNEP in association with the FAO and the **Joint Research Centre (JRC) of the European Commission**. The database has been adopted by Millennium Ecosystem Assessment as the baseline for 2000 and is being used by UNEP's World Conservation Monitoring Centre in its report on the state of the world's protected areas. In general terms, UNEP works to increase the ability of the UN system to address environmental problems such as **desertification**, and, more specifically, assists in negotiations of international agreements and provides support to them, such as the **United Nations Convention to Combat Desertification** and the **United Nations Convention on Biological Diversity**. Nearly half of the multilateral environmental agreements concluded since the organization's creation have been under UNEP auspices. UNEP provides secretariat services to a number of these agreements and acts as an implementing agency for projects carried out under the Global Environmental Facility and the Montreal Protocol Multilateral Fund. In addition, UNEP has worked to set guidelines for various environmental concerns such as the management of **common-pool resources** and environmental impact assessments.

See also: AFRICOVER; Global Terrestrial Observing System; GLOBCOVER; Investments; Land Cover Classification System; Land Degradation; Scenario.

Further Reading

United Nations Environment Programme, Environment for Development [Online, January 2005], The UNEP Web Site, www.unep.org.

HELMUT GEIST

United Nations Food and Agriculture Organization. One of the largest specialized agencies of the United Nations (UN), both in membership by states and its staff, which has multiple roles in providing services to governments through the compilation of data and a variety of information and support services in the fields of **agriculture**, agricultural commodities, fisheries, **forestry**, and **food security**. Located in Rome, Italy, the UN Food and Agriculture Organization (FAO) had a predecessor entity in the International Institute of Agriculture in Rome and was created at the same time as the United Nations (in 1945). The FAO, drawing upon its networks and technical staff, is also a source of advice to governments on agricultural policy and planning as well as administrative and legal planning. Through its field program, it gives assistance to a large number of developing countries. At the time of its creation, the primary focus of the FAO was on the improvement of the production and **consumption** of food and other agricultural products. This was fully in line with the concepts and strategies of the **Green Revolution**. In the course of the 1980s and 1990s, the FAO was criticized for being top-heavy, centralist, and out of touch with the mainstream of creative thinking and analysis. Thus, the FAO was not a crucial factor in the considerations of the follow-up to the major UN conferences of the 1990s, which are the basis of today's international environment-development agenda. Only since about the World Food Summit in 1996, the FAO now has created the necessary preconditions for reinventing itself as an effective player in the multilateral system. For example, **AFRI-COVER** was developed as a project in response to a number of national requests for assistance to develop reliable and georeferenced information on natural resources. Its interpretation and classification are based on the FAO **Land Cover Classification System**, which was conceived by an international working group and was developed by several FAO services, with the support of a large number of national and international agencies. Likewise, the global data set of the **Global Land Cover Map of the Year 2000** has been published by the FAO in association with the **Joint Research Centre (JRC) of the European Commission** and the **United Nations Environment Programme**. The database has been adopted by **Millennium Ecosystem Assessment** as the baseline for 2000.

See also: Global Land Cover Network; Global Terrestrial Observing System; Land-Use Planning; Land-Use Policies; Proximate Causes.

Further Reading
Food and Agriculture Organization of the United Nations, Helping to Build a World without Hunger [Online, January 2005], The FAO Web Site, www.fao.org.

HELMUT GEIST

United States of America. The land-cover history of the United States is largely a result of the nation's mid-latitude location; physical resources; the federal government's role in land dispersal, reservation, and management; the dominant technologies at the time of settlement, and society's demand for land-based resources. Initial **land-cover changes** subsequently were modified by increased population, **globalization** of the economy, maturation of

Wheat farming area near Goldendale in Klickitat County, Washington, in August 1939 (with abandoned farm in foreground and occupied farm beyond).

PHOTO: Dorothea Lange. [Courtesy Library of Congress]

the nation's settlement system, and development of new technologies that have resulted in the exploitation of different resources and **modification** of old ones. Regions with poor innate suitability, which are unable to respond to increases in the intensity of land use that is associated with greater technical inputs, were either bypassed by settlers and businesses or temporarily exploited and then allowed to revert to a "natural" land cover.

The Natural Resource Base

The major natural land-cover patterns were the result of temperature, precipitation, elevation gradients, and the flow of energy and moisture from the Pacific Ocean, Gulf of Mexico, and Atlantic that primarily followed the flow of the mid-latitude westerly winds. In general, temperatures decline south to north and precipitation declines east to west. These gradients are modified by three north-to-south mountain systems: the Appalachians in the east, Rocky Mountains in the interior west, and Pacific ranges along and near the west coast. These **mountains** block and channel air masses from the west and allow air from the south and north to penetrate to the continent's interior. The general land-cover result was moist forests east of the Mississippi River and on the windward slopes of mountains; **grasslands** between the

Mississippi River and Rocky Mountains; and arid land grasses and shrubs from the Rocky Mountains to the Sierra Nevada and Cascade Mountain ranges.

Presettlement

The land cover of the United States prior to European settlement is subject to debate. The noted biogeographer A. Küchler mapped potential natural vegetation of the lands now representing the conterminous forty-eight United States. Potential natural vegetation was defined as the **vegetation** that would cover the land if there were no disturbances from man or nature. Essentially, A. Küchler's map represents climax vegetation, an ecological condition that was unlikely in many areas due to natural disturbances and land uses associated with indigenous people. Still, the map provides a starting point for understanding U.S. land change: prior to European settlement, 50 percent of the conterminous United States was forested, grasslands covered 24 percent, shrublands covered 18 percent, and open woodlands (grassland-forest savannas) covered nearly 8 percent of the land.

Clearly, vegetation at the time of settlement had been modified by both indigenous people and by natural events. For example, hurricanes and other wind storms frequently damaged southeastern and northeastern forests, creating gaps and modifying successional status. In the west, **fires** changed the composition of forests and created a varied quilt of different seral stages. There is also evidence that indigenous populations and economies were sufficient to have a significant local to regional impact on the natural vegetation of the United States. Indigenous populations were relatively high in 1500 and then declined following the spread of infectious diseases. As a result, the impact of indigenous people due to land uses associated with their farming, grazing, hunting, and urban activities was likely to be highest in 1500 and then was reduced.

Development Frontier

The primary development frontier in the United States was east to west with secondary frontiers moving south to north from New Spain (Mexico) during the seventeenth and eighteenth centuries and a mining frontier that moved east from the California gold fields through the mountainous west during the second half of the nineteenth century. The most important land-cover changes were the result of U.S. expansion westward from the original thirteen states. The historical geography of U.S. land-cover changes follows four epochs of change that were the result of the time of initial land cover; settlement patterns; the predominant technologies of the time, especially transportation and energy technologies; the resources that attracted settlement; and the prevailing government policies associated with land.

Epoch I: 1620–1820

The forests east of the Mississippi River and south of the Great Lakes were settled by farmers who cleared their land with axes. Preferred areas of farming often included fields abandoned by indigenous people. The primary settlement motivation was to acquire farmland, and the primary resource was good soil. Most land was converted to fields and **pasture** with some small, developed areas in villages and incipient cities. In the South, global demands for **cotton** and **tobacco** were a **driving force**, while the northern crops of **wheat** and corn were produced for the domestic grain and meat markets. Before 1776, the British tried to prevent settlement west of the Appalachians, which confined virtually all land **conversion** to the eastern areas. After the American Revolution, land that was not within the boundaries of a state was considered to be in the public domain until statehood. The primary federal government role was to survey and sell the land to farmers and speculators as quickly as possible.

Epoch II: 1820–1870

Between 1820 and 1870, land conversion accelerated because of the nation's increased population and the application of **Industrial Revolution** technology to land conversion. The application of steam power to shipping and early railroads increased the magnitude of land conversion. The nation's forests were converted rapidly as the nation's larger population demanded additional farmland, wood for heating and construction, and fuel for steam engines, which consumed prodigious quantities of wood. Most forest conversion was located near water bodies, because waterpower sites were the primary industrial energy source and the inland waterway system of canals and rivers supported most long-distance transportation. The primary federal role continued to be the surveying and disposition of land as quickly as possible.

Epoch III: 1870–1920

By 1870, the settlement frontier had reached the forests of the northern states and the grasslands of the **Great Plains**. Improvements in steam technology, the availability of cheap steel, and commercial development of electricity allowed new resources to be developed. Coal became the dominant energy source, and areas with it had rapid land conversion. The demands for wood increased because of the nation's growing population, settlement of the Great Plains where lumber had to be imported from forested regions, and use of wood for mine supports and railroad ties.

The northern forests were commercially harvested and abandoned when forestry companies chose to relinquish land ownership and move to western mountains that had areas of virgin forests. The eastern half of the grassland interior was converted from natural grass to cropland and pastures, while the arid western half became a mosaic of dryland **wheat** fields and natural rangeland depending upon soil quality and crop prices. Farther west, the arid, intermountain west remained in natural grass and shrubs except where irrigation farming was available.

The band of needle leaf evergreen trees that stretched from Maine to Minnesota and along the slopes of the Appalachian Mountains and Cumberland Plateau began to be commercially logged. Where there was adequate soil, farmers moved in as the forestry companies left, but in the north the soil was often infertile and secondary forests began to grow, and in the south the slopes were often so steep that after a few seasons of farming, most of the good soil had eroded. By 1920, virtually all of the old-growth forest of the eastern United States had been harvested. Drainage of **wetlands** and conversion to farmland was significant in the glaciated portions of the Midwest.

In response to public concerns about environmental **conservation** and timber shortages, and settler reluctance to acquire lands that could not be profitably farmed, the federal government began to reserve lands to create national forests, national parks, and federal grazing lands. All of these were in the mountainous and arid west. These lands had minimal land conversion from their original cover, even when commercial **forestry** and grazing were permitted. After 1900, the federal government began to purchase private lands in the hilly areas of the east where farming had failed or forests had been cut and not yet regrown.

Epoch IV: 1920–Present

By 1920, the U.S. settlement pattern and infrastructure system was established. One of the two major conversions was the return of trees to cleared land that had been cleared for farms or commercial forestry. This occurred throughout the states that bordered the Gulf of Mexico, Atlantic Ocean, and Great Lakes as well as substantial parts of the Appalachian and Ozark highlands. Florida, the exception, had massive drainage and forest clearance during this epoch. The second major transformation was the growth in urban areas. After 1920, the population of the nation grew rapidly, and the automobile allowed a lower-density settlement pattern. Meanwhile, the general increase in per capita income led to larger residential lots and houses. The result was that more land per capita was converted to developed land covers (e.g., houses, residential lots, streets and highways, and parking lots), and the nation's developed area increased faster than the population.

During the 1930s, the federal government purchased degraded farmland in much of the Great Plains and converted the land to national grasslands. By the end of the twentieth century, the federal government owned about 28 percent of the United States. With the exception of urban offices and military bases, most of this land has the same land cover that it did in 1620.

During the 1950s and 1960s, the nation made a series of substantial one-time land conversions to transportation, reservoirs, and **urbanization**. The construction of the federal interstate highway system converted more than 700,000 hectares to highways and intersections that often became development nodes leading to further conversion to developed covers. States and cities also were responsible for substantial street and highway development. Construction of reservoirs had begun during the Industrial Revolution, but approximately 60 percent of the nation's reservoir storage capacity was constructed during the 1950s and 1960s. Most of these are located in the humid East, but most of the large ones are west of the Mississippi River. After 1930, up to 2 million farm ponds were constructed, with the greatest concentration in the eastern Great Plains and Southeast. Most of these are less than one hectare in size. Before 1980, significant areas of hardwood wetlands in the Mississippi alluvial plain and other river bottoms in the South were drained and converted to cropland.

After 1970, the nation's population growth, and associated conversion to developed covers, slowed in the traditional growth core of the northeastern and midwestern states and shifted to the warmer southern states, especially from California to Florida and north to North Carolina. Secondary population growth occurred in rural areas within commuting distance of cities and in amenity regions associated with oceans, lakes and reservoirs, mountains, and warm areas.

The Major Transformations

Visually, the primary land-cover transition was the decline in **forest** from 50 percent to 29 percent of the nation's land cover. Partially compensating for the loss of trees to farmland and developed areas were trees in areas historically devoid of forests. These trees were generally the result of fire prevention, the planting of farm windbreaks, and settlements of all sizes from hamlets to conurbations where trees were planted. Virtually all of the nation's humid and sub-humid grassland had been converted to crops and pasture covers. West of the 100th meridian, areas with good soil and **access** to water were typically converted to irrigated cropland.

Contemporary Land Cover

The documentation of the contemporary land cover of the United States is somewhat incomplete. A 2000 land-cover map of the nation is being developed and will provide the most complete land-cover description of all fifty states ever produced. A 1992 map of the land cover of the conterminous United States (excluding Alaska and Hawaii) reveals that forests cover over 29 percent of the land, cropland and pasture now occupy 26 percent, shrublands cover 18 percent, and grasslands cover 16 percent. The developed footprint of the conterminous United States is just over 2 percent of the area. Wetlands cover 4 percent, barren lands comprise 2 percent, and fresh water covers nearly 2 percent. The types and extent of land cover in the conterminous United States reflect the resource potential based on the inherent biotic and abiotic features of different regions of the country; and the influence of transportation systems, government policy, and shifting social and economic conditions.

See also: **Agricultural Revolution; Change Detection; Colonization; Corporate Strategies; Desertification; Economic Growth; Economic Restructuring; Extensification; Food Crops; Hot Spots of Land-Cover Change; Industrialization; Land Degradation; LANDFIRE Project; Land Reform; Land Rights; Land Tenure; Land-Use History; Leapfrogging; Mineral Extraction; Pristine Myth; Proximate Causes; Public Policy; Reforestation; River Basin; Suburbanization; Technological-Scientific Revolution; United States Geological Survey (USGS) Program; Urban-Rural Fringe; Urban Sprawl; Tobacco; Yellowstone Model.**

Further Reading

Borchert, John R., "American Metropolitan Evolution," *Geographical Review* 57 (1967): 301–32; Conzen, Michael, ed., *The Making of the American Landscape*, Boston: Unwin Hyman, 1990; Denevan, William M., "The Pristine Myth: The Landscape of the Americas in 1492," *Annals of the Association of American Geographers* 82 (1992): 369–85; Dilsaver, Lary M., and Craig E. Colten, eds., *The American Environment: Interpretations of Past Geographies*, Lanham, MD: Rowman and Littlefield Publishers, Inc., 1992; Graf, William L., "Dam Nation: A Geographic Census of American Dams and Their Large-Scale Hydrologic Impacts," *Water Resources Research* 35 (1999): 1305–11; Hart, John

Fraser, "Half a Century of Cropland Change," *The Geographical Review* 91 (2001): 525–43; Hart, John Fraser, "Loss and Abandonment of Cleared Farm Land in the Eastern United States," *Annals of the Association of American Geographers* 58 (1968): 417–40; Küchler, A. W., *Potential Natural Vegetation of the Conterminous United States* (Special Publication no. 36), Washington, DC: American Geographical Society, 1964; McIlwraith, Thomas F., and Edward K. Muller, eds., *North America: The Historical Geography of a Changing Continent*, 2nd ed., New York: Rowman & Littlefield Publishers, Inc., 2001; Prince, Hugh, *Wetlands of the American Midwest: A Historical Geography of Changing Attitudes*, Chicago: University of Chicago Press, 1997; Ramankutty, Navin, and Jonathan A. Foley, "Estimating Historical Changes in Land Cover: North American Croplands from 1850 to 1992," *Global Ecology and Biogeography* 8 (1999): 381–96; Vogelmann, James E., Howard, S. M., Yang, L., Larson, C. R., Wylie, B. K., and N. Van Driel, "Completion of the 1990's National Land Cover Data Set for the Conterminous United States from Landsat Thematic Mapper Data and Ancillary Data Sources," *Photogrammetric Engineering and Remote Sensing* 67 (2001): 650–62; Whitney, Gordon G., *From Coastal Wilderness to Fruited Plain: A History of Environmental Change in Temperate North America 1500 to the Present*, Cambridge, UK: Cambridge University Press, 1994; Williams, Michael, *Americans and Their Forests: A Historical Geography*, Cambridge, UK: Cambridge University Press, 1989.

DARRELL NAPTON, THOMAS R. LOVELAND

United States Geological Survey (USGS) Program. The U.S. Geological Survey (USGS) plays a significant land-cover research and development role. A noteworthy contribution occurred in the early 1970s when James Anderson, USGS chief geographer, led the development of "A Land Use and Land Cover Classification System for Use with Remote Sensor Data." This system established a hierarchical classification scheme and intellectual framework for using the then-new **Landsat** imagery, as well as **aerial photography**, for mapping land use and land cover. Each level of the classification hierarchy was matched with the information potential of different **scales** of remotely sensed data (i.e., Level I and Landsat multispectral scanner data, and Level II and high altitude aerial photography). The framework developed by James Anderson and colleagues was used to produce a twenty-one-class national land-cover map based on manual interpretation of 1970s aerial photographs and classified images. The map, often referred to as LUDA (land-use data analysis), was the first comprehensive land-cover map of the **United States of America**.

The USGS continues to produce land-cover maps of the United States. In 2000, the National Land Cover Database (NLCD) was completed for the conterminous United States. The NLCD was produced using a modified Anderson classification scheme, circa-1992 Landsat Thematic Mapper (TM) imagery, ancillary data sets, and computer-assisted classification methods. Development of a new NLCD covering all fifty states began in 2001 using circa-2000 Landsat 7 Enhanced Thematic Mapper data and improved classification methods that were designed to improve land-cover map accuracy and consistency.

Two additional nationwide land-cover mapping activities are underway in the USGS. The Gap Analysis Project produces natural **vegetation** maps using the same 1992 and 2000 Landsat data used to produce the NLCD maps.

The **LANDFIRE project** is an effort to produce detailed vegetation maps corresponding to wildfire fuels. Like NLCD and the Gap Analysis Project, LANDFIRE uses the same circa-2000 Landsat data. The three 2000 USGS national land-cover and vegetation-mapping projects (NLCD, Gap Analysis, and LANDFIRE) share data and are closely coordinated in order to ensure map consistency.

The USGS has been a leader in global-scale land-cover mapping. It led the development of the first validated **IGBP-DIS 1-km global land cover data set**. USGS land-cover specialists continue to investigate large-area land-cover topics and are contributors to the **Global Land Cover Map of the Year 2000** (GLC-2000).

USGS land-cover research spans a wide range of topics and spatial and temporal scales. Studies of contemporary U.S. land-cover trends (1972–2000) and paleo-ecological investigations of historical vegetation ranges are also underway. Research on large-area mapping and accuracy assessment strategies is part of the USGS core land-cover research program.

See also: Center for International Earth Science Information Network; Earth Resources Observation Systems (EROS) Data Center; Remote Sensing; Sahel Land Cover.

Further Reading

Anderson, James R., Hardy, E. E., Roach, J. T., and R. E. Witmer, *A Land Use and Land Cover Classification System for Use with Remote Sensor Data*, U.S. Geological Survey Professional Paper 964, Reston, VA: U.S. Geological Survey, 1976; United States Geological Survey, Earth Resources Observation Systems (EROS) Data Center, Land Cover Characterization Program [Online, November 2004], The USGS Land Cover Web Site, landcover.usgs.gov.

THOMAS R. LOVELAND

Urban Sprawl. A catch-all phrase that typically refers to low-density urban development that may also be characterized by a lack of contiguity, compactness, or centrality. No universally agreed-upon definition of sprawl exists. While some define sprawl as anything negative associated with urban form and others define it in terms of market failures, the most useful definitions are in terms of quantifiable aspects of an urban land-use pattern. This approach then allows further investigation of trends, causes, and consequences of this pattern. The literature ranges from very simple operational definitions, such as population per urban acre, to multidimensional definitions that seek to capture a variety of pattern aspects, including density, contiguity, concentration, centrality, mixed use, and proximity. Measurement of sprawl is limited by data availability and complicated by the dynamic process of land-use change, which implies that sprawl is not a discrete outcome, but rather a matter of degree and that sprawl may be a transient phenomenon.

Many of the factors that are hypothesized to cause sprawl are the same factors that have been linked to **suburbanization**, including cheap transportation dominated by automobiles and trucks; central-city fiscal and social problems; advances in information technologies; rising household incomes; and lifestyle preferences for new houses, large lots, and open space. A variety of external costs are commonly associated with sprawl, including overconsumption of natural resource lands, increased traffic congestion, and the added costs of providing more roads, utility lines, and other public services.

See also: **Extensification; Land Privatization; Leapfrogging; Urbanization; Urban-Rural Fringe.**

Further Reading

Breuckner, Jan, "Urban Sprawl: Diagnosis and Remedies," *International Regional Science Review* 23 (2000): 160–71; Galster, George, Hanson, Royce, Ratcliffe, Michael R., Wolman, Harold, Coleman, Stephen, and Jason Freihage, "Wrestling Sprawl to the Ground: Defining and Measuring an Elusive Concept," *Housing Policy Debate* 12 (2001): 681–717.

ELENA G. IRWIN

Urbanization. Urban built-up areas make up a small proportion of global land covers—2 to 3 percent by most estimates—yet their influence on surrounding landscapes is profound. This is because urban areas are centers of population and economic activity that draw on their hinterlands for renewable and nonrenewable resources such as water, food, energy, and raw materials for manufacturing. The urban-rural linkage thus becomes a critical element in understanding how urban areas impact land-use and **land-cover change**.

Coming up with a consistent definition of what constitutes an urban land-cover type is not as simple as it might at first appear. There is in fact a

Skyline of mid-New York in 1931.
PHOTO: William Frange. [Courtesy Library of Congress]

gradient from wild areas to rural agricultural lands to peri-urban areas to suburban and then urban. Traditionally researchers treated as urban anything that fell within a city's administrative borders or that met a certain population density threshold. However, using administrative boundaries is clearly deficient, since in many cases urban built-up lands spill over a city's limits, and in other instances city boundaries may have been drawn in such a way as to include large tracts of relatively undeveloped land. In recent years, **remote sensing** scientists, demographers, and other researchers have sought to develop more consistent definitions for what constitutes "urban" land cover.

The Global Urban-Rural Mapping Project (GRUMP) of the **Center for International Earth Science Information Network** utilized a combination of data sources—including the Gridded Population of the World, nighttime lights, and point estimates for city population—to develop a globally consistent urban extents database. Utilizing these data, it is possible to identify the patterns in geographic location of urban areas, including the kinds of land-cover classes in which urban areas tend to be located. Globally, the urban areas tend to be located in **coastal zones**, where 10.2 percent of the land is urban, followed by cultivated areas (6.8 percent), islands (4.7 percent), and in close proximity to inland waters (3.2 percent). Drylands, forests, and mountainous areas are all under 2 percent urban land cover. According to the study, 46.2 percent of the world's population resides in urban areas, and 40 percent of the urban population resides in cities under 500,000 in size, with a quarter residing in cities between 1 and 5 million in size. Some 6.7 percent of the world's population resides in "megacities" of 10 million or more. These megacities are generally located in the coastal zone, on islands, or near inland waters.

The processes and patterns of urbanization are different in developed and developing countries. In most developed countries, large cities arose during the **Industrial Revolution** as economies of scale were generated by the concentration of capital and labor in urban areas. Mills would locate near sources of raw materials and cheap power (rivers or coal deposits), and people flocked to the mills from the countryside, creating urban agglomerations. In 1950, seven of the ten largest cities were in the industrialized world. In the latter part of the twentieth century, the economies of major cities became less dependent on heavy industry and more tied to the service sector. In the Americas, **suburbanization**, reliance on cars rather than public transportation, and the decentralization of jobs led to progressive low-density development on the fringes. In **Europe** and **Japan**, the same phenomenon has occurred though to a lesser extent due to shortages of land, the protection of farmland, and greater concentration of development near public transportation.

Developing country cities have grown very rapidly in the past fifty years such that today, seven out of the ten largest urban agglomerations are in the developing world, with Mexico City, Mumbai, and Saõ Paulo following after Tokyo as the second, third, and fourth largest cities in the world. While industrial activity played a significant role in the development of Asian and Latin American cities (less so in Africa), the recent explosion of growth is also related to a combination of rapid population growth, lack of economic opportunities in rural areas, rural landlessness, political instability, and droughts and other natural disasters. This, coupled with a lack of **land-use planning**, weak zoning, and insufficient urban infrastructure (water and sanitation), has led to a deteriorating quality of life in many developing country urban areas, especially in slum or squatter settlements. As the poorest residents often reside in previously undeveloped lands on steep slopes or in low-lying flood-prone areas, they have an elevated risk of being affected by natural hazards.

These problems point to a major difference in the environmental outcomes of urbanization in developed and developing countries. Over a period of 150 years, urban managers in industrialized nations have developed mechanisms to mitigate the impact of household and industrial **consumption** and wastes on surrounding ecosystems through water and sanitation systems, sewerage treatment, automobile and industrial air pollutant emissions standards, environmental impact assessments for new development, zoning codes, and waste disposal regulations. Even so, many developed countries face problems of "brownfields"—former industrial zones that are difficult to redevelop due to buildup of hazardous wastes—as well as periodic air and water quality emergencies. Therefore, in light of rapid urban growth and a comparative lack of resources, it is not surprising that urban managers in developing countries struggle to meet the water, sanitation, and waste disposal needs of city dwellers.

From an ecological perspective, the intensity of the transformation of the landscape in urban areas has certain repercussions for ecosystem processes such as **nutrient cycling** and hydrology. These have not been as well studied as ecology in natural areas, largely because scientists have tended to see these areas as human-transformed landscapes of little ecological interest. This is beginning to change, and there are new studies examining urban ecosystems, including long-term ecological research sites in the U.S. cities of Baltimore, Maryland, and Phoenix, Arizona. Generally, urban areas are places of intensified nutrient cycling, as nutrients such as plant and animal food products are shipped to cities and consumed, and resulting wastes often end up in rivers, lakes, and coastal zones, affecting aquatic ecosystems. Urban areas also have a dramatic impact on hydrology from two perspectives. On the one hand, impervious surfaces result in more dramatic spikes of runoff during stormwater discharge, affecting flooding in downstream areas. On the other, water demand in urban areas often results in the diversion of water from distant areas, which affects ecosystems in both the source (rural) and destination (urban) areas.

Urban areas can also have significant impact on local microclimates through such phenomena as the urban heat island effect. Urban heat islands are created by the relatively low **albedos** (reflectivity) of urban land surfaces, and the relative absence of vegetative cover. **Vegetation** absorbs energy in the visible near infrared band of the spectrum, converting it to plant growth through photosynthesis. By contrast, paved surfaces absorb the sun's energy and re-radiate it after the sun has set. Research in Atlanta, Georgia (United States) found that the growing season was increased by several days as a result of the heat island effect. Also, the elevated air temperatures in Atlanta create local convection that results in thunderstorms in downwind areas.

As stated above, just because urban areas make up only a small proportion of global land area does not mean that they can be discounted as an insignificant contributor to land-use and land-cover change. On the contrary, if one traces urban-rural linkages, it becomes clear that the "**ecological footprint**" of urban areas is several orders of magnitude larger than the urbanized area itself. Urban centers rely on their hinterlands for water, food, fuel, and natural resources of all types. One study estimated that the cities of Baltic Europe appropriate for their resource consumption and waste assimilation an area of forest, agricultural, marine, and wetland ecosystems that is 565–1,130 times larger than the area of the cities themselves.

Cities also expropriate water from large distances. In some cases, such as in the New York City water supply districts, this may have the effect of conserving large areas of land that might otherwise be developed. In others, such as in Los Angeles, the expropriation may have a negative impact on the ecology of natural water bodies such as Mono Lake and the Colorado River, where much of the water is drained before it crosses the Mexican border.

Urban area energy demands may also have significant impacts on land cover in distant areas. Many rural agricultural areas and natural ecosystems have been submerged under the reservoirs created by large hydroelectric dams, such as the Three Gorges Dam in **China** and the Aswan High Dam in Egypt. At 500 kilometers in length, Lake Nasser, formed by the Aswan High Dam, is the largest artificial lake in the world. Mining, oil drilling, and other land-based energy extraction activities make up small but important portions of global land cover. Many African cities are still heavily dependent on **biomass** fuels such as wood and charcoal for household heating and cooking, which can have a significant impact on **deforestation** in surrounding areas.

A significant impact of urban areas on the hinterlands is the need for sinks (dumping sites) for the great volumes of waste they produce. Urban wastes can be solid, liquid, or gaseous. Problems of solid waste disposal have become increasingly problematic as landfills have reached their maximum capacities. Cities generally have few solutions other than to truck their waste at great cost to distant landfills, or to incinerate it. Owing to shortages of space for landfills, European and a number of North American cities have increasingly opted for incineration. Problems arise from emissions of toxic fumes, especially dioxins, lead, and mercury, and then the disposal of toxic ash.

Atmospheric pollutants can have a significant impact on land use and land cover. Carl Folke and colleagues estimate that the carbon dioxide emissions of major cities alone would consume 95 percent of the carbon absorptive capacity of the world's forests. Another major problem is air pollutant impacts on natural vegetation and crops. Acid rain has been a documented problem that damages **forest** lands in North America and eastern Europe. William Chameides and colleagues examine the impact of ground-level ozone on crop production in three large "metro-agro-plexes"—areas of intense agricultural and industrial production in North America, northern Europe, and China/Japan—and find that it may be reducing global crop production by a few percent. This impact is likely to intensify due to projected increases in nitrogen oxides, an important precursor to ozone.

Urbanization pathways lead to different impacts on rural landscapes in the developed and developing worlds. In the developed world, large-scale urban agglomerations and extended peri-urban settlements fragment the landscapes of such large areas that various ecosystem processes are threatened. Ecosystem fragmentation, however, in peri-urban areas may be offset by urban-led demands for **conservation** and recreational land uses.

See also: **Climate Impacts; Economic Restructuring; Extensification; Hot Spots of Land-Cover Change; Human Immunodeficiency Virus (HIV)/Acquired Immunodeficiency Syndrome (AIDS); Industrialization; International Human Dimensions Programme on Global Environmental Change; Investments; Land Reform; Land Rent; Leapfrogging; Malaria; Mineral Extraction; People at Risk; Population Dynamics; Reforestation; Suburbanization; Urban-Rural Fringe; Urban Sprawl.**

Further Reading

Chameides, William L., Kasibhatla, Prasad S., Yienger, James, and Hiram Levy, "Growth of Continental-Scale Metro-Agro-Plexes, Regional Ozone Pollution, and World Food Production," *Science* 264 (1994): 74–7; Elvidge, Christopher D., Sutton, Paul C., Wagner, Thomas W., Ryzner, Rhonda, Vogelmann, James E., Goetz, Scott J., Smith, Andrew J., et al., "Urbanization," in *Land Change Science: Observing, Monitoring and Understanding Trajectories of Change on the Earth's Surface* (Remote Sensing and Digital Image Processing Series no. 6), eds. Garik Gutman, Anthony C. Janetos, Chris O. Justice, Emilio F. Moran, John F. Mustard, Ronald R. Rindfuss, David Skole, B. L. Turner, II, and Mark Cochrane, Dordrecht, NL, Boston, and London: Kluwer Academic Publishers, 2004, pp. 315–28; de Sherbinin, Alex, Balk, Deborah, Yager, Karina, Jaiteh, Malanding, Pozzi, Francesca, Giri, Chandra, and Antoinette Wannebo, Social Science Applications of Remote Sensing: A CIESIN Thematic Guide [Online, December 2004], The CIESIN Web Site, sedac.ciesin.columbia.edu/tg; Folke, Carl, Jansson, Asa, Larsson, Jonas, and Robert Costanza, "Ecosystem Appropriation by Cities," *Ambio* 26 (1997): 167–72; Urbanization and Global Environmental Change [Online October 2005], The UGEC Web Site, co.hunter.cuny.edu/%7Eshodges/UGEC/index.htm.

ALEXANDER DE SHERBININ

Urban-Rural Fringe. Semi-rural areas that lie beyond built-up, urbanized areas, but that fall within the influence of urban areas. These areas consist of a mix of urban and rural functions that are spatially interspersed. A variety of terms have been used to describe these areas, including *peri-urban, exurban, urban fringe, rural-urban fringe, urban-rural fringe, urban periphery, extended metropolitan regions*, and *desakota* (meaning literally a village city).

While urban-rural settlement patterns vary by country, region, and time period, they are often characterized as low-density, scattered urban development that consists of largely residential and sometimes industrial land uses; existing rural towns often with new development on the periphery; and intervening rural land uses. Due to the low density and fragmented nature of this development, the impact on population redistribution to these areas is far less than the impact on land **conversion**.

Population growth in urban-rural areas—a process sometimes referred to as "exurbanization" or "counterurbanization"—is largely the result of **urbanization** pressures. In some ways, urban-rural fringe areas are the "new suburbs" associated with expanding urban areas. However, they differ from their suburban counterparts in ways other than just their location vis-à-vis the urban core. Rural economic activities, such as **agriculture** and other natural resource-based industries, are still a substantial contributor to the local economy. A mix of rural residents, for example, farmers and residents of rural towns and new residents moving outward from urban areas live in these areas. Government controls, including land-use controls and locally provided public services, are often much less extensive than in the built-up urban areas.

Developed countries witnessed a "rural renaissance" or "population turn-around" in the 1970s—faster population growth in nonmetropolitan versus metropolitan areas. While these trends waned in the 1980s and appeared to reverse themselves in certain countries, the 1990s saw renewed growth in semi-rural areas of many developed countries. Analyses of longer-term trends support the hypothesis that counterurbanization has been a dominant trend for the past several decades in many developed countries, including **Canada**, the **United States of America**, Great Britain, and other countries in **Europe**.

Similar trends in some developing countries have been documented as well. For example, Victor Sit and Chun Yang studied recent growth patterns in the Pearl River delta of **China** since the economic reforms of 1978 and found that the predominant growth pattern has been a more equal level of urbanization across the region coupled with a declining importance of the primate regional city. Even in sub-Saharan Africa, where economic growth is largely absent, increased demand for new, higher-quality residential development has spurred peri-urban growth in some metropolitan areas.

Low transportation costs, due to the lost cost of automobiles and an extensive roads network; new manufacturing technologies; advances in communications technologies; and rising household incomes are some of the primary **driving forces** that have lessened the advantages of urban agglomeration and led to urban-rural fringe growth. In addition, increased demand for environmental amenities and the **economic restructuring** of rural areas of the United States, Canada, and other developed countries have transformed some rural areas into recreation and tourism-based places with high-valued environmental amenities. In China and other Asian countries, economic **globalization** has led to growth that favors smaller cities and semi-rural areas due to production technologies that require large tracts of land and use information technologies that reduce the need for urban proximity. Peri-urban residential growth in some developing countries has also been tied to increasing incomes of a growing professional class and remittances from migrants.

See also: **Economic Growth; Leapfrogging; Remittances Landscape; Suburbanization; Urban Sprawl.**

Further Reading

Briggs, John, and Ian E. A. Yeboah, "Structural Adjustment and the Contemporary Sub-Saharan African City," *Area* 33 (2001): 18–26; Champion, Anthony G., "Urban and Regional Demographic Trends in the Developed World," *Urban Studies* 29 (1992): 461–82; Daniels, Tom, *When City and Country Collide*, Washington, DC: Island Press, 1999; Nelson, Arthur C., "Characterizing Exurbia," *Journal of Planning Literature* 6 (1992): 350–68; Sit, Victor F. S., and Chun Yang, "Foreign Investment-Induced Exo-Urbanization in the Pearl River Delta, China," *Urban Studies* 34 (1997): 647–77.

E L E N A   G .   I R W I N

V–Y

Validation. A process that measures the consistency between theory and data, which is used frequently to test models (see figure titled "The position of validation"). If a theory is consistent with accurate and comprehensive data, then the theory is said to be valid. In the context of **modeling**, Edward Rykiel Jr. defines validation as "a demonstration that a model within its domain of applicability possesses a satisfactory range of accuracy consistent with the intended application of the model." In order to test the validity of any particular theory, the theory must be translated into a form such that it can be compared directly to data. For example, if a scientist wants to test the theory that agricultural market forces drive land change, then the scientist could express this theory in terms of a model that generates a prediction in the form of a map. Validation is the process of measuring the agreement between the model prediction and independent data. If there is a good match between the model prediction and independent data, then the method used to make the prediction is said to be valid. Scientists must respect rigorous definitions of "model prediction," "independent data," and "good match."

In order to distinguish between model prediction and independent data, it is crucial to distinguish between model calibration and model validation. Calibration is the process of creating a model such that it is consistent with some subset of the entire available data set. It is essential that the available data set be split into two separate subsets, called calibration data and validation data. The modeler uses the calibration data and theory to create the model. The validation data are reserved for the validation procedure and must not be used to help to create the model. After the model is calibrated using the calibration data, the model generates a prediction that is then compared to the validation data. The word "prediction" refers to the fact that the model writes the output before the model sees the validation data. There are two common techniques to separate calibration data from validation data. The first method is separation through time, where calibration data consist of information at or before some specific time, and validation data consist of information after that time. Under this first arrangement, the validation procedure tests whether the model is valid as it predicts across time. The second common method is separation through space, where calibration data are from one location and the validation data are from some other location. Under this second arrangement, the validation procedure tests whether the model is valid as it extrapolates across space.

Much of the confusion about validation derives from lack of distinction between goodness-of-fit of calibration and goodness-of-fit of validation. The former refers to how well a model can fit a mathematical form to existing data during the calibration phase, where the data are used to help define

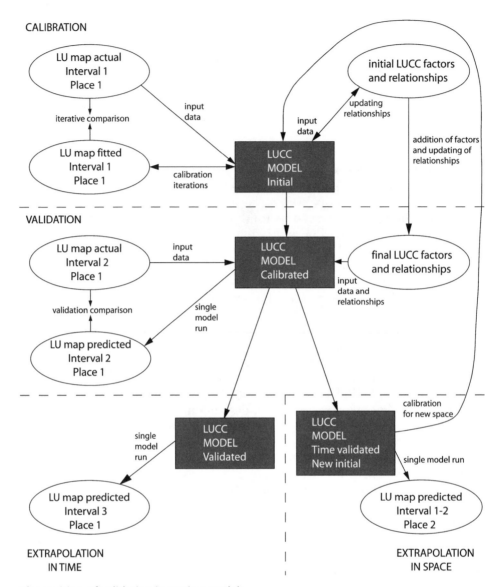

The position of validation in testing models.

the mathematical form and the parameters. For example, an R-squared value in regression is a measure of goodness-of-fit of calibration, if all the data are used to create the regression line. This type of calibration is referred to as model fitting, not as model prediction. Model prediction occurs when the scientist uses the fitted model to attempt to predict the validation data. When scientists use the terms "modeled" and "simulated," it is not clear whether they mean "fitted" or "predicted"; therefore, it is not clear whether the goodness-of-fit refers to calibration or validation. An elaborate fitting procedure can attain a high goodness-of-fit for calibration; however, such a fit does not necessarily indicate that the model will have a good fit for validation. High-speed computers can induce modelers to over-fit a model, so the model becomes calibrated to both the signal and the noise in the calibration data; so the fit of the calibration can be high. When an over-fitted model extrapolates to data that were not used for calibration, the resulting measure of goodness-of-fit of validation is likely to be less than the goodness-of-fit of calibration. In order to assure an appropriate measure of a model's predictive power, one must separate the information used for calibration from the information for validation. This can be difficult for places where availability of information is limited and for models that demand substantial data for calibration.

The validation exercise assumes that the scientist already has knowledge of the time and place to validate the model, because proper validation requires accurate validation data. However, usually the major purpose of the model is to understand a phenomenon for a time or place that is unknown, for which data do not exist. Therefore, in order to capitalize on the validation exercise, the scientist should translate the measurement of validation into a statement about confidence in a model as it extrapolates into times and places that are unknown. Extrapolation is the process whereby a model makes predictions beyond the extent of the calibration and validation data. If there is a good fit between the model prediction and the validation data, then the model might be able to make accurate extrapolations to other spatial and temporal extents. If the characteristics and mechanisms of the other spatial and temporal extents are similar to the characteristics and mechanisms that existed during the calibration and validation phases, then the model should be able to extrapolate to other extents with a level of accuracy similar to the performance in the validation. If the model fails to attain a good fit in the validation, then one should have little confidence in the model's ability to extrapolate accurately to other extents. A high goodness-of-fit of validation is a necessary condition for a scientist to have confidence in a model's predictions to other spatial and temporal extents. If the mechanisms that the model simulates are consistent during the calibration, validation, and extrapolation phases, then a high goodness-of-fit in the validation phase is also a sufficient condition for a scientist to have confidence in a model's extrapolations to other spatial and temporal extents. Obviously, we will not know *a priori* whether the mechanisms during the calibration and validation phases will continue into the extrapolation, because we can never know something until we have empirical data. Alas, if empirical data existed, there would not be much reason to make a model prediction. Ultimately, separation of calibration from validation, and objective measurement of validation are important techniques that a scientist should use in order to measure the level of trust one should have in a model as it is used to help one learn about what one does not yet know.

There are no agreed criteria among scientists concerning either the method or the level of what is considered a "good match" between validation data and a model prediction. A reasonable minimum criterion would be that the agreement between the validation data and the prediction from a scientist's model should be better than the agreement between the validation data and the prediction from a null model. The prediction from a null model is the naive prediction that one would make if one were not to create any analytical model. For example, suppose the purpose of the model were to predict the change in a land-cover map from an initial time to a subsequent time. The calibration data consist of information at or before the initial time. The validation of the model would compare the independent validation map of the subsequent time to the prediction map of the subsequent time from the scientist's model. The prediction map of the subsequent time from the null model would be exactly the map of the initial time. In other words, the null model would predict no change between the initial time and the subsequent time. In many cases, the agreement between the independent validation map and the prediction map from a null model can be large, because the only disagreement between those two maps is the change between the two points in time, which can be small relative to the study area. It is not advisable to use a map of randomness as the prediction of the null model, because a naive prediction would be pure persistence rather than randomness.

Validation of a land-change model usually requires the comparison between a validation map and a prediction map. Hence, there is a need to measure objectively the agreement between any pair of maps in order to determine which among the many possible prediction maps has the closest agreement with the validation map. There are numerous mathematical methods to compare the patterns in maps. The methods can be separated into those that apply to a real variable and those that apply to a categorical variable.

The usual first step in comparing two maps of a real variable is to plot the values in a Y versus X space where the vertical axis shows the prediction values and the horizontal axis shows the validation data. Each point in the plot denotes an element on the map, such as a **pixel** or a polygon. The closer the points are to the $Y = X$ line, the better the fit between the validation data and the prediction of the scientist's model. This fit should be compared to the fit between the validation data and the prediction of the null model. Two standard measurements of how tightly the values are clustered around the $Y = X$ line are the root mean square error (RMSE) and the mean absolute error (MAE).

A common first step in computing the agreement between a validation map and a prediction map of a categorical variable is to calculate the percentage of pixels classified correctly. This statistic should be compared to the percentage correct in the comparison between the validation data and the prediction from the null model. The percentage correct is the most common statistic reported because it is simple to compute and is relatively intuitive to interpret. The disadvantage is that it fails to capture patterns that are immediately obvious to the human eye. There are many other statistics available that compute **pattern metrics** to compare the patterns in two maps. Whatever statistic is chosen, the scientist should be cognizant of two important components in the comparison of the pattern between two maps of a common categorical variable: comparison in terms of quantity of each category and comparison in terms of location of each category. Comparison in terms of quantity considers whether the proportion of each category on one map is similar to the proportion of the corresponding category on the other map. Comparison in terms of location considers whether the position of each category on one map is similar to the position of the corresponding category on the other map. Robert Gilmore Pontius Jr. describes how to budget the agreement and disagreement for these two components.

In some cases, the validation data are of a categorical variable, and the prediction from the model is of a real variable. For example, the validation data may consist of pixels in which 0 indicates land persistence and 1 indicates land change, whereas the prediction from the model is a real number that indicates probability of land change. The scientist would like to know whether high probabilities of land change are concentrated at locations of true land change according to the Boolean validation map. For this case, there are a variety of statistical measures of agreement. One of the most appropriate ones is the relative operating characteristic (ROC).

None of the statistical measurements above match perfectly the human eye's ability to recognize patterns. Therefore, all of the methods of statistical map comparison should be complemented by a visual assessment to determine whether the selected statistic is measuring the characteristic that the scientist thinks is important. However, visual assessment can be influenced dramatically by subjective aspects of map production, such as selection of the color palette. Therefore, objective statistical measurement is essential to maintaining scientific rigor.

Any measurement to compare two maps can be extremely sensitive to the **scale** of the analysis. Therefore, it is advisable to compute the measurement at various scales to examine the degree to which the results are sensitive to changes in scale. Increasingly, scientists are creating methods to examine how results are sensitive to scale. Robert Gilmore Pontius Jr. shows how to compute the components of agreement and disagreement in terms of quantity and location at multiple resolutions. He and colleagues describe how to compute the ROC for multiple resolutions and how to compute the entire cross-tabulation matrix at multiple resolutions. They further show how to measure the value of information at nested levels of stratification structure, and how changes in categorical scale influence the measurement of map agreement, while Kasper Kok and colleagues perform their validation at multiple scales. Multiple-scale analysis allows the statistics to conform more closely to the patterns and clusters that the human eye sees.

Multiple-scale validation is also important because it allows the scientist to see whether the model makes predictions at scales that are relevant to the purpose of the model. Many models are calibrated and make predictions at fine resolutions that match the available data. For example, satellite data are commonly available in the form of thirty-meter pixels. However, the relevant questions concerning land change may occur at coarser resolutions. For example, many global climate models operate at scales of 1 degree longitude by 1 degree latitude. Scientists are sometimes reluctant to change the scale of the raw data because any adjustments introduce additional artifacts into the data. Multiple resolution validation allows a scientist to see whether the thirty-meter resolution model performs sufficiently accurately at a resolution that is relevant to the purpose of the model.

When there is lack of fit between the data and the prediction, then the scientist must consider the quality of both the model and the data. The scientist should improve the quality of whichever is worse. If the data quality is high, then the scientist will create a better model by changing the model to make it more consistent with the data. If the data quality is low, then the scientist should focus on getting better data; otherwise the scientist might be induced to make an inaccurate model by redesigning the model to be consistent with inaccurate data.

There is tremendous desire on the part of some scientists to label a model as valid or not. It is not particularly useful to attempt to crown a model as valid or to condemn a model as invalid. It is more useful to state carefully the degree to which a model is valid. Validation should measure the performance of a model in a manner that enables the scientist to know the level of trust that one should put in the model. Useful validation should also give the modeler information necessary to improve the model.

See also: **Agent-Based Model; Change Detection; Remote Sensing.**

Further Reading

Hagen, Alex, "Fuzzy Set Approach to Assessing Similarity of Categorical Maps," *International Journal of Geographic Information Science* 17 (2003): 235–49; Jolliffe, Ian T., and David B. Stephenson, eds., *Forecast Verification*, West Sussex, UK: Wiley, 2003; Kok, Kasper, Farrow, Andrew, Veldkamp, Tom A., and Peter Verburg, "The Need for Multi-Scale Validation in Spatial Land Use Models," *Agriculture, Ecosystems and Environment* 85 (2001): 223–38; Oreskes, Naomi, Shrader-Frechette, Kristin, and Kenneth Belitz, "Verification, Validation, and Confirmation of Numerical Models in the Earth," *Science* 263 (1994): 641–46; Pontius, Robert Gilmore, Jr., "Quantification Error Versus Location Error in Comparison of Categorical Maps," *Photogrammetric Engineering & Remote Sensing* 66 (2000): 1011–66; Pontius, Robert Gilmore, Jr., "Statistical Methods to Partition Effects of Quantity and Location during Comparison of Categorical Maps at Multiple Resolutions," *Photogrammetric Engineering & Remote Sensing* 68 (2002): 1041–49; Pontius, Robert Gilmore, Jr., Agrawal, Aditya, and Diana Huffaker, "Estimating the Uncertainty of Land-Cover Extrapolations while Constructing a Raster Map from Tabular Data," *Journal of Geographical Systems* 5 (2003): 253–73; Pontius, Robert Gilmore, Jr., and Kiran Batchu, "Using the Relative Operating Characteristic to Quantify Certainty in Prediction of Location of Land Cover Change in India," *Transactions in GIS* 7 (2003): 467–84; Pontius, Robert Gilmore, Jr., Huffaker, Diana, and Kevin Denman, "Useful Techniques of Validation for Spatially-Explicit Land-Change Models," *Ecological Modelling* 179 (2005): 445–61; Pontius, Robert Gilmore, Jr., and Jeffrey Malanson, "Comparison of the Structure and Accuracy of Two Land Change Models," *International Journal of Geographical Information Science* 19 (2004): 243–65; Pontius, Robert Gilmore, Jr., and Nicholas Malizia, "Effect of Category Aggregation on Map Comparison," *Lecture Notes in Computer Science* 3234 (2004): 251–68; Pontius, Robert Gilmore, Jr., and Laura Schneider, "Land-Use Change Model Validation by a ROC Method for the Ipswich Watershed, Massachusetts, USA," *Agriculture, Ecosystems & Environment* 85 (2001): 239–48; Pontius, Robert Gilmore, Jr., and Beth Suedmeyer, "Components of Agreement in Categorical Maps at Multiple Resolutions," in *Remote Sensing and GIS Accuracy Assessment*, eds. Ross S. Lunetta and John G. Lyon, Boca Raton, FL: CRC Press, 2004, pp. 233–51; Ritters, Kurt H., O'Neill, Robert V., Hunsaker, Carolyn T., Wickham, James D., Yankee, Dennis H., Timmins, Sidey P., Jones, K. Bruce, and Barbara L. Jackson, "A Factor Analysis of Landscape Pattern and Structure Metrics," *Landscape Ecology* 10 (1995): 23–39; Rykiel Jr., Edward J., "Testing Ecological Models: The Meaning of Validation," *Ecological Modelling* 90 (1996) 229–44.

ROBERT GILMORE PONTIUS JR., KASPER KOK

Vegetation. Green mantle of plants that covers the land surface, consisting of assemblages of plant species that, depending on the prevailing ecological conditions of each site or region, range from a small number of species in a simple spatial and structural arrangement, to very complex and highly diverse arrangements. Vegetation is conspicuous; it is absent only in areas that are too dry or cold to sustain it. The flora of a region are the complete range of plant species found there. The world's natural or semi-natural vegetation includes about 400,000 plant species.

The kinds of organisms constituting vegetation are those belonging to the *Plantae* kingdom, of which vascular plants are the most prominent. Vascular plants are those that contain vessels, functioning as a support structure and as a means of distributing the water and nutrients uptaken by the roots to the various plant tissues. Vascular plants can range from giant trees to tiny herbs and include vegetative forms such as woody shrubs, vines, and tall herbs with non-woody stems. Some trees may live for up to a thousand years, while annual herbs may live for only a few months and produce seed to establish future generations. Reproduction can be by seed or by vegetative means. Vascular plants include angiosperms, gymnosperms, ferns, and a few other groups. Angiosperms are, by far, the group with the larger number of species, which reflects their evolutionary success, and include trees, shrubs, and herbs. Gymnosperms are composed of tree and shrub species, whereas ferns are mostly herbaceous.

Each vegetation type is characterized by the growth form of its dominant plants (the largest, most abundant, characteristic plants). Examples of growth forms are annual herbs, broadleaf evergreen trees, or drought deciduous shrubs. Vegetation is also characterized by the architecture of its canopy layers, which together with the prevailing growth forms contributes to its

Desert vegetation in Yuma County, Arizona, March 1942.
PHOTO: Russell Lee. [Courtesy Library of Congress]

outer appearance, or physiognomy. A vegetation type that extends over a large region is called a formation. The prevailing climatic conditions determine the length of the growing season, which corresponds to that part of the year when the biophysical processes that make plants grow and reproduce are active; canopy phenology or the timing of leaf production and loss generally tracks this period. The length of the growing season and the total amount of foliage displayed determine the productivity of the system. Perennial plant species that never lose their leaves are called evergreens, whereas those that lose their leaves by the end of the growing season are called deciduous.

Plants need a substrate, from which they uptake mineral nutrients and water. They also need light and temperatures within their range of tolerance (which may vary widely). Biotic interactions, such as competition or consumption by herbivores, constitute additional constraints to their establishment, growth, and reproduction. Vegetation type distribution and abundance are controlled by the interaction of many biotic and abiotic environmental factors, such as the length of the dry season, soil nitrogen availability, or degree of shading from the canopy of taller plants. For each plant species there are favorable ranges of intensities of each factor, marginal ranges where unfavorable conditions cause stress, and ranges of intolerance along environmental gradients. The interaction of these gradients determines the potential ecological range of each vegetation type.

Temporal change is ubiquitous in vegetation. Regular changes, correlated with seasons, include the sequence of annual or other periodic variations that are part of the normal cycle of growth. These phenological changes do not correspond to abrupt changes in species composition. A different type of change is that resulting from sequential replacement of the dominant set of plant species by a different set. These sequential replacements, known as plant succession, correspond to the unfolding of the life history of those plants having propagules available at the site. In theory, grasses establish first, and then longer-lived shrub species dominate, followed by several types of trees, culminating in an equilibrium community dominated by shade-tolerant species. However, for a given set of environmental conditions the sequence of establishments is a function of initial floristic composition, differential dispersal, growth, and survival. The occurrence of disturbances, such as land clearing and **fire**, can interfere with succession, promoting alternative successional pathways and maintaining vegetation at a given successional stage. For example, savanna systems in sub-Saharan Africa are maintained by the periodic burning of woody vegetation, which would otherwise establish and become dominant.

Given appropriate availability of water, temperature, and light, plants synthesize their food materials through photosynthesis: the process by which chlorophyll traps light energy to produce sugars from carbon dioxide and water. Carbon dioxide is drawn from the atmosphere through pores in leaves and stems—the stomata—which are open when water is available for transpiration, and are shut otherwise, halting the process. Transpiration is the process by which leaf temperature is decreased, as a large amount of energy is required to convert liquid water to vapor, resulting in evaporative cooling. Plant green tissues are thus directly involved in transfers of energy, carbon, and water between the atmosphere and terrestrial ecosystems. The primary products of photosynthesis are converted into carbohydrates, fats, and proteins through other plant biochemical processes, and the surplus that is not consumed in plant biochemical activities (respiration) is accumulated in seeds, fruits, and vegetative tissues. Mass accumulation by photosynthesis in ecosystems is called primary production. **Biomass** accumulation also depends on the ability of some plant species to fixate nitrogen in their roots by a symbiotic association with specific bacteria. Nitrogen is an essential nutrient needed by all plants in relatively large quantities and is not available from the bedrock underlying soil formation. Once the nitrogen-fixing plants incorporate atmospheric nitrogen into their own system, it can be made available to other plants through decay. The leaves, stems, and trunks that constitute plant canopies also play a physical role in the energy budget of their ecosystems because, depending on their architecture and density, canopies intercept light and water, also affecting wind patterns. The water trapped in plant canopies is evaporated back into the atmosphere and thus is not used to replenish soil reservoirs; similarly in some tropical forest ecosystems almost no photosynthetically active light reaches the ground. Vegetation has a major role in ecological and biogeochemical processes: it modulates **atmosphere-land interlinkages**, produces the energy that feeds all organisms in the trophic chain, protects and improves soil quality, and is the matrix where many other terrestrial species live. Photosynthesis has been identified as the most important chemical process in the biosphere, because it provides energy used by all living organisms. It also supplied the energy now stored in the fossil fuels used in industrial societies.

Given its structural, functional, and ecological roles, vegetation is commonly used to denominate terrestrial ecosystems or **biomes**, in association with climate. For example, one broad classification of the world biomes is that of tropical forests, temperate forests, cold forests, forest/shrub/grassland combinations, **grassland**, tundra, and sparse or no vegetation.

Even though **climate impacts** on vegetation have been known for a long time, awareness of vegetative **feedbacks** on climate is recent. In addition to evaporation and transpiration of water, reflection and absorption of incoming radiation, and transfer of momentum to/from the lower boundary of the atmosphere, recent studies have shown that the relative amounts of vegetation and the spatial patterns of vegetated/bare areas can affect rainfall patterns. Vegetation also plays a central role in the **carbon cycle**. Several studies have shown that in the last two decades there has been a net carbon dioxide uptake in the northern and tropical biomes (despite tropical deforestation). This is a consequence of anthropogenic activities that increase atmospheric carbon dioxide concentrations, which in turn drive higher photosynthetic activity and net primary productivity in some terrestrial ecosystems. The role of vegetation as a carbon sink, however, depends on the transfer of carbon to forms with long residence time (e.g., wood), which in turn depends on appropriate land-management practices. Vegetation is also a source of carbon, and eventually all carbon fixed by photosynthesis is returned to the atmosphere. However, the return rate can vary from almost instantaneous (**deforestation** and biomass-burning) to slow (heterotrophic respiration by decomposers). The net carbon dioxide release due to land-use change during the 1980s (mainly tropical deforestation and **conversion** of natural vegetation to **agriculture**) has been estimated as 0.6–2.5 Pg C/yr.

Humans have used and manipulated vegetation cover since ancient times. Activities such as harvesting, hunting, sheltering, food growing, and **cattle ranching** involve various, ever more sophisticated processes of land clearing and intervention. There is virtually no vegetation type on earth that has not been somewhat influenced by anthropogenic activities, and in many parts of the developed world there are no remains of the potential natural vegetation types. Concomitantly, increases of population density in developing countries are associated with **agricultural intensification** and occupation of increasingly hazardous areas. Clearing of natural and semi-natural vegetation increases the risk of problems associated with **land degradation**, such as **soil erosion** and nutrient depletion from soils, which may determine a decrease in the capability of the environment to support adequate levels of food production while hampering other environmentally dependent mechanisms of raising income.

Vegetation has always been central in land-cover mapping. Its role in ecosystems and its ability to function as an indicator of the prevailing environmental conditions determined that different land-cover and land-use classification schemes always used vegetation as a main reference (presence/absence, condition, type, or use). The most traditional classification schemes rely on vegetation physiognomy or on its use by humans. For example, **forest** is a vegetation type dominated by tall woody plants (above six meters), and scrub is dominated by woody plants less than six meters tall, whereas classes such as irrigated agriculture or dryland agriculture use related classes. More recently, the use of **remote sensing** for mapping and **monitoring** land cover led to the development of new land-cover classification systems, such as the **Land Cover Classification System** (LCCS) used by the **United Nations Food and Agriculture Organization** (FAO).

See also: **Boreal Zone; Coastal Zone; Conservation; Disequilibrium Dynamics; Ecological Colonization; Fallow; Food Security; Holocene; Human Appropriation of Net Primary Production; Mountains; Nutrient Cycle; Pasture; Regrowth; Savannization; Secondary Vegetation; Taiga; Tropical Humid Forest; Tropical Dry Forest; Wetlands.**

Further Reading

Aber, John D., and Jerry M. Melillo, *Terrestrial Ecosystems*, 2nd ed., Oxford, UK: Academic Press, 2001; Breckle, Siegmar-Walter, *Walter's Vegetation of the Earth*, Berlin: Springer, 2001.

MARIA JOSÉ P. DE VASCONCELOS

Vulnerability. The extent to which changes both within and outside an entity—acting alone or in combination—predispose it to harms over time. A concept considered capable of bringing global change research to focus, vulnerability is a subject of diverse interpretations bandied about that are fueled by differences from many sources. A diversity of terminology has been enlisted to define and clarify its polyphony of meaning and implications (evidenced by its plural form "vulnerabilities") and its proper operationalization (including measurement), ranging from vocabularies from hazard, risk, and **food security** literatures where the study of vulnerability finds its root, to terms from systems ecology. Vulnerability defined in these terms varies widely, including a synonym for **poverty**, marginalization, fragility, or brittleness; a function of different combinations of exposure, sensitivity, resistance/robustness, irreversibility, coping, and so on; a relative state along a continuum from **resilience** to susceptibility; and an error term to account for residual consequences after adaptation has been taken. Some of these terms tend to illuminate and strain against more precise definitions at once, owing simultaneously to semantic baggages from mother disciplines, positivist overtones, and their equally (if not more) complex meanings. Consequently, the very term "vulnerability" itself has become vulnerable to misuse, even abuse.

Ongoing debate surrounding vulnerability points to a wide variety of contributing factors operating at different **scales** and in different contexts, which may be woven into different "explanatory frameworks" falling along several analytical axes. These variant frameworks have significant bearing on alternative answers to three interrelated questions: what (who) is vulnerable, and to what? How does a vulnerable state come about? Where are the solutions?

The "what" question highlights differences in understanding the connections between social and biophysical dimensions of vulnerability, and the extent to which exposure alone can be used to reliably measure vulnerability. The problem of specifying what is vulnerable often arises when we consider the variegated forms vulnerable conditions take: degradation, marginalization, deprivation, instability, etc. "Vulnerable" in one condition does not necessarily mean vulnerable in the other. Although vulnerability is frequently used to describe attributes of different systems at different levels, its mechanisms and implications in each setting are usually different. Vulnerability of parts does not necessarily translate into vulnerability of the whole, and vice versa (e.g., artificial flooding in flood control). The reliability of using exposure (meaning increased contact with environmental hazards) as a surrogate indicator of vulnerabilities also needs to be considered here. In a given place, vulnerability to one type of hazard does not mean vulnerability to the other; then how about multi-hazards? Besides, exposure unit is difficult to define due to the notorious boundary problem that impedes separation between **endogenous factors** and extraneous or **exogenous factors**. Cascades of relations often render upstream causes of exposure difficult to trace. Moreover, the biophysical and social dimensions of vulnerability rarely correspond in a given place as the concept of exposure might suggest: the most vulnerable people do not necessarily live in the most vulnerable regions.

The "how" question underscores the local and historical configuration of vulnerability and its structural shaping forces. One definitional approach is to take exposure to a harmful event as given and search for explanations of differential consequences. In this vein, social vulnerability is seen to have two sides: an external side of shocks and stresses to which a system is subject (exposure) and an internal side—a lack of coping capability (a function of resistance, and the rate of recovery) that is largely determined by socio-economic structure anchored in **property** relations (e.g., architecture of entitlements). Those focusing on "social construction of vulnerability" even take exposure formally out of the definition of vulnerability. Some structural explanations emphasize configuration of contextual factors of a geographic place in shaping the **syndrome** of vulnerabilities, while others see vulnerability of complexity as intrinsic to technological societies; yet others, seeing the need to distinguish complexity (large number of fully distinct elements and their spatial distribution) and complicatedness (many interlocking but structurally similar parts), attribute vulnerability to the very fabric of modern industrial societies that, in contrast to the evolutionary tendency of biological systems to become complex, have replaced structural diversity and complexity with structurally similar parts (e.g., **monocultures**) arranged in very complicated patterns. It is this complicatedness that builds industrial efficiency that also creates vulnerability to failure in that the more centralized structures have the ability to transfer both positive and negative repercussions over border areas and lower the margins of tolerance. When temporal dimension is brought in, adaptability is emphasized, where two research paths diverge: one projective, interpreting vulnerability as a residual variable to be measured by the adverse consequences remaining after the processes of adaptation has taken place; another reconstructive, attempting to use historical analogs to profile local constraints that limit the capacity to adapt. The distinction between chronical and current vulnerability as well as the notion of "windows of vulnerability" are all relevant here; the latter sees heightened vulnerability arising from conjunction of circumstances. Special note should be made about the interactions among **institutions**, nature, and society with the ultimate "pathology" unfolding in the forms of institutional rigidity, ecological brittleness (lower margin of tolerance), and resulting societal dependencies and vulnerabilities.

The "where" question brings to the fore the differences in the identification of "winners and losers," the values placed on the differential consequences and their **tradeoffs**, the distributions of responsibilities based on appraisal of these distributed consequences, and the level of uncertainties inherent to all of the above. Clearly, the burdens of vulnerabilities are not distributed evenly across places and social scales, hence the notion of differential vulnerabilities. Tangible or intangible "harms" is a value-laden term that means different things to different people, as studies of discourse of social vulnerability illustrate. Also, harms that are certain to happen are different from potential harms to be avoided. Irreversible loss of land and culture causes much more harm to people displaced from a dam building than the potential harm avoided by the dam. For an individual household, vulnerability to job loss is very different from vulnerability to loss of entitlement to a social safety net. Lessening vulnerability of a social system is not necessarily equivalent to the enhanced level of security for the social groupings and the ecological systems residing within its boundary, whose delineation is a matter of uncertainties. For technological infrastructures, equal amounts of protection from hazardous events do not ensure equal capacities to recover nor equal distribution of benefits for different groups of end users, contingent upon such systemic attributes as criticality, substitutability, connectedness, and dependency. Difficulties of stratifying degrees of vulnerabilities across social and spatial scale are exacerbated by the often-neglected normative dimension of vulnerability. Reducing vulnerability to changes does not mean maintaining status quo, or defending a vulnerable (read as rigid, unfair, etc.) establishment, in the same sense that the sustainability (of biosphere) should not be used to justify sustainability of everything. In addition, redundancy and diversity/diversification (in both the cultural and biological sense) that are seen as conducive to sustainability have to be achieved at the expense of efficiency. All these issues will enter into policy debate and bear on the critical question of "for whom or for what are the priorities in designing responses?"

The three questions posed above all point to the importance of scale of observation and the level of analysis in understanding vulnerability in its finer grain. Both need to be fine-tuned to correspond to the appropriate level where policy leverages can be applied to achieve their goals. A rarely clarified question is how vulnerability is linked with sustainability. Arguably, vulnerability is a variable state essentially intrinsic to all evolving, complex systems, while sustainability is not. Tough tradeoffs must be made between higher and lower levels, and between global and local interests in sustainability transition, often meaning (if not prescribing) increased level of vulnerability of different systemic units to changes at different junctures.

See also: **Biodiversity; Cascades; Driving Forces; Feedback; Hot Spot Identification; Human Immunodeficiency Virus (HIV)/Acquired Immunodeficiency Syndrome (AIDS); People at Risk; Proximate Causes; Threshold.**

Further Reading
Turner, B. L., II, Kasperson, Roger E., Matson, Pamela A., McCarthy, James J., Corell, Robert W., Christensen, Lindsey, Eckley, Noelle, et al., "A Framework for Vulnerability Analysis in Sustainability Science," *Proceedings of the National Academy of Sciences of the United States* 100 (2003): 8074–9.

KE CHEN

Water-Land Interlinkages. On its transit overland from the **watershed** to the **coastal zone**, water is exposed to the properties of the terrestrial surface, which, in conjunction with land-use and **land-cover change**, are important determinants for both water quantity (i.e., sufficient supply of freshwater to support human and natural systems) and water quality (i.e., suitability of supply for an intended use). It is well recognized that direct land changes such as **deforestation** and **reforestation**, cropland change (such as **agricultural intensification**), **mineral extraction**, and **urbanization**—but also indirect effects stemming from underlying **driving forces**—hold considerable potential to significantly modify or even disrupt hydrological cycles.

Overview
The recent intensification of agriculture, in particular, has its largest impacts on both freshwater and marine ecosystems, which are and continue to be greatly eutrophied by high rates of nitrogen and phosphorus release from agricultural fields. Aquatic nutrient **eutrophication** triggers the growth of blue-green algae that renders water unpalatable and increases the growth of weeds. It increases the turbidity of water and links to several other types of impacts: it can lead to shifts in the structure of food chains, including fish killings, outbreaks of nuisance species, loss of **biodiversity**, and negative

Flooding of low-lying areas by extreme high tides at Holland Cliffs Shore on mid-Patuxent River, Maryland, in fall 1993.

PHOTO: Mary Hollinger. CREDITS: NOAA Central Library and the Office of the CIO/ HPCC (America's coastline collection).

health impacts. For example, water shortages already exist in many regions of the world (with more than one billion people without adequate drinking water, and thus consequences), and 90 percent of the infectious diseases in developing countries are transmitted from polluted water.

It has early been estimated that, on a global scale, withdrawals (i.e., water removed from a source and used for human needs) from streams, rivers, and aquifers, combined with instream flow requirements (altogether totaling 6,780 cubic kilometers per year), already account for more than 50 percent of total accessible runoff. Claims have arisen that the human impact on the terrestrial water cycle during the last fifty years (actually dating back to about 4,000 years with water engineering in association with cropland expansion) has likely exceeded natural forcings of continental aquatic systems in many parts of the world.

Several interconnected properties intervene into the relationships shaping the impact of land-use/cover change upon freshwater hydrology. First, the life history of water on a landscape encompasses its appearance already in precipitation until its exit to the ocean. Indeed, any land-use decision very often turns out to be a water decision, be it the **conversion** of coastal marshes into **agriculture**, settlements or industrial zones, or the **modification** of farmland through the development of freshwater resources for irrigation farming. In a broader earth system perspective, responses to these influences are already discernible to reverberate through the hydrological cycle, which go well beyond the direct human appropriation of freshwater and coastal zones for drinking, agriculture, and industry.

Various **tradeoffs** need to be addressed between the potential benefits of land-use/cover change and the potentially negative consequences upon the hydrological cycle. Clearly, agricultural ecosystems have become incredibly good at producing food, and it has been the irrigation of croplands that

Surrounded by desert and steppe, the Aral Sea (Kazakhstan, Uzbekistan) is still the fourth largest lake in the world (since the development of agricultural irrigation projects in the late 1950s, the sea has shrunk drastically and two islands can be seen emerging in July 1985).

PHOTO: Space Shuttle. CREDITS: Image Science and Analysis Laboratory, NASA-Johnson Space Center, 18 Mar. 2005 (Earth from Space—Image Information). http://earth.jsc.nasa.gov/sseop.efs

contributed enormously to **food security**. Currently, around 40 percent of all agricultural production comes from irrigated areas, and global food production foreseeably becomes largely dependent on artificial irrigation systems. On the other hand, current agricultural practices involve deliberately maintaining ecosystems in a highly simplified, disturbed, and nutrient-rich state—with threats to biodiversity and the supply of water for food production. Especially in irrigated dryland zones of the world, the negative hydrological consequences of land-use/cover change are most pressing. They illustrate modern or industrial society's capacity to transform large coupled human-environment systems rapidly such as in the case of the Aral Sea basin (see image above), or the collapse even of ancient societies. In any case, and not only in drylands, the increased yields of food production have environmental costs that cannot be ignored, especially if the rates of nitrogen and phosphorus triple, and the amount of irrigated land doubles in the coming decades. Not included in tradeoff considerations so far are the large social costs associated with social disruptions due to dam construction (and population displacement) and other large-scale water infrastructure constructions, particularly along transnational rivers in dryland zones of the world.

Many insights into the hydrological consequences of land-use/cover change stem from the experimental manipulation of land cover at rather small spatial, observable scales such as research plots, hill slopes, and small catchment areas (e.g., 100–1,000 hectares). These manipulations prove that human activities can modify or disrupt interception losses by different plant species, soil infiltration, storm runoff, water yields, flood peaks, evapotranspiration rates, concentration of water-quality constituents, snow accumulation, and snow melting. However, extrapolating from such studies to larger systems such as **river basins** is confounded by the diversity of **land-use systems** as well as hydrological systems.

There is an unequal distribution of water resources or hydrological systems, for which the demand likewise varies greatly. Regions in which water quantity is particularly stressed by human demands are western North America; areas bordering the Sahara Desert; the Arabian peninsula; and several densely populated zones in Asia, namely India, Pakistan, and northeastern **China**.

The dominance of different processes changes at different **scales**. For example, land-use/cover changes in the upstream of a catchment may have a different impact on hydrology than changes downstream; and processes of interception, infiltration, and storage dominate at the plot scale; while channel processes assume a greater role with increasing catchment size. Only few examples exist for controlled long-term studies of the impacts of permanent land conversions at multiple scales such as **forest** to agriculture or agriculture to urban cover.

The impact of land-use/cover change also varies in terms of time scale. On the scale of catchments, changes usually occur at irregular-time intervals, while crop planting, drainage for afforestation, and other changes at the plot scale occur fairly regularly. Furthermore, river and lake quality can be restored in quite a short time, while destroyed biodiversity can take several thousands of years to recover to the original condition. Especially in the case of sediments, even after a "beneficial" land-use change, there will be still enough sediment in the system from prior human-induced or natural erosion that would lead to increased sediment loads in the rivers. In most of the cases, the time taken to contaminate a system is only a fraction of the time required to later clean up the same system. Remedies against high loadings of pathogens as a result of large population centers, for example, can be effective within less than one year, while eutrophication and micro-pollutants may contaminate the system for up to 100 years. Agrochemicals, in particular, may have a large impact over a long period of time. Also, mining and other sources of suspended load may have an impact on the ecosystem over many years.

Identifying a linkage between land-use and hydrological change often implies isolating the impact of biophysical forces and climatic variability, in particular. There is some reason to assume that climate change and long-term climatic fluctuations are particularly inherent to extreme weather events such as flooding and droughts. Likewise, human interventions at the microscale appear easily possible and are well documented, but the change in flood peaks, sediment load, and base flow at the large scale becomes much stronger when dominated by natural processes. Also, distinguishing the impact of land-use/cover change on hydrology from the impact of climatic variability is more difficult at the catchment scale than at the plot level.

Forest Zones

In **forest** zones, land clearance through deforestation—with the subsequent loss of the surface organic layer and decline in soil organic matter—increases overland stream flow through decreased evapotranspiration as well as raindrop detachment of soil particles, sheet erosion, rill erosion, gullying, and downstream sedimentation, though the latter processes are often episodic. The hydrological consequences following deforestation, however, are highly variable, and depend upon a wide array of host factors. At the watershed scale, forest clearing generally results in a significant increase in annual water yield, but generalization across different streamflow response measures remains difficult. On deforested slopes in particularly steep terrain, rates of erosion are maximized if the terrain is subsequently subjected to intensive cropping, **fire**, or both. Barren wastelands, unable to support any **vegetation**, can be the end result of this process, and the foothills north of Mexico City are an example of this. In forest zones where either revegetation occurs rapidly or **secondary vegetation** and **regrowth** are part of the land-use cycle, the effects of forest clearing on hydrological cycles are transient and less extreme. It has been noted though that geological conditions can often override the effects of land-use/cover change.

Reforestation (or afforestation) is seen to help reverse the hydrological responses to deforestation. Complete reversal of trends, however, depends upon the restoration of both vegetation and soil properties that had been characteristic of native forests within a particular climate on a particular parent material. In temperate climates, for example, hydrological processes are restored quite gradually, given the slow speed of soil development there. It is estimated that old-growth forests help to completely recover water flows to original levels only after about 150 years past disturbance. There appears to be a clear link between forests and the quality of water, a much more sporadic link between forests and the availability of water quantity, and a variable link only between forests and the constancy of flow at the catchment level.

At the scale of river basins, modeling studies for the Amazon basin suggest that complete conversion of the rainforest to degraded **pasture** would cause annually a decrease of 26 percent in mean precipitation, 30 percent in evapotranspiration, and 18 percent in runoff; while other studies have produced similar, if less dramatic, responses to deforestation. It has been shown that changes in riparian forest and vegetation have a major impact on the instream biota as well as on the pollutants entering the stream or river stretch.

There is little scientific evidence for the largest, most damaging flood events being caused by deforestation at the global scale. Studies in **South America**, **South Africa**, and in the **Himalayas** indicate that the increase in infiltration capacity of forested lands over non-forested lands is insufficient to influence major downstream flooding events. Rather, the intensity, amount, and spatial distribution of rainfall appear to be the key elements determining the extent and magnitude of damage caused by such disasters, with local geology, land use, and topography being important, concomitant factors.

Croplands

Agriculture is the largest consumer of water by humans worldwide (nearly 85 percent of total human consumptive use), and land-use practices associated with agricultural intensification have been identified to exert an impact on hydrological cycles, especially where improper cultivation techniques were applied in environments with high natural variability. In the **United States of America**—where per capita withdrawals were about 1,700 cubic meters in 1995—crop irrigation accounts for more than 40 percent of the withdrawn freshwater on average, but the figure is as high as 85 percent in California, where agriculture accounts for only 3 percent of the state's economic production. In Africa—where reported water uses range from approximately 600–800 m³/person/year in Egypt, Libya, the Sudan, and some other countries to under 20 m³/person/year in the poorest countries of the continent—as much as 90 percent (or more) of reported water uses go to agriculture. As a matter of fact, large countries that produce grain in **monocultures**—such as **Canada**, the United States, Argentina, and Australia—all have significantly higher per capita water uses than average. In another group of large countries with extensive irrigation—such as India and China—agricultural water use numbers approximately 90 percent.

The central arguments for water-land interlinkages in croplands are that, first, those agricultural practices that can retard **soil erosion** are practices that will also increase infiltration, thus reducing and delaying surface runoff (eroded soils absorb 87 percent less water through infiltration than uneroded soils); and, second, that most problems such as **salinization** are not problems at all with rainfed crops, because the soils are naturally washed away, for instance. In the U.S. southern **Great Plains**, for example, the paucity of farm-level conservation strategies—combined with a period of extreme drought and the Great Depression—caused the Dust Bowl of the 1930s, during which two to twelve inches of topsoil were removed by wind and water erosion and hundreds of thousands of farming households were deprived of their **economic livelihood**. Most of the region eventually recovered from the Dust Bowl episode. For example, changes in streamflows and sediment yields were observed and simulated for several watersheds in Texas, and both observational data and modeling results indicated a significant decline in erosion and **siltation** (reservoir sedimentation) during the period 1910–1984—attributable to the combined effects of conversion from rural to urban land, changes in agricultural crops (i.e., replacement of **cotton** by **wheat** and sorghum), and the implementation of soil and water **conservation** measures beginning in the 1940s (i.e., terracing, contour plowing, strip cropping, and no-till cropping).

In the high-intensity agricultural production zones across the world, the use of commercial fertilizers—containing primarily nitrogen, phosphorus, and potassium—bear several effects on water quality, extending into stream chemistry across both watershed zones and coastal systems. Indeed, there is a direct and quantitative link between the amounts of nitrogen in the major rivers of the world and the magnitude of agricultural nitrogen inputs to their watersheds. In particular, small but severely disturbed agricultural areas such as animal feedlots export about 100 times more nitrogen and phosphorus than other types of agricultural land use. One of the major (unintended) consequences of land-use intensification has been the contamination of shallow groundwater with nitrates, especially due to an explosive growth in fertilizer application following the end of World War II. Nitrate concentrations of 10–20 mg NO_3-N/liter are frequently observed in most shallow aquifers of agricultural areas in North America, for example. Due to the tendency of nitrate concentrations greater than ten milligrams to cause methemoglobinemia in infants and to form carcinogenic nitrosamines in the human intestine, these waters are undrinkable. In the United States, about 40 percent of water is deemed unfit for drinking or recreational use due to contamination by microorganisms, pesticides, and fertilizers; and more than 76 million Americans are infected (and 5,000 die per year) as a result of pathogenic *Escherichia coli* (E. coli) and related foodborne pathogens. In recent decades, more U.S. livestock production systems have moved closer to urban areas, contaminating water and food with manure there.

It has been noted that the effect of decreasing forest cover and increasing agriculture is not linear. As agriculture expands to more than 70 percent of land cover, the transformation of the last remaining normal landscape traps for NO_3 to agriculture—such as **wetlands** and riparian forests—makes NO_3 concentrations in streams rise exponentially. For example, in an estimation of the biogeochemical effects of land-use change in the Choptank basin of the Chesapeake Bay area in the eastern United States over the last 150 years, it was found that conversion of forest to agriculture in the eighteenth and nineteenth centuries increased nitrogen and phosphorus by a factor of two, but application of fertilizers in the twentieth century resulted in about a fivefold increase in nitrogen. Modeling results suggest that not all NO_3 in the groundwater may appear in the base flow, with hydric watershed soils driving denitrification by as much as 80 percent below the expected concentrations based on land use.

Drylands

Irrigation farming, or the consumptive and nonrecoverable use of water by irrigated crops, is a major component of the water balance at many scales, and the availability of freshwater is a key factor for intensification and expansion of agriculture in drylands, in particular. The hydraulic control on previously natural river systems—with damming, water extraction, and redirection of flows as the most important engineering works—alters the behavior of rivers immediately and exerts an array of impacts reverberating through the coupled land-water system.

Soil salinization is one of the problems associated with irrigation farming. It has been estimated that about half of all existing irrigated soils worldwide are adversely affected by salinization, and that the amount of world agricultural land destroyed by salinized soils is 10 million hectares. Another problem associated with crop irrigation is waterlogging, which means that—in the absence of adequate drainage—water levels rise in the upper soil levels, including the plant root zone, and crop growth is impaired. Such irrigated fields are sometimes called "wet deserts" because they are rendered unproductive. In India, for example, waterlogging adversely affects 8.5 million hectares of cropland and results in the loss of as much as 2 million hectares to grain every year.

In most developed countries, the total area of irrigated land has been relatively stable over the last decade, but irrigation has remained a major offstream use of both surface and groundwater resources. In the western United States, for example, groundwater withdrawals for irrigation have been among the most significant impacts of agricultural activities, contributing to dramatic increases in evapotranspiration, excessive declines in water tables, surface subsidence, and soil salinization. Following recoverage from the Dust Bowl episode, changing regional and national economies have promoted substantial pumping of groundwater for crop irrigation. Globally, groundwater aquifers provide an estimated 23 percent of water per year available for sustainable use—with approximately 60 percent of the water intended for crop irrigation never reaching the crop due to water losses during pumping and transport—while the United States relies disproportionately on water pumped from aquifers (i.e., 65 percent). The capacity of the large Ogalla aquifer, for example, which underlies parts of Nebraska, South Dakota, Colorado, Kansas, Oklahoma, New Mexico, and Texas, has decreased 33 percent since about 1950, and water withdrawal is three times faster than its recharge rate. This continues to trigger decreasing water levels and subsequent abandonment of agricultural land, thus raising new concerns about the long-term sustainability of cultivation in the wider region.

The abandonment of irrigated agricultural land—due to soil and water degradation (but also as a consequence of shifting priorities in water allocation)—has been identified as a common feature of land-use irrigation zones of arid and semiarid regions across the world. In the United States, about 150,000 hectares of agricultural land have already been abandoned solely because of high pumping costs. Despite **land stability** in irrigation farming, depleted freshwater resources and land-use legacies have been playing together in the western United States at some locations in producing an abundance of opportunistic shrubs and non-native annual plants as well as lower species diversity than lands that were never cultivated.

In contrast, developing countries that hold large dryland zones—such as Central Asia or Middle Eastern countries along the Euphrates River—have been experiencing an explosive growth in irrigation over the last decade, often driven by huge government-funded water engineering projects. As in developed countries, groundwater withdrawal, declining water tables, and subsequent **land abandonment** are common. In the agriculturally productive Chenaran plain in northeastern Iran, for example, the water table has been declining by 2.8 meters annually since the late 1990s, likewise in Guanajuato in Mexico by as much as 3.3 meters per year. In countries like Turkey and Syria, a pattern of reallocation of land and water use has been typical, that is, irrigated lands along river bottoms and floodplains got abandoned for upland sites, due to a myriad of factors including groundwater depletion, reallocation for surface water, salinization, and waterlogging. In Central Asia, the impacts of land-use/cover change upon the hydrological cycle in low-lying ecosystems over the last 300 years are linked to a typical pathway of water and **land-use transition**. The transition more or less paraphrases those impacts that are typical for the transformation from a predominantly rural mode to a largely industrialized mode of farming and society.

Large hydraulic structures for river regulation are often designed to support the extension of irrigation farming—besides their many other purposes, e.g., for electricity generation, domestic water supply, and flood control. Examples are dams, reservoirs, diversions, levees, artificial channels, and detention/retention ponds. Only 23 percent of the flow in 139 of the largest rivers in the Northern Hemisphere remains unaffected by reservoirs, and the equivalent of 40 percent of the total global runoff to the oceans is intercepted by large dams. There is growing evidence that land-use changes associated with the introduction of hydraulic control of especially large river systems in drylands contribute to rapid water degradation, disruption of hydrological cycles, and the partly irreversible collapse even of regional human-environment systems such as the Aral Sea basin through processes of **desertification**. There is further evidence that some of the responses reverberate through the coupled land-water cycle, strongly impacting hydrological conditions through changes in the partitioning of incoming solar radiation between evapotranspiration and sensible heat, which in turn affect the amount of water that runs off into riverine systems or infiltrates into soil.

In the Lake Chad basin of northern Africa, for example, long-term decreases in lake area, lake level, and river discharges were primarily attributed to climatic variations (rainfall has been declining since the 1960s), but increases in water losses from rapidly growing irrigation explained a large proportion of the variation. Likewise, a decline in potential evapotranspiration in a regional-scale water development project of arid, southeastern Turkey could be attributed to increasing irrigated land area, playing together with decreasing wind speed and increasing atmospheric humidity. From a **paleo perspective**, records for both regions prove that pumping for crop irrigation means the mining of fossil water, because groundwater reservoirs have not been fully recharged for thousands of years. Worldwide, the Nile and Syr-Darya rivers are among the most heavily regulated rivers, partly demonstrating drastic changes in coupled land-water cycles. The Nile, after the erection of the Aswan High Dam in 1968, shows reduced overall discharge, truncated peak flows, higher lower flows, and a seasonal shift in the timing of the natural hydrograph. In the case of the Syr-Darya River, the progressive losses of discharge are associated with expanded water use for irrigation and the contraction of the Aral Sea. In the Aral Sea basin, water from the two rivers entering the inland sea was diverted from the 1960s onward, and led to an enormous decrease in the area and volume of the sea over the following thirty-five years, that is, from 6.7×10^4 km^2 to 3.2×10^4 km^2 and 1,064 km^2 to 310 km^2, respectively. Most studies anticipate its complete disappearance within the next twenty-five years. Associated with the hydrological consequences of land-use/cover change in the sea basin have been an array of ecological influences reverberating in the coupled land-water-atmosphere complex (e.g., surface temperature and local climate changes) as well as an array of social influences on economic livelihoods (e.g., decline in agricultural productivity due to salinization and water logging) and human health.

Owing to impoundments, still waters (i.e., the standing stock of river channel water) have increased worldwide by more than 700 percent relative to the natural state. The consequences of such water aging for associated material transport are to trap a substantial proportion of the incoming suspended sediments and to modify the concentration of dissolved components of nitrogen, phosphorus, and silicon. It has been estimated that the current registered 45,000 largest reservoirs (i.e., greater than 0.5 km^3 storage capacity) trap nearly 30 percent of global sediment flux destined for the ocean, and that this estimate of sediment retention rises further with inclusion of about 800,000 smaller impoundments—and, of course, with continued dam construction.

Urban Zones

The ramifications of land transformation toward urban cover and its impact on the hydrological cycle have been qualitatively described for various stages of urbanization. They include decreases in transpiration from loss of vegetation, decreases in infiltration due to decreased perviousness associated with urban development (streets, roofs, sidewalks, parking lots, etc.), increases in storm runoff volumes, increases in flood peaks, declines in water quality from discharges of sanitary wastes to local streams and rivers, and reductions in baseflow.

Both empirical analyses and modeling results suggest that urbanization and **suburbanization** increase stream flow through increased runoff (and thus flood potential), but also decrease water quality when the amount of impervious surface in a watershed exceeds 10–15 percent of the total land cover. In catchments, the degree of impervious areas in catchments is often directly related to the size of floods. Several studies point to the fact that increased nitrate-nitrogen exports across river basins (such as that of the Mississippi River) can be associated with the amount of developed land there. Agricultural intensification and the disposal of human waste in septic systems are clearly the principal causes of elevated NO_3 in groundwater, and contribute to the eutrophication of aquatic ecosystems, besides various consequences for **human health**.

See also: **Atmosphere-Land Interlinkages; Climate Impacts; Decision-Making; Degradation Narrative; Ecosystem Services; Environmental Change; LandSHIFT Model; Nutrient Cycle; People at Risk; Vulnerability.**

Further Reading

Eshleman, Keith N., "Hydrological Consequences of Land Use Change: A Review of the State-of-the-Science," in *Ecosystems and Land Use Change* (Geophysical Monograph no. 153), eds. Ruth DeFries, Gregory Asner, and Richard Houghton, Washington, DC: American Geophysical Union, 2004, pp. 13–29; Gleick, Peter H., "Water Use," *Annual Review of Environment and Resources* 28 (2003): 275–314; Kabat, Pavel, Claussen, Martin, Dirmeyer, Paul A., Gash, John H. C., Bravo de Guenni, Lelys, Meybeck, Michel, Pielke Sr., Roger A., Vörösmarty, Charles J., Hutjes, Ronald W. A., and Sabine Lütkemeier, eds., *Vegetation, Water, Humans and the Climate: A New Perspective on an Interactive System*, Berlin: Springer, 2004; Meybeck, Michel, and Charles Vörösmarty, "Human-Driven Changes to Continental Aquatic Systems," in *Global Change and the Earth System: A Planet under Pressure* (The IGBP Series), eds. Will Steffen, Angelina Sanderson, Peter D. Tyson, Jill Jäger, Pamela A. Matson, Berrien Moore III, Frank Oldfield et al., Berlin: Springer, 2004, pp. 112–3; Mustard, John F., and Thomas R. Fisher, "Land Use and Hydrology," in *Land Change Science: Observing, Monitoring and Understanding Trajectories of Change on the Earth's Surface* (Remote Sensing and Digital Image Processing no. 6), eds. Garik Gutman, Anthony C. Janetos, Christopher O. Justice, Emilio M. Moran, John F. Mustard, Ronald R. Rindfuss, David Skole, B.L. Turner II, and Mark A. Cochrane,

Dordrecht, NL: Kluwer Academic Publishers, 2004, pp. 257–76; Pimentel, David, Berger, Bonnie, Filiberto, David, Newton, Michelle, Wolfe, Benjamin, Karabinakis, Elizabeth, Clark, Steven, Poon, Elain, Abbett, Elizabeth, and Sudha Nandagopal, "Water Resources: Agricultural and Environmental Issues," *BioScience* 54 (2004): 909–18.

HELMUT GEIST

Watershed. Boundary defining an area where all water flows into a single drainage basin. In Britain, a watershed traditionally refers to the divide between catchments, where water falling on one side of the divide will flow into a particular **river basin**, and water on the other side will flow into a different river basin. A typical watershed divide is generally located along a **mountain** range or topographical high point. In North America, watershed commonly refers to the area within this boundary, and is defined as the total area where all the water falling within that land flows into the same drainage basin, or the region contributing to the supply of a waterbody. In this latter sense, watershed is used interchangeably with "catchment" or "drainage basin." A watershed can be defined at a range of **scales**, depending on the resolution of interest. The largest watersheds, for example, are the global watersheds that separate land areas draining into the Atlantic, Pacific, and Indian oceans. These watersheds can be further subdivided into continental watersheds separating drainage into various oceans. Watersheds can be defined at progressively finer scales, such as the drainage area for a single river or stream.

As noted by David Thomas and Andrew Goudie, the watershed is the basic unit of landscape hydrology. While watersheds are usually drawn based on map elevation contours, actual drainage is likely to differ due to disparities between surface and subsurface flow. The hydrological cycle acting within a watershed unit drives processes of runoff, **soil erosion**, sedimentation, and evapotranspiration. These processes are influenced by land use within a watershed unit. Dense **vegetation**, for example, will increase water infiltration and decrease runoff. Removal of vegetation, or the presence of urban areas, is associated with increased overland flow and increased risk of **soil erosion** and flooding. Similarly, watershed topology drives climate, waterflow, and sediment transport, and thus influences the natural and potential land covers and land uses for a given catchment region.

Characteristics and processes of land cover and land use are driven, to a large extent, by hydrological systems; this makes the watershed an important unit of study for land-use/cover analyses. As noted by William Cunningham and colleagues, a watershed is a natural boundary that provides an excellent basis for analysis of land-cover processes; land and water systems can be integrated easily, downstream effects can be conceptualized, and the watershed provides an excellent unit for management planning. The nested nature of watersheds (i.e., the ability to select a well-defined watershed unit at a variety of scales) means that various resolutions of land-cover processes can be assessed, understood, and integrated using the same conceptual unit of analysis. Dennis Soden and Brent Steel add that watersheds are relatively well-defined, rational boundaries that can be easily understood. As a result, the watershed has emerged as a popular unit of analysis for land-cover and land-use research, management, and planning.

See also: Eutrophication; Land-Use Planning; Mineral Extraction; Mountains; Reforestation; Regrowth; Tradeoffs; Water-Land Interlinkages.

Further Reading

Cunningham, William P., Ball, Terence, Cooper, Terence, Gorham, Eville, Hepworth, Malcolm T., and Alfred A. Marcus, *Environmental Encyclopedia*, Detroit, MI: Gale Research Inc., 1994; Snelder, Ton H., and Barry J. F. Biggs, "Multiscale River Environment Classification for Water Resources Management," *Journal of the American Water Resources Association* 38 (2002): 1225–39; Soden, Dennis L., and Brent Steel, *Handbook of Global Environmental Policy and Administration*, New York: Marcel Dekker, 1999; Thomas, David S. G., and Andrew Goudie, eds., *The Dictionary of Physical Geography*, 3rd ed., Oxford, UK: Blackwell Publishing, 2000.

LEA BERRANG FORD

West-Central Europe.   Western part of a region that is often conceived of as central Europe in a geographic context, comprising the countries of Austria, Belgium, Germany, Luxembourg, the Netherlands, and Switzerland. The region falls within the ecological zone of temperate nemoral climate and is thus characterized by warm summers that hold most of the annual precipitation, and by cool, but not extremely cold, winters. In the coastal lowlands in the north, particularly Belgium, the Netherlands, and northern Germany, the climate is milder and more humid; toward the east the climate is becoming slightly continental. Accordingly, temperate deciduous and mixed **forests** constitute the zonal potential natural **vegetation**. Deviations with respect to climate and potential vegetation occur mainly in the southern part of west-central Europe, which is dominated by the Alps, which reach an elevation above 4,000 meters. With respect to soils, podzols prevail in the northern part, while cambisols and luvisols dominate most of the rest of the region.

West-central Europe covers an area of 550,000 km^2 and accounts for about 11 percent of the total European territory (excluding **Russia** and the Ukraine). In 2000, the total population of west-central Europe was about 124 million, which is about one-quarter of the European total. The west-central European countries are among the economically most advanced, highly industrialized countries in the world. Taken together, they produced about one-third of the total European gross domestic product (GDP) in 2001. The share of urban population in west-central Europe amounts to 85 percent, significantly above the European average (75 percent). West-central Europe includes some of the most densely populated regions in **Europe**. A belt of highly urbanized and industrialized land stretches from Belgium and the Netherlands (average population density above 350 persons/km^2) through the German industrial zone of the *Ruhrgebiet* to southern Germany, and continues south of the Alps into northern Italy.

Land Use and Landscape Patterns

The west-central European landscape is heterogeneous and includes large urban/industrial areas, regions of very intensive and specialized agricultural production, and extensively used landscapes dominated by traditional mixed agriculture, woodlands, and **pastures**. The area used for human settlement, transport infrastructure, and industrial production occupies approximately 15 percent of the total land area, but may exceed 30 percent in the densely populated regions in the north. About half of this urban fabric is actually built-up land covered by artificial surface. Significant areas, predominantly in eastern Germany and the *Ruhrgebiet*, are characterized by vast, often abandoned open pit coal mines and the corresponding industries. Almost half of the total area of west-central Europe is agricultural land. More than 60 percent of this area is used for rainfed crop production; the rest is **grassland**. Agricultural yields are 30–40 percent above the European average due to favorable conditions and highly intensive production. Roughly 30 percent of the total area is covered with woodlands and forests. The wooded areas are concentrated in the hilly and mountainous regions in the southern and central part of the region, while the northern lowlands are largely deforested.

The northern part of west-central Europe is characterized by fertile—in its western parts highly urbanized—coastal lowlands stretching from Belgium and the Netherlands into northeastern Germany. A significant share of this area, particularly in the Netherlands, was gained by coastal land reclamation and drainage of **peatland** during the last centuries. Cattle farming on grasslands prevails in the marshier areas of the **coastal zone**, whereas large areas further inland and eastward are devoted to intensive high-input crop production, mostly grain and rapeseed, but also root crops such as sugarbeets. Especially the sparsely populated eastern part of Germany is characterized by huge farm operations and an agro-industrial landscape that is a heritage of collective farming of the former German Democratic Republic. The central part of west-central Europe features low **mountain** ranges with mixed farming and woodlands, but also extensive fertile lowlands with intensive crop production. The southern part is dominated by the mountainous regions of the Alps. Mont Blanc in the French Alps is Europe's highest peak and reaches an elevation of 4,807 meters. The Alps are comparatively sparsely populated (an average of 75 inhabitants/km^2) and cover a total territory of approximately 192,000 square kilometers, with the major parts situated in Austria (30 percent), northern Italy (27 percent), southeastern France (21 percent), and Switzerland (14 percent). Major land-use types in the Alpine regions are grassland-based cattle farming and **forestry**. Population and infrastructure are concentrated in the major valleys. Non-urban areas in these valleys are mostly used for intensive, mixed agriculture. The Alps are one of Europe's main tourist regions (370 million overnight stays per year). Alpine tourism is an important economic factor in the region and contributes roughly 10–15 percent to total GDP in Austria and Switzerland. However, the impact of tourism, and particularly winter tourism, on Alpine ecosystems and land use there is considerable. Among the most crucial environmental issues related to alpine mass tourism are the large volume of traffic, construction activities, waste and sewage generation, mechanical production of artificial snow, and skiing.

LAND-USE HISTORY

Europe has a long history of intensive agricultural land use. As early as in the fifth century BC, west-central Europe had lost a large part of its forest cover to land clearing for **agriculture**. The expansion of agricultural area in this region reached its all-time high during the thirteenth century when medieval land clearing had reduced forest cover to less than 20 percent. Overpopulation of favored areas and large-scale **deforestation** contributed to considerable **soil erosion** and deterioration of agricultural land in the Middle Ages. In combination with deteriorating climate and a number of years with particularly bad harvests, these **land degradation** tendencies resulted in a downward trend in population that culminated around 1350 when the bubonic plague (Black Death) wiped out 25–35 percent of west-central Europe's population. This drastic and sudden population decline resulted in the large-scale abandonment of human settlements, and consequently triggered major, comparatively rapid changes in land use and land cover. Cropland and grain production receded, and the land was increasingly used for grazing. Forests re-advanced into the fields, and woodland area more than doubled. In the time period between 1450 and the beginning of **industrialization**, the agricultural area gradually expanded again as population and food demand grew. By 1800, woodland was again reduced to less than 30 percent and remained only to the minimum extent that it was necessary to provide for timber and firewood, respectively on habitats due to soil conditions or morphology not worth the effort of cultivation.

At the eve of industrialization, most of west-central Europe's agricultural land was farmed in cereals (e.g., rye), mostly in three-course rotation with **fallow**. In most parts, animals were only subsidiary to provide draft power and manure. Animal-rearing with subordinate crop farming only prevailed in the alpine regions in the south and in the marshy lowlands in some parts of northern west-central Europe. Agriculture was most advanced in Belgium and the Netherlands, where root crops and leguminous fodder crops were introduced early into the crop rotation and allowed for more livestock and higher yields. Beginning in the eighteenth century, this modern type of agriculture gradually expanded all over central Europe and allowed for a doubling of agricultural output during the nineteenth century.

After World War II, the most significant driver of land-use change in west-central Europe was **agricultural intensification**. Supported by agricultural policies, including massive national and European Union subsidies, large-scale mechanization and the use of artificial fertilizers and pesticides were introduced within only two decades. This resulted in, among other things, highly specialized agro-industrial landscapes in large parts of central Europe and multiplied output and productivity. The consequences were increasing farm and field size, a reduction of crop diversity, the introduction of **monocultures**, and the spatial separation of arable farming and livestock husbandry. Agriculture intensified on the most productive lands, while less-favored areas were increasingly abandoned. One of the outcomes of these processes was a steady decrease of the area of agricultural land and a gradual process of **reforestation** in many central European regions, for example, in the Alps. The total area of agricultural land, above all grassland, has decreased by roughly 15 percent in central Europe during the last fifty years.

In the mid-1980s, the subsidized agricultural overproduction became a severe economic burden. Instead of its former focus on fostering production, agricultural policy and subsidization schemes were reformed to include a number of measures to encourage farmers to adopt more extensive and less-polluting farming practices. As a consequence, inputs such as artificial fertilizer began to level off and even decrease. A significant proportion of the region's cropland (10–15 percent) was taken out of production and managed within programs for temporary or permanent set-aside land. Organic farming experienced a significant growth, and the number of certified organic farms in west-central Europe has almost tripled since 1993. More than 1.2 million hectares or approximately 5 percent of the total agricultural area was farmed organically in 2002, the share being highest in Austria (9 percent) and Switzerland (7 percent). Another measure to reduce grain production with significant spatial impact was the promotion of the cultivation of so-called "energy crops," which caused the share of rapeseed and sunflower of total cropland to increase from less than 5 percent in the 1980s to 20 percent in 2002.

Crucial Current Issues

At present, hardly any natural or undisturbed areas can still be found in the central European landscape. Cultural landscapes in which traditional small-scale farming remained in place are currently considered to be among the most diverse and valuable landscapes. This includes, above all, regions in the mountainous and pre-alpine regions of Austria, Germany, and Switzerland, where richly structured agrarian landscapes with mixed farming remained; these were regions with high-alpine grassland farming or extensively used heath lands in the north. These diverse landscapes still cover approximately 5–15 percent of central European territory and are regarded as being of high cultural and ecological value. However, these regions are particularly threatened either by intensification or **land abandonment** and reforestation.

Among the most intensively discussed issues in central European land use are the high intensity of agricultural land use, which results in negative environmental impacts; the increasing reforestation and depopulation in marginally productive regions; the adverse effects of mass tourism on alpine landscapes; and the persistent growth of areas covered by settlements, industry, and transport infrastructure. Land consumed by urbanization and infrastructure is ever increasing. For example, in Germany each year about 0.11 percent of the total area (5 m^2/inhabitant/year) is converted into built-up land.

See also: **Agricultural Revolution; Agrodiversity; East-Central Europe; Land Rights; Land-Use Policies; Mineral Extraction; Urbanization; Western Europe.**

Further Reading

Brouwer, Floor M., Thomas, A. J., and M. J. Chadwick, *Land Use Changes in Europe: Processes of Change, Environmental Transformations and Future Patterns*, Dordrecht, NL: Kluwer Academic Publishers, 1991; European Environmental Agency [Online, August 2004], The EEA Web Site, www.eea.eu.int/products; International Commission for the Protection of the Alps [Online, August 2004], The CIPRA Web Site, www.cipra.org; Pounds, Norman J. G., *An Historical Geography of Europe*, Cambridge, UK: Cambridge University Press, 1990; United Nations Food and Agricultural Organization [Online, August 2004], The FAO Web Site, www.fao.org.

FRIDOLIN KRAUSMANN

Western Europe. Countries of the pre-2004 European Union (EU) with the addition of Iceland, Norway, and Switzerland. This definition is used by the **United Nations Food and Agriculture Organization** (FAO) in compiling its land-use statistics, but it is to some extent arbitrary. The notion of western Europe has been fluid, and its limits are variously defined. In much of the second half of the twentieth century, it was delineated with reference to "Eastern Europe," with the boundary along the Iron Curtain. Prior to the Cold War, it would have been used to describe a small area including France, Ireland, and the United Kingdom (UK) together with the Low Countries (Belgium, Luxembourg, and the Netherlands). Even this narrower definition, however, would have depended on the treatment afforded to Norway and the Iberian countries of Spain and Portugal, and to the recognition of "northern" and "southern" Europe.

Most of the land cover of western Europe has been modified by human activity. Few if any areas are free from human influences. Even the high **mountains** have been affected to some degree by air pollution from power stations and other sources. Over most of the lowlands and intermediate altitudes, the original **vegetation** has been completely replaced by crops or severely modified by the grazing of farm animals.

Land Cover

In the absence of humans, most of western Europe would have been clothed in **forest**. Little remains of this forest in its natural condition. In the maritime lands of Ireland, Scotland, and Denmark, the forest extent had fallen to under 10 percent by modern times. Once removed, **regrowth** or forest regeneration in the moist and exposed climate was difficult, especially if peat formed on deforested ground. In the more continental interior, the forest survived more extensively, but not without human **modification**. According to land-use statistics of international agencies, the relative extent of forest "undisturbed by man" ranges from zero (in several countries) to a maximum of 17 percent in Sweden: parts of today's forests have been planted, while most of the remainder has been affected by some form of management. Perhaps the fate and state of the forest epitomize the degree of human impact on land cover. Over sizable areas, "artificial" land cover, in the form of built-up areas with sealed soil surfaces, has been created; and over much larger areas, near-**monocultures** of intensively produced crops have resulted in drastic reductions in **biodiversity**.

CORINE land cover (CLC) information from the **Coordination of Information on the Environment (CORINE) database** exists for much of the region. Based primarily on satellite imagery, the CORINE project quantifies land cover at a scale of 1:100,000, using a classification system of forty-four categories that can be grouped hierarchically into larger classes. Maintained by the European Environment Agency, the database can provide aggregations for various spatial units such as countries, EU statistical areas, biogeographical provinces, and river catchments. Gaps in coverage mean that a breakdown for the whole of western Europe as here defined cannot be achieved, but it is clear that semi-natural habitats are restricted to a very small proportion of the land area.

The first table (below) summarizes the pattern of variation in land cover across two transects, from maritime Ireland to continental Austria, and from the southwest to the northeast.

In each of the countries included in the table, the proportion of artificial surfaces is under 10 percent. Though western Europe is usually regarded as heavily urbanized, the urban area is still small and in all except a few countries is very low. The relative extent of arable land is also small. Even when "complexes" and permanent crops are included, it remains under 50 percent in each of the countries. "Forest and transitional woodland" forms the most variable category. It covers more than three-quarters of the land area of Finland but less than one-tenth in Ireland and the Netherlands. One of the most distinctive features of the maritime periphery is the low forest extent compared with the continental interior: this extent, however, is much greater than it was a hundred years ago.

Trends

The dynamics of land cover and land use in western Europe are complex in detail, and different trends are discernible at different time **scales**. Over the long sweep of history, the forest has retreated and agricultural land has expanded. Over shorter time scales, different trends have dominated. Over the last 100–200 years, a **forest transition** has occurred and forests have expanded significantly, while the agricultural area has contracted. The second table in this section summarizes FAO data for western Europe: these data are associated with some problems, and indeed FAO discontinued the publication of statistics on forests and woodland in this form after 1995. The table is included here simply as an indicative summary of trends: (total) agricultural area and arable area contracted by 14 and 16 percent respectively between 1961 and 2001, while the forest area increased by 12 percent. The third table in this section focuses on the 1990s and includes more reliable forest data.

Land cover across western European transects

	Ireland	France	Austria	Portugal	The Netherlands	Finland
Artificial surfaces	1.2	3.8	1.8	1.4	9.5	0.5
Arable land	5.9	25.8	13.6	17.7	20.9	6.3
Vineyards, fruit trees, and olive groves	0	2.6	0.7	13.9	0.3	0.1
Pastures	56.7	14.7	11.0	0.1	31.0	0
Complexes etc.^	6.3	19.3	10.9	13.9	16.8	0.4
Forest and transitional woodland	6.1	27.3	44.4	41.3*	8.0	78.5
Natural grass and moors	6.8	3.9	8.8	8.5	2.0	1.9
Bogs	14.3	0.3	0.2	0.3	1.1	2.6
Bare and water surfaces	3.1	2.4	8.7	2.9	10.3	9.8

^ Complex cultivation patterns, and land principally under cultivation but with significant areas of cultivate land;
* including agroforestry.
Source: Based on Corine Land Cover 90 Version 12/2000, European Environment Agency, Copenhagen.

One of the biggest changes in western European land use and landscapes in the twentieth century has been forest expansion, although that process has operated more strikingly in the peripheries of individual countries than in their cores. In recent years, this process has been especially marked in the maritime periphery, as the third table indicates. Forest extents on the maritime fringes are low, but the rate of expansion is high. In countries in the maritime fringe, most of the expansion is through forest plantations. In the more continental parts of the region, it is mostly through natural regeneration. This process is especially marked in alpine regions, where the retreat or **disintensification** of agriculture has been followed by the spread of woodland over alpine **pastures**.

Trends in agricultural and forest areas in western Europe (ha × 1,000)

	Agricultural Land	Arable Land	Forest and Woodland
1961	170,012	89,021	108,957
1971	162,215	81,617	116,612
1981	157,767	80,193	120,538
1991	150,561	78,548	122,437
2001	145,755	75,151	*

*Statistics discontinued after 1995.
Source: Based on FAO FAOSTATS.

Land is coming out of agriculture especially on the more extensive margins, though some is also being lost through **urbanization** at the intensive margin. There is some evidence that arable land is becoming increasingly concentrated in the areas with more favorable soils and climate, such as the Paris basin.

These trends of forest expansion and agricultural contraction are accompanied by other less evident but equally significant processes. For many centuries, the forest resource in western Europe dwindled in volume as well as in area. In recent times, however, this process has been reversed. Timber removals are now equivalent to less than two-thirds of the annual increment, with the result that the volume of standing timber is increasing. While the goal of sustainability might indicate that it is desirable that removals not exceed growth, the aging of the forest may bring its own problems.

Trends within agriculture are different. Here the dominant process has been **agricultural/intensification** of the more productive areas. This intensification has led to dramatic improvements in yields, but these have been achieved at an environmental cost. Semi-natural habitats have been converted to cropland as drainage has been improved, field boundaries have been removed, and grazings plowed up. The result has been a loss of biodiversity and an opening and simplification of the ecosystem and landscape. The

Change in forest and agricultural areas in western Europe in the 1990s (%)

	Ireland	France	Austria	Portugal	The Netherlands	Finland
Forest cover (annual)	3.0	0.4	0.2	1.7	0.3	n.s.
Agricultural area (decennial)	−21.9	−2.8	−3.1	+4.5	−2.9	−7.7

Forest cover: annual change 1990–2000; n.s. not significant; agricultural area: change 1990–2000.
Sources: Forest cover: FAO 2001; Global Forest Resources Assessment 2000, Main Report; *FAO Forestry Paper 140*; Rome: FAO; Agricultural area: FAOSTAT.

process has probably been most severe in the core areas of eastern England, northern France, and the Low Countries. It is now slowing and perhaps in some cases being reversed: net loss of hedge length in England, for example, stabilized in the 1990s, and habitat re-creation has been carried out in some areas under agri-environmental programs.

Driving Forces

Historically, demography or human **population dynamics** was a key driver. High population densities provided a ready market for food and encouraged agricultural expansion and intensification. Over the last 100–150 years, however, there has been a partial decoupling of demographic trends from land-use change. On the one hand, transport improvements, internationalization, and **globalization** have ensured that local populations are not dependent on local food production. On the other hand, most but not all countries in western Europe have sought to shield their domestic agriculture from the effects of cheap imports, and thus the extent of decoupling has been less than it might otherwise have been. The recent stabilization of population might be expected to lead to a lessening of dynamism of land use, but again the relationship has been complex. While urban populations have stabilized, urban areas are still expanding, in part because of increasing numbers of households. Even if the urban area is still relatively very small, the local environmental effects can be large, especially in the more heavily urbanized areas such as the Randstat area of Holland and the London area, where formerly separate settlements now merge together in **urban sprawl**.

A weakening of the effects of traditional drivers such as population growth would in any case now be expected, because the influences of these drivers are now mediated through complex sets of **land-use policies** and plans. These are especially prominent in two spheres—the regulation of the expansion of urban land uses over rural land, and support for agriculture. In the absence of such plans and policies, the character of land cover and use in western Europe would be significantly different.

The Role of the State

The significance of government intervention in land use is unmistakable. In the nineteenth century, the state became a powerful entity across most of western Europe, and since then it has involved itself in a widening range of areas and activities that relate to land use. These include rural welfare, resource **conservation** and strategic interests, and environmental management. The supra-state, in the form of the European Union, has of course also become a prominent actor, associated in particular with **agriculture**, but increasingly also involved in other areas.

The EU's Common Agricultural Policy (CAP) has been a major influence, but national policies in Iceland, Norway, and Switzerland have had comparable effects. In essence, agricultural policies weakened the direct relationship between producer and consumer of food. For reasons of national self-sufficiency and rural welfare, farm prices have been supported at levels well above those of the world market. Production has accordingly been higher than it would have been in the absence of these policies, as have intensity of management and extent of area. The influence of wars and food shortages shaped interventionist policies, but arguably they were maintained long after the specter of scarcities had receded. The inevitable consequence has been a tendency toward overproduction of food relative to regional demand, and the high cost of production of that food meant that it could be exported only if subsidized. By the 1990s, the existence of a problem had been recognized, but an acceptable solution was elusive. A push to decouple support measures from production began tentatively during the 1990s, with a more radical step in 2004. In place of production-related subsidies, farmers will receive direct support. As long as they manage their land in accordance with defined environmental criteria, they will receive payments regardless of their volume and type of production. This change in approach could have consequences in terms of land use and land cover that are both profound and spatially variable. One **scenario** is that agricultural production will decline in the more marginal areas, while continuing to increase through further intensification in the areas that are environmentally more favorable. On the other hand, environmental regulation, imposed by the state in response to changing societal demands, may limit the spatial concentration to which such a trend could lead. Already, pollution loadings from intensive livestock units and fertilizer applications are high in some areas.

Another manifestation of the role of the state is its attempts to curb urban expansion. Greenbelt and similar policies have been pursued for more than half a century. Their effects, however, have been partial: urban expansion has continued apace. In theory, the increasing emphasis now placed on brownfield development should help to curb expansion. Nevertheless, with an apparent problem of overproduction of food, it will be difficult to prevent the **conversion** of further areas of agricultural land into urban uses.

Overview

The role of land in relation to economy and society in western Europe has been transformed. Today, land-based industries account for only tiny proportions of national gross domestic products (GDPs) and employment. The historical role of land in producing food and wood has weakened: memories of physical scarcities are fast receding. The role of land in providing a pleasant environment and the setting for pleasant lifestyles has grown as the productive role has weakened. Increasingly, farms are being purchased as residential rather than productive units. In time, this trend is likely to have its own impact on land use and landscape. The effect will probably be to hasten the rate of agricultural contraction. It may not accelerate the rate of forest expansion, but it will probably lead to new patterns and styles of tree planting, perhaps reminiscent of those of country estates of the past. In other words, a trend of agricultural area may be sustained as the environmental Kuznets curve suggests, along with an inverse-Kuznets curve of forest area. As yet, however, there is little indication that the top of a Kuznets-type curve of urban area has been reached. Perhaps it is unfortunate that the scope for complementarity between the agricultural and urban trends is so limited. Historically, cities grew up in fertile plains: subsequent expansion inevitably eats into the stock of fertile agricultural land that is most likely to be the scene of further intensification.

See also: **Agricultural Revolution, Consumption; East-Central Europe; Economic Restructuring; Europe; Forestry; Hot Spots of Land-Cover Change; Peatlands; Public Policy; Reforestation; West-Central Europe.**

Further Reading

Kronert, Reinhold, Baudry, Jacques, Bowler, Ian, and Annette Reenberg, eds., *Land Use Changes and Their Environmental Impact in Rural Areas in Europe*, New York: Parthenon, 1999; Kuusela, Kullervo, *Forest Resources in Europe 1950–1990*, Cambridge, UK: Cambridge University Press, 1994; Tracey, Michael, *Government and Agriculture in Western Europe 1880–1980*, New York: Harvester Wheatsheaf, 1989; United Nations Economic Commission for Europe and Food and Agriculture Organization, *Forest Resources of Europe, CIS, North America, Australia, Japan and New Zealand*, New York: United Nations, 2000.

ALEXANDER S. MATHER

Wetlands.   Since about the second half of the twentieth century, a substitute for terms such as "swamp", "marsh", and "bog". Wetlands include a wide variety of habitat types, such as **peatlands**, mires, fens, and mangroves. Most wetlands are natural, but some wetlands are artificial, such as farm ponds, reservoirs, sewage farms, gravel pits, and salt pans. Once regarded as unhealthy areas or wastelands of little value, wetlands are being recognized as one of the most biologically productive of all ecosystems. The quite recent awareness of wetlands' functional role and their increasing disappearance in farming landscapes have involved the need to inventory them for **conservation** or restoration purposes. However, the inventories of wetlands are very often debated, insofar as the definitions of these ecosystems are numerous and the interests of users and managers for these areas are greatly conflicting.

Marsh vegetation on the mid-Patuxent River, Maryland, September 1996.
PHOTO: Mary Hollinger. CREDITS: NOAA Central Library and the Office of the CIO/HPCC (America's Coastlines Collection).

Definitions

Most wetlands are transition areas between terrestrial and aquatic ecosystems. There is no single, universal, indisputable definition for the term wetland because the gradation between totally dry and totally wet environments is continuous but also because the reasons for defining wetland vary. This explains why wetlands have been defined in various ways, and a multitude of definitions have been proposed.

L. Cowardin and colleagues define wetlands as transitional lands between terrestrial and aquatic systems where the water table is usually at or near the surface, or the land is covered by shallow water. With regard to this classification, a wetland must have one or more of the following attributes: at least periodically, the land supports predominantly hydrophytes; the substrate is predominantly undrained hydric soil; and the substrate is nonsoil and is saturated with water or covered by shallow water at a given time during the season of growth each year.

In this classification, the limit between wetland and upland is defined according to the three following boundaries: the boundary between land with predominantly hydrophytic cover and land with predominantly mesophytic or xerophytic cover; the boundary between predominantly hydric soil and predominantly nonhydric soil; or in the case of wetlands without vegetation or soil, the boundary between flooded or saturated land at a given time each year and land that is not in these hydrologic conditions. The limits between wetland and deepwater systems are also defined. Although the boundary between wetland and deepwater systems is important for inventory purposes, it is unusually at issue in conflicts of standardization and is not mentioned completely in the regulatory definitions of wetland.

The classification by L. Cowardin and colleagues relies on the use of soils, **vegetation**, and hydrology factors for wetland identification. It introduces the concepts of hydrophytes and hydric soils and of predominance both for hydrophytes or undrained hydric soils in wetland areas. In addition, this classification includes some areas that lack vascular plants or soils. These concepts have been later developed in wetland delineation manuals. For example, the U.S. Army Corps of Engineers defines wetlands as areas that are inundated or saturated by surface or groundwater at a frequency and duration sufficient to support a prevalence of vegetation typically adapted for life in saturated soil conditions.

In most current approaches to the mapping and assessment of wetlands, the identification variables used to determine wetland presence are hydrophytic vegetation, hydric soil, and hydrodynamics. The concept of wetland comprises a number of characteristics, including the elevation of the water table with regard to the ground surface, waterlogging, duration of surface water, soil types that form under temporarily or permanently saturated conditions, and various kinds of plants and animals that are adapted to a wet environment. In other terms, wetland is defined as land where the water table is at, near, or above the land surface long enough each year to allow hydric soils formation and to support hydrophytes' growth. The single characteristic that is common to all wetlands is the presence of more soil moisture than necessary to support the growth of most plants. In certain types of wetlands, soils and vegetation are poorly developed or even absent because of important and frequent fluctuations of water levels, turbidity rates, or very high concentration levels of salts or other substances in the water or substrate.

The use of scientific principles for wetland definition contributes to stabilize both the discrimination of wetlands from the other ecosystems and the determination of wetland boundaries for the purposes of inventory, assessment, and management. It also helps to rationalize the application of regulations. In addition, legal implications need to be considered in the application of any wetland definition, because specifying the exact boundary of a wetland that present and projected future users will accept is not an easy task in these frequently conflicting and sometimes controversial areas.

Wetland Functions

Wetlands perform many vital functions, including three main categories: hydrological, biogeochemical, and ecological functions. Hydrologic functions concern regulation of water regime through water storage, flood mitigation, groundwater recharge, and discharge. These functions reduce the amplitude of flooding peaks downstream and maintain basic flow rates by buffering flow distributions. Biogeochemical functions encompass the transformation and cycling of elements, the accumulation of inorganic sediments and peats, and the retention and removal of dissolved substances from surface waters. These functions improve water quality, retain nutrients and other elements, and affect aquatic and atmospheric chemistry. In particular, wetlands play an important role in nitrogen and carbon storage. They can have high rates of denitrification and nitrogen fixation, contributing to control non-point source pollution. Wetlands are major natural sources of reduced gases such as methane and sulfur compounds. Wetlands act also as sediment sinks, in particular in floodplains. But they also contribute, with **rice** fields, to be the largest emitters of methane to the atmosphere due to the anoxic conditions occurring in their flooded soils and their high primary production, totaling roughly 40 percent of the methane input to the troposphere. Ecological functions include habitat maintenance and food web support. This latter includes maintenance of wetland plant communities that provide food and habitat for waterfowl and other animals, thus maintaining biodiversity. Some wetlands are among the most productive environments of the world.

Benefits and Values

Wetlands provide ecological services and economic benefits such as water supply, food supply, biodiversity, and recreation. Most of the sources used to inventory wetlands provide some information on their values and benefits. However, this information frequently takes the form of a summarization of the **biodiversity** values and human use, with few quantitative or economic data included. The exceptions are the productivity of artificial wetlands, rice paddies, Salinas, and fish ponds. The benefits and values of all wetlands for biodiversity and human uses are often highlighted at a global scale. Information is particularly detailed for mangroves, where values and benefits include flood reduction, coastal protection, sediment accumulation, and fish nurseries. Similar descriptions are available for **peatlands**. Protected areas are also valued by people for various reasons, including **conservation**, fishing, and tourism. For example, in **Europe** there is a focus on the values of protected areas, especially as feeding or breeding habitat for birds.

Wetland Inventories

Wetlands are distributed worldwide, from tropical to Arctic countries, but the real extent of wetland is not yet precisely determined. Numerous detailed and precise wetland inventories are conducted on a local scale, and national inventories are developed in a number of countries, either through national wetland policies or implementation of wetland priorities within national environmental action plans and national biodiversity strategies. The global areal extent of wetlands at the earth surface has been estimated by several institutions, but they are uncertain because it is difficult to identify and classify wetlands on a global scale. For example, the World Conservation Monitoring Centre has estimated that they roughly cover some 5.7 millions km^2, which is about 6 percent of the earth's land surface. Based on the wetland inventory in each region of the **Ramsar Convention on Wetlands**, supplemented by a review of regional and international inventories, a global review of wetland resources and priorities for wetland inventory by C. Finlayson and A. Spiers estimated that the area of wetlands worldwide ranges from 7.48 million km^2—the "best" minimum global estimates—to 12.79 million km^2, which corresponds to the amount of the regional minimum estimates. These variations highlight major discrepancies in the areal estimates, which can be attributed to many factors, such as differences in the definition of wetlands, gaps and inaccuracies in the information, the techniques used to collect and interpret the basic data, and the scale of the analyses.

Wetland areas are often underestimated for a number of reasons. Some of the main omissions are due to lack of **access** (e.g., Siberian or Brazilian wetlands), or to their dispersion. For example, small riparian wetlands are generally not inventoried because they are inserted in agricultural landscapes in the bottomlands of the headwater catchments or **watersheds**, scattered in the rural landscape. They are often neglected in national and regional wetland inventories although they strongly influence hydrology, water quality, and biodiversity. Some inventories are carried out by groups with narrow topical interests like habitat maintenance, thus excluding parts of wetland areas. In forested areas, wetlands are also underestimated and difficult to determine, especially in those areas where the water table is below the surface.

Moreover, the areal extent of wetlands is being modified as a result of land-use changes. As a result, once a large inventory is set up, the areal distribution must be monitored and revised periodically while existing ones are updated and new local or regional inventories are created.

Globally, there is very little information available or even attempts made to evaluate the loss of wetlands on a systematic basis, except in North America and **western Europe**. In the **United States of America**, most of the wetland loss occurred during the first half of the twentieth century. Between the first phases of European **colonization** and the 1980s, about half of the wetland area in the conterminous states was converted to other land uses, mainly by drainage, for agricultural purposes. Federal intervention by various acts of swampland and ambitious drainage projects of the U.S. Army Corps of

Engineers (USACE) promoted wetland conversion, and this resulted in the creation of productive croplands that today constitute an important part of the agricultural resource base of the United States. **Conversion** has been distributed very unevenly. The conversion, which had exceeded 80 percent in a number of states by the end of the 1980s, was highest where conversion was both feasible and advantageous, or where wetlands were of limited extent but were located on the most favorable areas for agriculture and population growth. In Europe, loss of wetlands dramatically increased from the second half of the century with **agricultural intensification**. The loss of wetlands worldwide has been estimated at 50 percent of those that existed in the beginning of the twentieth century. This evaluation underestimates the amount of wetland loss, because it includes inland wetlands and possibly mangroves, but not marine wetlands and large estuaries. Since the 1950s, tropical and sub-tropical wetlands, particularly peatlands, swamp forests, and mangroves, have increasingly been converted to cultivated areas.

The driving factors of variations in areal extent of wetlands have both natural and anthropogenic origins, but human factors are predominant, with **agriculture** considered the principal cause of wetland loss worldwide. By 1985, it was estimated that 56–65 percent of available wetland had been drained for intensive agriculture in North America and Europe, 27 percent in Asia, 6 percent in **South America**, and 2 percent in Africa. These figures are probably underestimated, as in the last thirty years **deforestation** has strongly increased the loss of wetlands, mostly in developing countries.

Moreover, the loss of wetlands is the most visible impact of human use on these ecosystems, but parts of existing ones have been degraded, leading to the loss or, at least, to the decrease of their functional efficiency. The Ramsar site database provides a regularly updated but still uneven and incomplete analysis of threats to wetlands. It indicates that 84 percent of Ramsar-listed wetlands had been eliminated or were threatened by ecological change. The most widespread threats were drainage for agriculture, pollution, settlements and **urbanization**, and hunting.

Consequences of Wetland Loss and Degradation

Loss and degradation of wetlands cause important and sometimes irreversible damage to the provision of **ecosystem services**. For example, the issue of water allocation and distribution is linked to the rate and extent of wetland loss and degradation. Planning actions on the rivers like regulation and on associated natural waterbodies and wetlands have impacts on water supply in diminishing underground water reserves and increasing **salinization**, but also on biodiversity in decreasing fish stocks because of degraded habitat and impeded migration.

The conversion of wetlands for agriculture by drainage and transformation of these areas into **grasslands** and/or dry cropland and urban areas leads to major changes in hydrology, vegetation, soil characteristics, and therefore biogeochemical cycles. When wetlands are totally drained, they are excluded from wetland inventories, but it is necessary to determine the areal extent of lost wetlands, because the influence of human activity can be assessed and these areas can be seen as "potential" wetlands that can, under certain circumstances, be rehabilitated and become functional again.

Wetland degradation is more subtle, because it remains as wetland, but is functionally altered, which is not always easy to assess. For example, grazing without hydrologic control results in changes in vegetation and consequent changes in biodiversity. Partial water management for the purpose of reducing the frequency and extent of flooding either for agricultural purposes or fishing leads to wetland degradation of biogeochemical functions and possibly removes carbon in important quantities through grazing and/or peat harvesting.

Wetland areas are influenced by onsite factors like drainage, but also by offsite or upland areas where deforestation or intensive cultivation generally lead to enhanced runoff, **soil erosion**, sedimentation, and increasing pollutant fluxes to aquatic systems. From the perspective of land use, wetlands are drinking-water and irrigation-water sources, and a direct source of food and grazing grounds. Uplands surrounding wetlands must be taken into account because of their controls on hydrology and nutrient supply, which affect wetland functions.

For existing wetland inventories, which are often incomplete, attention must be given to the inventory of priority wetland habitats, focusing on those with the greatest risk of destruction and degradation, and those for which there is little or no information. It appears necessary to consider wetlands that have not been included in wetland inventories, but also to improve knowledge of wetlands that have been better mapped than others, but where serious inconsistencies and few areal estimates exist; more comprehensive inventory is required in that case. Priority wetland habitats designated in the review by C. Finlayson and A. Spiers are the following:

1. Sea grasses in South Asia, the South Pacific, South America, and some parts of Africa that are under increasing threat from pollution, coastal development, destructive fishing practices, recreational use, etc.
2. Salt marshes and coastal flats, because they are under increasing threat worldwide, particularly in Africa, Asia, and Oceania due to increasing coastal development; mangroves, to better determine their loss that is occurring at a dramatic rate in many parts of Africa, Southeast Asia, and Oceania through deforestation, land reclamation, and development for aquaculture.
3. Arid-zone wetlands, which are important when considering the increasing population pressures and water supply, although poorly inventoried.
4. Peatlands, which are well mapped in comparison with other wetland types, are threatened by drainage for agriculture and afforestation in Europe, Asia, and North America in particular, despite their importance as a global carbon sink and economic resource, and are poorly known in tropical regions such as Southeast Asia.
5. Small riparian wetlands can be included in that summary, as they are generally not inventoried despite their important functional role.

Besides natural wetlands, artificial wetlands that can partially compensate for the loss and degradation of natural wetlands have to be taken into account in inventories of priority wetland habitats: reservoirs, dams, salinas, paddies, and aquaculture ponds are increasingly important in many regions, particularly in Asia, Africa, and the neotropics, where they can provide habitat for wildlife and many values and benefits.

In developing countries, land-use changes are increasingly eliminating wetlands. In addition, given the expected increase in human population, mostly in developing countries, the pressure to convert wetlands is expected to increase even more with growing food requirements. Demographic pressure and related agricultural and urban land use could be the most important factors affecting wetland distribution changes in the future and could dominate some of the expected changes in sea level, temperature, and carbon dioxide concentration variations associated with possible climate changes.

In response to the extent and rate of wetland loss and degradation, numerous scientific and conservation programs involved in wetlands currently recommend rehabilitation of degraded wetlands whenever possible so that their structure and functions can be restored. They also outline the need for stronger protective measures for still-remaining functional wetlands. The definition of priority conservation actions should ensure a wise use of these areas.

See also: Atmosphere-Land Interlinkages; Carbon Cycle; Coastal Zone; Cumulative Change; Extensification; GLOBWETLANDS; Land Degradation; Land Rehabilitation; Malaria; Modification; Monitoring; River Basin; Taiga; Water-Land Interlinkages.

Further Reading

Birkett, C., et al., Global Wetland Distribution and Functional Characterization: Trace Gases and the Hydrologic Cycle [Online, January 2005], The Global Analysis and Integrated Modeling Web Site, gaim.unh.edu; Cowardin, L. M., Carter, V., Golet, F., and E. LaRoe, Classification of Wetlands and Deepwater Habitats of the United States [Online, January 2005], The U.S. Fish and Wildlife Service Web Site, www.nwi.fws.gov; Finlayson, C. M., and A. G. Spiers, eds., Global Review of Wetland Resources and Priorities for Wetland Inventory [Online, January 2005], The Wetlands Web Site, www.wetlands.org; Mitsch, W., and J. Gosselink, *Wetlands*, New York: Van Nostrand-Reinhold, 1993; National Research Council, Committee of Wetlands, Wetlands: Characteristics and Boundaries [Online, January 2005], The National Academy Press Web Site, www.nap.edu; U.S. Army Corp of Engineers, *Wetlands Delineation Manual* (Technical Report Y-87-1), Washington, DC: Department of the Army, 1987.

LAURENCE HUBERT-MOY

Wheat. Next to **rice**, wheat is the most important cereal worldwide and provides almost 20 percent of all human food energy. Wheat (or *triticum*) is used for making bread, pastries, crackers, biscuits, macaroni, spaghetti, bulgur, couscous, and breakfast food, as a brewing ingredient, and for alcohol production and livestock feed.

Domestication of wheat started around 7500–6500 BC in the Middle East. In prehistoric times, wheat was grown in Greece, Persia, Egypt, and throughout **Europe**. It spread into **China** and southwest Asia and was finally introduced by the Spanish to the New World in the sixteenth century.

Mule-drawn combine in wheat field of Walla Walla County, Washington, July 1941.
PHOTO: Russell Lee. [Courtesy Library of Congress]

Wheat is adapted to a wide range of environments. It is grown from the equator to the Arctic Circle and from sea level up to 4,500 meters high. Bread and durum wheat are the most important wheats and are grown on about 90 percent and 10 percent, respectively, of the total area cultivated with wheat. Durum wheat is less frost-resistant and requires warmer temperatures than many bread wheats and is grown mainly in Mediterranean environments. Bread wheat is more variable with about 25,000 varieties grown across a wide range of conditions worldwide.

Wheat can resist frost; the forms that have the ability to survive low temperatures in winter are known as winter wheat. However, about two-thirds of the total area devoted to wheat (i.e., 220 million hectares in 2004) is cultivated with spring wheat. As opposed to spring wheat, winter wheat requires vernalization at cold temperatures to reach the generative stage. For both spring and winter wheat, phasic development depends on temperature and, as for most cultivars, also on day length. The degree of sensitivity to cold temperatures and day length varies with cultivar and is an important parameter characterizing the agro-ecological adaptation. Wheat has a C_3 photosynthetic pathway. Optimum growth temperatures are between 10 and 24°C. Growth stops at above 35°C, and temperatures above 40°C induce irreversible heat stresses. Unlike spring wheat, most winter wheat is grown under rainfed conditions.

Historic changes in wheat area and production are closely related to population growth and economic development. During the last three centuries, the demand for food steadily increased with the rising number of people and the industrial development in many countries. Expansion of cropland including wheat land was for a long time the most effective way to increase food production but has eventually reached the limits set by available land area, other land uses, and environmental reasons. Global wheat harvested area has more than doubled within the last hundred years and reached its maximum extension around the 1980s. A reduction in wheat-growing area has been observed in recent decades due to abandonment of cropland in mainly developed countries. Presently, about 15 percent of all arable land is used for wheat production. Most wheat is grown in the European Union, China, India, and the **United States of America**, accounting together for more than 55 percent of the global wheat production. Other important wheat-growing countries are Australia, **Canada**, and **Russia**. In the first half of the twentieth century, wheat was produced mainly in developed countries; but during the second half of the century, production gradually shifted toward developing countries, which presently produce almost 50 percent of all wheat.

In recent decades, raising productivity per unit land has become the most important factor in meeting the growing demand for food. Average global wheat yields were about 2.9 t ha^{-1} in 2004 and have increased by almost 200 percent since 1961. Large differences among countries, however, are evident, which is due mainly to differences in climatic and socio-economic conditions. High yields are obtained in countries of the European Union: they often exceed 8 t ha^{-1}, for example, in the Netherlands, United Kingdom, Germany, and France. Yields have increased with a high rate since the **Green Revolution** in the 1960s. Advances in technology through improved varieties and crop management, including the use of fertilizers and biocides, have nearly tripled wheat yields in many countries within a few decades. Increasing wheat yields has in some regions, for example the European Union, resulted in overproduction and abandonment of cropland. However, supply is still insufficient for many countries in the developing world, and further increase in the demand for wheat due to population and **economic growth** is expected for the future. Since cropland expansion is limited, increases in production have to be realized via higher yields through further development of production technologies. The long-term effects on the environment of intensive crop production and the use of genetically modified organisms remain unclear but are of increasing concern.

Future productivity of wheat will also depend on changes in climatic conditions and atmospheric carbon dioxide (CO_2) concentration as projected by the **Intergovernmental Panel on Climate Change**. Experimental evidence and **modeling** studies suggest that doubling the present CO_2 concentration may increase wheat productivity by 20–30 percent, which may ameliorate some of the negative effects of increasing temperature and reduced rainfall. Effects of climate changes are projected to vary depending on the region. Wheat may extend into colder regions in the North and South due to increasing temperatures and disappear from warmer regions in the South due to increasing stresses from drought and heat. However, relationships that will determine future changes in yields and the spatial distribution of wheat are complex and not well understood. Impacts of climate change will also depend on the **vulnerability** and more specifically on the adaptive capacity of the agricultural systems, which again will depend on the socio-economic conditions of the farms, regions, and countries.

See also: **Agricultural Intensification; Agricultural Revolution; Cash Crops; Consumption; Driving Forces; Economic Restructuring; Food Crops; Great Plains; Land Abandonment; Land Degradation; Mountains; Population Dynamics; Turnover.**

Further Reading

Ewert, Frank, Rounsevell, Mark D. A., Reginster, Isabelle, Metzger, Marc, and Rik Leemans, "Future Scenarios of Agricultural Land Use in Europe: I: Estimating Changes in Productivity," *Agriculture, Ecosystems & Environment* 107 (2005): 101–16; Parry, Martin L., Rosenzweig, Cynthia, Iglesias, Ana, Livermore, Matthew, and Günther Fischer, "Effects of Climate Change on Global Food Production under SRES Emissions and Socio-Economic Scenarios," *Global Environmental Change* 14 (2004): 53–67; Ramankutty, Navin, and Jonathan A. Foley, "Estimating Historical Changes in Global Land Cover: Croplands from 1700 to 1992," *Global Biogeochemical Cycles* 13 (1999): 997–1027; Rounsevell, Mark D. A., Ewert, Frank, Reginster, Isabelle, Leemans, Rik, and Timothy R. Carter, "Future Scenarios of Agricultural Land Use in Europe: II: Projecting Changes in Cropland and Grassland," *Agriculture, Ecosystems & Environment* 107 (2005): 117–35; United Nations Food and Agriculture Organization [Online, February 2005], The FAO Web Site, www.fao.org; Van Ginkel, M., and R. L. Villareal, "Triticum L," in *Plant Resources of South-East Asia Handbook 10: Cereals*, eds. G. J. H. Grubben and Soetjipto Partohardjono, Prosea Foundation, 1996, pp. 137–43.

FRANK EWERT

World Agroforestry Centre. One of the international agricultural research institutions supported by the **Consultative Group on International Agricultural Research**, also known as the International Centre for Research in Agroforestry (ICRAF). Taking a broad view of the economic, social, cultural, and environmental roles of trees on farms and at the landscape/**watershed** level, ICRAF seeks to improve livelihoods for poor people in developing countries while enhancing sustainability of land use, local ecosystems, and the global environment. The Centre's headquarters are in Nairobi, Kenya. Nearly 200 professional staff work in twenty-nine developing countries within seven regional programs in Africa, Asia, and **Latin America**. ICRAF works in partnership with farmers, national scientists, policymakers, and others to conduct **agroforestry** research and development, strengthen the capacity of partners, communicate useful information about agroforestry, and provide scientific leadership in the field of integrated natural resource management.

See also: **Agrodiversity; Alternatives to Slash-and-Burn Programme; Biodiversity; Ecosystem Services; Forestry; Land Rehabilitation; Land-Use Policies; Natural Resources; Participatory Land Management; Poverty; Sustainable Land Use; Water-Land Interlinkages.**

Further Reading

World Agroforestry Centre, Transforming Lives and Landscapes [Online, November 2004], The World Agroforestry Centre Web Site, www.worldagroforestrycentre.org.

THOMAS P. TOMICH

Yellowstone Model. Legislation to create the world's first national park—Yellowstone—in Wyoming was passed on March 1, 1872. It was the starting point for national governments to start annexing and managing "natural" areas for various objectives (e.g., nature protection and recreation), and has loomed large in the psyche of conservationists and park administrators ever since. Yellowstone was promoted as a wilderness: an area untouched by humans, an area where nature remained undisturbed. The wilderness concept that has evolved in North America is deeply entrenched (e.g., in 1964, the **United States of America** passed the Wilderness Act) but is not as strong elsewhere. The strict definition of wilderness as a place untouched by people is strongly disputed on the basis of archaeological, biological, and paleo-ecological evidence. Retrospectives on the establishment of Yellowstone National Park illustrate that the area was considered undisturbed because Indians were ignored entirely. In one sense, then, the Yellowstone Model of **conservation** comprises large areas of wilderness being annexed by national governments, usually to protect nature and/or to encourage recreation. When the Yellowstone Model is applied to the design of a nature reserve, there is often a strong assumption of exclusion. People are evicted from the land they live on, or their **access** to natural resources (e.g., fuel and medicinal plants collection) is restricted. In the worst-case **scenarios**, areas are controlled by guards, they may be fenced off, and access is granted only to scientists and/or tourists. This practice is also colloquially known as "fortress conservation" or the "fences-and-fines approach." National parks were established using the Yellowstone Model in the first part of the twentieth century throughout the world (e.g., Southern [now Nahuel Huapi, Argentina], 1902; Udjon Kulun [Indonesia], 1915; and Krüger [**South Africa**], 1926). But the Yellowstone Model has led to other types of protected areas (PAs)—not simply large areas that exclude indigenous people—being established. The International Union for Conservation of Nature and Natural Resources (IUCN) currently recognizes seven PA categories. Therefore, while strict protectionism is one clear outcome of the Yellowstone Model, the growth in PAs has been another. At the Fifth World Parks Congress (Durban, South Africa, 2003), IUCN reported 102,102 PAs covering 18.8 million square kilometers. This equates to 11.5 percent of the earth's surface under some kind of protection, an area greater than that under permanent arable crops. Statistics show that the number and area of PAs is growing exponentially. The land-use and **land-cover change** implications of the Yellowstone Model are found within and outside park boundaries. Access regulations in many types of protected areas can allow relatively small areas in parks to recover to near-natural land covers. However, the communities that have been evicted from the parks or have had their access restricted often settle around park perimeters. The areas around their settlements often become degraded, and illegal access to natural resources within the PAs continues.

See also: **Arena of Land Conflict; Public Policy; Tragedy of Enclosure.**

Further Reading

Gomez Pompa, A., and A. Kaus, "Taming the Wilderness Myth," *BioScience* 42 (1992): 271–9; Stevens, Stan, "The Legacy of Yellowstone," in *Conservation through Cultural Survival*, ed. Stan Stevens, Washington, DC: Island Press, 1997, pp. 13–32.

ANDREW MILLINGTON

Selected Bibliography

Abrol, Yash P., Satpal Sangwan, and Mithilesh K. Tiwari, eds. *Land Use: Historical Perspectives: Focus on Indo-Gangetic Plains.* New Delhi, India: Allied Publishers, 2002.

Bičík, Ivan, Pavel Chromý, Vit Jančá, and Helena Janů, eds. *Land Use/Land Cover Changes in the Period of Globalization.* Prague, Czech Republic: Charles University Faculty of Science, International Geographical Union Commission on Land-Use and Land-Cover Change, 2002.

Blaikie, Piers M., and Harold Brookfield. *Land Degradation and Society.* London, New York: Routledge, 1987.

Bridge, Gavin. "Contested Terrain: Mining and the Environment," *Annual Review of Energy and Resources* 29 (2004): 205–59.

Bryant, Raymond L., and Sinéad Bailey. *Third World Political Ecology.* London, New York: Routledge, 1997.

DeFries, Ruth, Gregory Asner, and Richard Houghton, eds. *Ecosystems and Land Use Change.* Washington, DC: American Geophysical Union, 2004.

Foley, Jonathan A., DeFries, Ruth, Asner, Gregory P., Barford, Carol, Bonan, Gordon, Carpenter, Stephen R., Chapin, F. Stuart, Coe, Michael T., Daily, Gretchen C., Gibbs, Holly K., et al. "Global Consequences of Land Use." Science 309 (2005): 570–74.

Fox, Jefferson, Ronald Rindfuss, Stephen Walsh, and Vinod Mishra, eds. *People and the Environment: Approaches for Linking Household and Community Surveys to Remote Sensing and GIS.* Dordrecht, NL: Kluwer, 2003.

Geist, Helmut. *The Causes and Progression of Desertification.* Aldershot, UK, Burlington, VT: Ashgate, 2005.

Geist, Helmut J., and Éric F. Lambin. "Proximate Causes and Underlying Driving Forces of Tropical Deforestation." *BioScience* 52 (2002): 143–50.

Gutman, Garik, Anthony Janetos, Christopher Justice, Emilio Moran, John Mustard, Ronald Rindfuss, David Skole, B. L. Turner II, and Mark Cochrane, eds. *Land Change Science: Observing, Monitoring and Understanding Trajectories of Change on the Earth's Surface* (Remote Sensing and Digital Image Processing Series 6). Dordrecht, NL: Kluwer, 2004.

Haberl, Helmut, Matthis Wackernagel, and T. Wrbka, eds. "Land Use and Sustainability Indicators." *Land Use Policy* 21 (2004).

Himiyama, Yukio, Alexander Mather, Ivan Bičík, and Elena V. Milanova, eds. *Land Use/Cover Changes in Selected Regions in the World: Volumes I–IV.* Asahikawa, JP: International Geographical Union Commission on Land-Use and Land-Cover Change, 2001, 2002, 2005.

Johnston, R. J., Peter J. Taylor, and Michael Watts, eds. *Geographies of Global Change: Remapping the World in the Late Twentieth Century.* Oxford, UK: Blackwell, 2002.

Kasperson, Jeanne X., Roger E. Kasperson, and B. L. Turner II, eds. *Regions at Risk: Comparisons of Threatened Environments.* Tokyo: United Nations University Press, 1995.

Lambin, Éric, Baulies, Xavier, Bockstael, Nancy, Fischer, Günther, Krug, Thelma, Leemans, Rik, Moran, Emilio, et al. *Land-Use and Land-Cover Change (LUCC): Implementation Strategy* (IGBP Report 48, IHDP Report 10). Stockholm, SW: International Geosphere-Biosphere Programme, and Bonn, GE: International Human Dimensions of Global Environmental Change Programme, 1999.

Lambin, Éric, and Helmut Geist, eds. *Land Use and Land Cover Change: Local Processes, Global Impacts.* Berlin: Springer, 2006.

Lambin, Éric, Geist, Helmut, and Erika Lepers. "Dynamics of Land Use and Cover

Change in Tropical and Subtropical Regions." *Annual Review of Environment and Resources* 28 (2003): 205–41.

Lambin, Éric F., Turner, B. L. II, Geist, Helmut J., Agbola, Samuel B., Angelsen, Arild, Bruce, John W., Coomes, Oliver, et al. "The Causes of Land-Use and Land-Cover Change: Moving beyond the Myths." *Global Environmental Change* 11 (2001): 261–9.

Lepers, Erika, Lambin, Éric F., Janetos, Anthony C., DeFries, Ruth, Achard, Frédéric, Ramankutty, Navin, and Robert J. Scholes. "A Synthesis of Information on Rapid Land-Cover Change for the Period 1981–2000." *BioScience* 55 (2005): 115–24.

Liverman, Diana, Emilio Moran, Ronald Rindfuss, and Paul Stern, eds. *People and Pixels: Linking Remote Sensing and Social Science.* Washington, DC: National Academy Press, 1998.

Mather, Alexander S. "The Forest Transition." *Area* 24 (1992): 227–37.

Mather, Alexander S. "Forest Transition Theory and the Reforesting of Scotland." *Scottish Geographical Journal* 120 (2004): 83–98.

Meyer, William B., and B. L. Turner II, eds. *Changes in Land Use and Land Cover: A Global Perspective.* Cambridge, UK, New York, Melbourne, AU: Press Syndicate of the University of Cambridge, 1994.

Moran, Emilio F., and Elinor Ostrom, eds. *Seeing the Forest and the Trees: Human-Environment Interactions in Forest Ecosystems.* Cambridge, MA: MIT Press, 2005.

Morris, Dick, Freeland, Joanna, Hinchliffe, Steve, and Sandy Smith. *Changing Environments.* Chichester, UK: John Wiley & Sons, 2003.

Oldfield, Frank, *Environmental Change: Key Issues in and Alternative Perspectives,* Cambridge, UK: Cambridge University Press, 2005.

Neil, Roberts, ed. *The Changing Global Environment.* Oxford, UK: Blackwell, 1994.

Parker, Dawn C., Manson, Steve M., Janssen, M. A., Hoffmann, M. J., and Peter Deadman. "Multi-Agent System Models for the Simulation of Land-Use and Land-Cover Change: A Review." *Annals of the Association of American Geographers* 93 (2003): 316–40.

Ramankutty, Navin, Foley, Jonathan A., and N. J. Olejniczak. "People on the Land: Changes in Global Population and Croplands during the 20th Century." *Ambio* 31 (2002): 251–7.

Reynolds, James, and Mark Stafford Smith, eds. *Global Desertification: Do Humans Cause Deserts?* (Dahlem Workshop Report 88). Berlin: Dahlem University Press, 2002.

Rudel, Thomas K, Coomes, Oliver T., Moran Emilio, Achard, Frédéric, Angelsen, Arild, Xu, Jianchu, and Éric F. Lambin. "Forest Transitions: Towards a Global Understanding of Land Use Change." *Global Environmental Change* 15 (2005): 23–31.

Steffen, Will, Sanderson, Angelina, Tyson, Peter, Jäger, Jill, Matson, Pamela, Moore, Berrien III, Oldfield, Frank, et al. *Global Change and the Earth System: A Planet under Pressure.* Berlin: Springer, 2004.

Tri-Academy Panel. "Population and Land Use in India, China, and the United States: Context, Observations, and Findings." In *Growing Populations, Changing Landscapes: Studies from India, China, and the United States,* edited by Indian National Science Academy, Chinese Academy of Sciences, and U.S. National Academy of Sciences. Washington, DC: National Academy Press, 2001, pp. 9–72.

Turner, B. L., II. "Toward Integrated Land-Change Science: Advances in 1.5 Decades of Sustained International Research on Land-Use and Land-Cover Change." In *Challenges of a Changing Earth: Proceedings of the Global Change Open Science Conference, Amsterdam, The Netherlands, 10–13 July 2001* (The IGBP Series), edited by Will Steffen, Jill Jäger, David Carson, and Clare Bradshaw. Berlin: Springer, 2002, pp. 21–26.

Turner, B. L., II, Skole, David, Sanderson, Steven, Fischer, Günther, Fresco, Louise, and Rik Leemans. *Land-Use and Land-Cover Change Science/Research Plan* (IGBP Report 35, HDP Report 7). Stockholm, SW: International Geosphere-Biosphere Programme, and Geneva: Human Dimensions of Global Environmental Change Programme, 1995.

United Nations Environment Programme: *One Planet, Many People: Atlas of Our Changing Environment.* Nairobi: UNEP, 2005.

Veldkamp, Tomas, and Éric Lambin, eds. "Predicting Land-Use Change." *Agriculture Ecosystems and Environment* 85 (2001).

Verburg, Peter H., and Tom Veldkamp, eds. "Spatial Modeling to Explore Land Use Dynamics." *International Journal of Geographic Information Science* 19 (2005): 99–102.

Walsh, Stephen, and Kelley Crews-Meyer, eds. *Remote Sensing and GIS Applications for Linking People, Place, and Policy.* Dordrecht, NL: Kluwer, 2002.

Wisner, Ben, Blaikie, Piers M., Cannon, Terry, and Ian Davis. *At Risk: Natural Hazards, People's Vulnerability, and Disasters.* 7th ed. London: Routledge, 2003.

Index

Note: Page numbers in bold type indicate main encyclopedia entries.

List of Contributors

(in alphabetical order; * = co-authored entry; † = introductory essay)

Achard, Frédéric (Dr.)
Joint Research Centre of the European
    Commission
Institute for Environment and
    Sustainability
Italy
1. Hot Spots of Land-Cover Change*
2. Joint Research Centre (JRC) of the
    European Commission
3. Kyoto Protocol*
4. Tropical Ecosystem Environment
    Observations by Satellite (TREES)
    Project
5. Tropical Humid Forest*

Arneth, Almut (Dr.)
Lund University
Department of Physical Geography and
    Ecosystems Analysis
Sweden
1. Atmosphere-Land Interlinkages
2. Integrated Land Ecosystem-Atmosphere
    Processes Study
3. International Geosphere-Biosphere
    Programme

Baade, Jussi (Prof.)
Friedrich Schiller University Jena
Department of Geography
Germany
1. Carbon Sequestration
2. Salinization
3. Siltation
4. Soil Degradation
5. Soil Erosion

Bartholomé, Etienne (Dr.)
Joint Research Centre of the European
    Commission
Institute for Environment and Sustainability
Italy
1. Global Land Cover Map of the Year
    2000*

Belward, Alan (Dr.)
Joint Research Centre of the European
    Commission
Institute for Environment and Sustainability
Italy
1. Global Land Cover Map of the Year
    2000*

Berrang Ford, Lea
University of Guelph
Department of Population Medicine
Canada
1. African Trypanosomiasis
2. Human Immunodeficiency Virus (HIV)/
    Acquired Immunodeficiency Syndrome
    (AIDS)
3. Malaria
4. Watershed

Biggs, Reinette (Oonsie) (Dr.)
University of Wisconsin, Madison
Center for Limnology
United States of America
1. South Africa*
2. Southern Africa*

Bridge, Gavin (Prof.)
Syracuse University
Department of Geography
United States of America
1. Corporate Strategies
2. Economic Liberalization
3. Economic Restructuring
4. Investments
5. Mineral Extraction
6. Multiplier Effect

Brouwer, Floor (Prof.)
Agricultural Economics Research Institute
The Netherlands
1. Agriculture
2. Land Abandonment

Brown, Daniel G. (Prof.)
University of Michigan
School of Natural Resources and
    Environment
United States of America
1. Agent-Based Model
2. Auto-Correlation

Busch, Gerald (Dr.)
University of Kassel
Center for Environmental Systems Research
Germany
1. Carbon Cycle
2. Land Degradation

Chase, Thomas N. (Prof.)
University of Colorado
Cooperative Institute for Research in
    Environmental Sciences and
    Department of Geography
United States of America
1. Climate Impacts*

Chen, Ke (Dr.)
Stockholm Environment Institute
SUSTF International Secretariat
Sweden
1. Extensification
2. George Perkins Marsh Institute
3. Land Tenure
4. Land-Use System
5. Sauer, Carl Ortwin
6. Vulnerability

Crews-Meyer, Kelley A. (Prof.)
University of Texas
Department of Geography and the
    Environment
United States of America
1. Access*
2. Airborne Light Detection and Ranging*
3. Change Detection
4. Global Positioning System*
5. Pattern Metrics
6. Savannization*

Cunfer, Geoff (Prof.)
University of Saskatchewan
Department of History
Canada
1. Great Plains

de Sherbinin, Alexander (Dr.)
Columbia University
Center for International Earth Science
    Information Network
United States of America
1. Center for International Earth Science
    Information Network
2. Consumption
3. Population Dynamics
4. Urbanization

de Vasconcelos, Maria José P. (Prof.)
Tropical Research Institute
Cartography Centre
Portugal
1. Cellular Automaton
2. Fire
3. Vegetation

de Vries, Bert J. M. (Prof.)
Netherlands Environmental Assessment
    Agency
Global Sustainability and Climate
The Netherlands
1. Integrated Model to Assess the Global
    Environment*

Eickhout, Bas (Dr.)
Netherlands Environmental Assessment
    Agency
Global Sustainability and Climate
The Netherlands
1. Integrated Model to Assess the Global
    Environment*

Erb, Karlheinz (Dr.)
Klagenfurt University
Institute of Social Ecology
Austria
1. Biomass
2. Colonization
3. Ecological Colonization
4. Ecological Footprint
5. Turnover

Eva, Hugh (Dr.)
Joint Research Centre of the European
    Commission
Institute for Environment and Sustainability
Italy
1. Global Land Cover Map of the Year
    2000*

2. Hot Spots of Land-Cover Change*
3. Tropical Humid Forest*

Evans, Thomas P. (Prof.)
Indiana University
Department of Geography
United States of America
1. Center for the Study of Institutions, Population and Environmental Change
2. Decision-Making
3. Pixel
4. Secondary Vegetation

Ewert, Frank (Dr.)
Wageningen University
Plant Production Systems
The Netherlands
1. Green Revolution
2. Maize
3. Rice
4. Wheat

Fox, Jefferson (Dr.)
East-West Center
Program on Environment
United States of America
1. Fallow
2. Slash-and-Burn Agriculture
3. Southeast Asia, Mainland
4. Swidden Cultivation

Freitas, Helena (Prof.)
University of Coimbra
Department of Botany
Portugal
1. Biodiversity
2. Eutrophication
3. Grassland

Geist, Helmut (Prof.)
University of Aberdeen
Department of Geography
United Kingdom
1. *Agriculture, Ecosystems & Environment*
2. Agrodiversity
3. Agroforestry
4. Agro-Industry*
5. *Ambio, A Journal of the Human Environment*
6. Anthropological Center for Training and Research on Global Environmental Change
7. Biodiversity Novelties

8. Biome
9. BIOME 300 Project
10. *BioScience*
11. Blaikie, Piers
12. Boserup, Ester
13. Brookfield, Harold Chilingworth
14. Carson, Rachel Louise
15. Cassava
16. Center for International Forestry Research
17. Contract Farming
18. Conversion
19. Cultural Factors*
20. Deforestation
21. Desertification*
22. DIVERSITAS
23. Domestication
24. *Ecology and Society*
25. Endogenous Factor
26. *Environment*
27. Exogenous Factor
28. Exotic Species
29. Feedback
30. Forest Degradation
31. *Frontiers in Ecology and the Environment*
32. *Global Environmental Change*
33. Global Environmental Change and Food Systems (GECAFS) Project
34. Human Health
35. IKONOS
36. Initial Conditions
37. International Food Policy Research Institute
38. *Journal of Arid Environments*
39. *Journal of Land Use Science*
40. *Land Degradation & Development*
41. Land Degradation Assessment in Drylands (LADA) Project
42. Land Quality
43. Land Stability
44. Land-Ocean Interactions in the Coastal Zone (LOICZ) Project
45. Landsat
46. Land-Use Legacy
47. Land-Use Transition
48. Land-Use/Cover Change (LUCC) Project
49. Large-Scale Biosphere-Atmosphere Experiment in Amazonia
50. Maathai, Wangari M.
51. Marsh, George Perkins
52. Mediating Factor
53. Meta-Analysis

54. Millennium Ecosystem Assessment
55. Miombo
56. Mode of Interaction
57. Moderate Resolution Imaging Spectroradiometer
58. Modification
59. Monoculture
60. Non-Timber Forest Products
61. Past Global Changes (PAGES) Project
62. Population-Environment Research Network
63. Project on People, Land Management and Ecosystem Conservation
64. Species Extinction
65. Sustainable Land Use
66. Syndrome
67. Tobacco
68. Tragedy of Enclosure
69. United Nations Convention on Biological Diversity
70. United Nations Convention to Combat Desertification
71. United Nations Environment Programme
72. United Nations Food and Agriculture Organization
73. Water-Land Interlinkages

Geoghegan, Jacqueline M. (Prof.)
Clark University
Department of Economics
United States of America
1. Economic Growth
2. Pixelizing the Social
3. Socializing the Pixel

Göbel, Barbara (Dr.)
International Human Dimensions Programme on Global Environmental Change
Germany
1. Earth System Science Partnership
2. Globalization
3. International Human Dimensions Programme on Global Environmental Change
4. Perception

Goldman, Mara Jill
University of Wisconsin
Department of Geography
United States of America

1. Participatory Land Management*
2. Pastoral Mobility*

Grégoire, Jean-Marie (Dr.)
Joint Research Centre of the European Commission
Institute for Environment and Sustainability
Italy
1. Global Land Cover Map of the Year 2000*

Grimm, Heike (Dr.)
University of Erfurt
School of Public Policy
Germany
1. Agro-Industry*
2. Cultural Factors*
3. Public Policy

Gutman, Garik (Dr.)
NASA Land-Cover/Land-Use Change (LCLUC) Program
United States of America
1. NASA Land-Cover/Land-Use Change (LCLUC) Program
2. Remote Sensing

Haberl, Helmut (Prof.)
Klagenfurt University
Institute of Social Ecology
Austria
1. Cascades
2. Human Appropriation of Net Primary Production
3. Metabolism
4. Threshold

Haileselassie, Amare (Dr.)
Institute of Soil Science and Forest Nutrition
George August University
Germany
1. Nutrient Cycle*

Heistermann, Maik
University of Kassel
Center for Environmental Systems Research
Germany
1. LandSHIFT Model*
2. Nutrient Cycle*

Himiyama, Yukio (Prof.)
Hokkaido University of Education
Japan

1. International Geographical Union (IGU) Commission on Land-Use and Land-Cover Change
2. Japan

Homewood, Katherine (Prof.)
University College London
Department of Anthropology
United Kingdom
1. Disequilibrium Dynamics
2. Transhumance

Hostettler, Silvia
Ecole Polytechnique Fédérale de Lausanne
Laboratory of Urban Sociology
Switzerland
1. Participatory Geographic Information System
2. Remittances Landscape

Hubert-Moy, Laurence (Prof.)
University of Rennes 2
Department of Geography
France
1. Peatlands
2. Ramsar Convention on Wetlands
3. Wetlands

Hwang, Manik (Prof.)
Seoul National University
Department of Geography Education
South Korea
1. Coastal Zone

Irwin, Elena G. (Prof.)
Ohio State University
Department of Agricultural, Environmental, and Development Economics
United States of America
1. Leapfrogging
2. Suburbanization
3. Urban Sprawl
4. Urban-Rural Fringe

Jeleček, Leoš (Prof.)
Charles University
Department of Social Geography and Regional Development
Czech Republic
1. Agricultural Revolution
2. Cadastre

3. East-Central Europe
4. Industrial Revolution
5. Land Reform
6. Land Rent
7. Southeast Europe
8. Technological-Scientific Revolution

Jepson, Wendy (Prof.)
Texas A&M University
Department of Geography
United States of America
1. Biotechnology
2. Brazilian Cerrado

Klein Goldewijk, Kees (Prof.)
Netherlands Environmental Assessment Agency
Global Sustainability and Climate
The Netherlands
1. Center for Sustainability and the Global Environment
2. History Database of the Global Environment
3. Land-Use Change in a Global, Historical Perspective†
4. National Institute of Public Health and the Environment

Klepeis, Peter (Prof.)
Colgate University
Department of Geography
United States of America
1. Degradation Narrative*
2. Land-Use History*
3. Pristine Myth*
4. Southern Cone Region

Kok, Kasper (Dr.)
Wageningen University and Research Center
Department of Environmental Sciences
The Netherlands
1. Integrated Assessment
2. Scenario
3. Validation*

Krausmann, Fridolin (Dr.)
Klagenfurt University
Institute of Social Ecology
Austria
1. Industrialization
2. West-Central Europe

Laris, Paul (Prof.)
California State University
Department of Geography
United States of America
1. Degradation Narrative*
2. Land-Use History*
3. Pristine Myth*

Lawrence, Peter J. (Dr.)
University of Colorado
Cooperative Institute for Research in
    Environmental Sciences and Department
    of Geography
United States of America
1. Climate Impacts*

Li, Xiubin (Prof.)
Chinese Academy of Sciences
Institute of Geographic Sciences and
    Natural Resources Research
China
1. China
2. River Basin

Linke, Sophia
United Nations Food and Agriculture
    Organization
Environment and Natural Resources Service
Italy
1. Land Cover Classification System*

Loveland, Thomas R. (Prof.)
South Dakota State University
United States of America
1. Earth Resources Observation Systems
    (EROS) Data Center
2. IGBP-DIS Global 1-Km Land Cover
    Data Set
3. LANDFIRE Project
4. Sahel Land Cover
5. United States of America*
6. United States Geological Survey (USGS)
    Program

Lu, Qi (Prof.)
Chinese Academy of Sciences
Institute of Geographic Sciences and
    Natural Resources Research
China
1. Green Wall Project
2. Land Rehabilitation
3. Regrowth

Mather, Alexander S. (Prof.)
University of Aberdeen
Department of Geography
United Kingdom
1. Driving Forces
2. Europe
3. Forest
4. Forest Transition
5. Forestry
6. *Land Use Policy*
7. Land-Use Policies
8. Proximate Causes
9. Thünen, Johann Heinrich von
10. Western Europe

Mayaux, Philippe (Dr.)
Joint Research Centre of the European
    Commission
Institute for Environment and Sustainability
Italy
1. Global Land Cover Map of the Year
    2000*
2. Hot Spots of Land-Cover Change*
3. Tropical Humid Forest*

Millington, Andrew (Prof.)
Texas A&M University
Department of Geography
United States of America
1. Agenda 21
2. Conservation
3. Land-Cover Change†
4. South America
5. Yellowstone Model

Mollicone, Danilo (Dr.)
Joint Research Centre of the European
    Commission
Institute for Environment and Sustainability
Italy
1. Kyoto Protocol*

Morais, João M. F. (Prof.)
International Geosphere-Biosphere
    Programme
Sweden
1. Global Land Project

Moseley, William G. (Prof.)
Macalester College
Department of Geography
United States of America

1. Cotton
2. Food Security
3. Poverty

Müller, Daniel (Dr.)
Humboldt University
Institute for Agricultural Economics and
   Social Sciences
Germany
1. Aerial Photography
2. Cash Crops
3. Food Crops
4. Georeferencing
5. Hot Spot Identification*
6. Subsistence Agriculture

Munroe, Darla K. (Prof.)
Ohio State University
Department of Geography
United States of America
1. Hot Spot Identification*
2. Probit/Logit Model

Nagendra, Harini (Dr.)
Indiana University
Center for the Study of Institutions,
   Population, and Environmental Change
United States of America
1. Community Involvement
2. Decentralization
3. Forest Anomalies
4. Himalayas
5. Reforestation

Napton, Darrell (Prof.)
South Dakota State University
Department of Geography
United States of America
1. United States of America*

Neuenschwander, Amy L.
University of Texas
Department of Geography and the
   Environment
United States of America
1. Airborne Light Detection and Ranging*
2. Global Positioning System*

Norman, Amy L.
University of North Carolina
Carolina Population Center
United States of America

1. Access*
2. Savannization*

Oldfield, Frank (Prof.)
University of Liverpool
Department of Geography
United Kingdom
1. Anthropocene
2. Environmental Change
3. Holocene
4. Human Impact on Terrestrial Ecosystems
   (HITE) Initiative
5. Paleo Perspectives

Pacheco, Pablo (Dr.)
Center for International Forestry Research
Embrapa Amazônia Oriental
Brazil
1. Agricultural Frontier
2. Amazonia
3. Arc of Deforestation
4. Cattle Ranching
5. Economic Livelihood
6. Land Concentration
7. Land Fragmentation
8. Landholding
9. Pasture

Parker, Dawn Cassandra (Prof.)
George Mason University
Department of Geography
United States of America
1. Ecosystem Services
2. Integrated Model
3. Tragedy of the Commons

Parton, William J., Jr. (Dr.)
Colorado State University
Natural Resource Ecology Laboratory
United States of America
1. Century Ecosystem Model

Pischke, Frederik
United Nations Food and Agriculture
   Organization
Environment and Natural Resources Service
Italy
1. Global Land Cover Network*

Pontius, Robert Gilmore, Jr. (Prof.)
Clark University
Department of International Development,
   Community, and Environment

United States of America
1. Pattern to Process
2. Scale*
3. Transition Matrix
4. Validation*

Priess, Jörg A. (Prof.)
University of Kassel
Center for Environmental Systems Research
Germany
1. Agricultural Intensification
2. Disintensification
3. LandSHIFT Model*
4. Nutrient Cycle*

Robbins, Paul (Prof.)
University of Arizona
Department of Geography and Regional
    Development
United States of America
1. Institutions

Rounsevell, Mark (Prof.)
University of Louvain
Department of Geography
Belgium
1. Global Environmental Change
2. Greenhouse Gases
3. Intergovernmental Panel on Climate
    Change
4. Spatial Diffusion

Sánchez-Azofeifa, Arturo (Prof.)
University of Alberta
Department of Earth and Atmospheric
    Sciences
Canada
1. Canada
2. Tropical Dry Forest

Schmullius, Christiane (Prof.)
Friedrich Schiller University
Institute of Geography
Germany
1. Boreal Zone
2. ENVISAT ASAR
3. European Remote Sensing (ERS-1/-2)
    Satellites
4. Monitoring
5. Taiga

Scholes, Robert J. (Prof.)
CSIR Division of Water, Environment and
    Forest Technologies

South Africa
1. Global Observation of Forest and Land
    Cover Dynamics
2. Global Terrestrial Observing System
3. South Africa*
4. Southern Africa*

Sessa, Reuben (Dr.)
United Nations Food and Agriculture
    Organization
Environment and Natural Resources Service
Italy
1. AFRICOVER
2. Global Land Cover Network*
3. Land Cover Classification System*

Shvidenko, Anatoly (Prof.)
International Institute for Applied Systems
    Analysis
Forestry Program
Austria
1. International Institute for Applied
    Systems Analysis
2. Russia

Sikor, Thomas (Dr.)
Humboldt University
Institute for Agricultural Economics and
    Social Sciences
Germany
1. Common-Pool Resources
2. Land Privatization
3. Land Rights
4. Private Property*
5. Property*

Southworth, Jane (Prof.)
University of Florida
Department of Geography and Land Use
    and Environmental Change Institute
United States of America
1. Coffee
2. Continuous Data
3. Discrete Data
4. Thermal Band Analysis

Stahl, Johannes
Humboldt University
Institute for Agricultural Economics and
    Social Sciences
Germany
1. Private Property*
2. Property*

Stephenne, Nathalie (Dr.)
Joint Research Centre of the European
    Commission
Italy
1. Geographical Information System
2. Indigenous Knowledge
3. Land Evaluation
4. Land-Use Planning
5. Resilience
6. Sahelian Land Use (SALU) Model

Stibig, Hans-Jürgen (Dr.)
Joint Research Centre of the European
    Commission
Institute for Environment and Sustainability
Italy
1. Global Land Cover Map of the Year
    2000*
2. Hot Spots of Land-Cover Change*
3. Tropical Humid Forest*

Tansey, Kevin J. (Dr.)
University of Leicester
Department of Geography
United Kingdom
  1. Advanced Very High Resolution
      Radiometer
  2. Albedo
  3. Along Track Scanning Radiometers
      (ATSR) World Fire Atlas
  4. Coordination of Information on the
      Environment (CORINE) Database
  5. Global Burnt Area 2000
  6. GLOBCARBON
  7. GLOBCOVER
  8. GLOBICE
  9. GLOBSCAR
  10. GLOBWETLANDS
  11. JERS Radar Mosaics
  12. Medium Resolution Imaging
      Spectrometer
  13. Normalized Difference Vegetation
      Index
  14. SPOT Vegetation
  15. TERRA-ASTER

Tomich, Thomas P. (Dr.)
World Agroforestry Centre
Alternatives to Slash-and-Burn Programme
Kenya

1. Alternatives to Slash-and-Burn (ASB)
    Programme
2. Consultative Group on International
    Agricultural Research
3. Tradeoffs
4. World Agroforestry Centre

Turner, B. L., II (Prof.)
Clark University
George Perkins Marsh Institute
United States of America
1. Cumulative Change
2. Land Change as a Forcing Function in
    Global Environmental Change†
3. Systemic Change

Turner, Matthew D. (Prof.)
University of Wisconsin
Department of Geography
United States of America
1. Arena of Land Conflict
2. Desertification*
3. Participatory Land Management*
4. Pastoral Mobility*
5. Sudan-Sahel

Verburg, Peter H. (Prof.)
Wageningen University and Research
    Center
Department of Environmental Sciences
The Netherlands
1. Conversion of Land Use and Its Effects
    (CLUE) Model
2. Modeling
3. Scale*

Vogel, Coleen H. (Prof.)
University of the Witwatersrand
School of Geography, Archaeology and
    Environmental Studies
South Africa
1. People at Risk

Young, Kenneth R. (Prof.)
University of Texas
Department of Geography
United States of America
1. Andes
2. Latin America
3. Mountains

About the Editor

Helmut Geist is professor of human-environment at the University of Aberdeen in Scotland, United Kingdom. From 2000 until 2005, he was Executive Director of the IGBP-IHDP Land-Use/Cover Change (LUCC) project at LUCC's International Project Office at the University of Louvain in Louvain-la-Neuve, Belgium. His research interests include political ecology, sustainable land use, land-change theory, and human-driven environmental changes such as land degradation, deforestation, desertification, and urbanization. Based at universities in the United Kingdom, Belgium, Germany, and the United States, he was involved in the World Bank's Human Development Network on tobacco control (1998), in reviews of the European Commission's desertification research (2003), and in the United Nations Millennium Ecosystem Assessment (2004).